Advances in Oil and Gas Exploration & Production

Series editor

Rudy Swennen, Department of Earth and Environmental Sciences
K.U. Leuven, Heverlee, Belgium

The book series Advances in Oil and Gas Exploration & Production publishes scientific monographs on a broad range of topics concerning geophysical and geological research on conventional and unconventional oil and gas systems, and approaching those topics from both an exploration and a production standpoint. The series is intended to form a diverse library of reference works by describing the current state of research on selected themes, such as certain techniques used in the petroleum geoscience business or regional aspects. All books in the series are written and edited by leading experts actively engaged in the respective field.

The Advances in Oil and Gas Exploration & Production series includes both single and multi-authored books, as well as edited volumes. The Series Editor, Dr. Rudy Swennen (KU Leuven, Belgium), is currently accepting proposals and a proposal form can be obtained from our representative at Springer, Dr. Alexis Vizcaino (Alexis.Vizcaino@springer.com).

More information about this series at http://www.springer.com/series/15228

Neil Craigie

Principles of Elemental Chemostratigraphy

A Practical User Guide

Springer

Neil Craigie
Exploration Department
Saudi Aramco
Dhahran
Saudi Arabia

ISSN 2509-372X ISSN 2509-3738 (electronic)
Advances in Oil and Gas Exploration & Production
ISBN 978-3-319-89052-4 ISBN 978-3-319-71216-1 (eBook)
https://doi.org/10.1007/978-3-319-71216-1

Printed on acid-free paper

This Springer imprint is published by Springer Nature
The registered company is Springer International Publishing AG
The registered company address is: Gewerbestrasse 11, 6330 Cham, Switzerland

Preface

After spending several years studying towards B.Sc. (Hons) and M.Sc. degrees, I developed an interest in chemostratigraphy as I felt that 'true' innovation is most easily found at the borders of science. Chemostratigraphy may be loosely defined as a correlation technique involving the application of inorganic geochemical data, but this masks the fact that it involves elements of geochemistry, sedimentology and other disciplines. I was very fortunate to have been given the chance to work on a Ph.D. study in Chemostratigraphy at the University of Aberdeen from 1994 to 1998, under the supervision of Dr. Nigel Trewin and Dr. Malcolm Hole. At that time, very few publications existed on chemostratigraphy, and I gleaned much information through the painstaking process of trial and error.

After graduation, I worked as a geologist for an oil service company based in Croydon, Surrey, UK but I always yearned to apply the technique of chemostratigraphy in hydrocarbon reservoirs, rather than on the field outcrop sections I had been studying during my Ph.D. years. Finally, I got the opportunity to do this in the year 2000 when I joined Chemostrat Ltd., a small consultancy based in mid-Wales, UK. At that time, the company only employed four persons, but was in the process of expanding as they had signed an alliance agreement with Halliburton to develop a wellsite chemostratigraphy service. Not only was I able to work alongside Dr. Tim Pearce and Dr. Ken Ratcliffe (two of the Worlds leading experts on chemostratigraphy at that time) on conventional laboratory-based projects, I was also involved in the establishment of the first wellsite chemostratigraphy service. After working for Chemostrat Ltd. for 4 years, I was employed as a chemostratigrapher for Ichron Ltd. (2004–2012) and then Saudi Aramco (2012–present). I have watched chemostratigraphy grow in stature and popularity from 1994 to present, exemplified by a sharp increase in the number of publications from 2009. The level of interest is also shown by a rise in the number of companies offering chemostratigraphy as a service. In the late 1990s, for instance, nearly all chemostratigraphy projects were completed by only two oil service companies—Chemostrat Ltd. and IRES Ltd, both being based in the UK. At the time of writing, the list of companies undertaking Chemostratigraphy projects include Chemostrat Ltd., IRES Ltd., Weatherford, Halliburton, Schlumberger, CGG, RPS Ichron, Stratochem and others.

Despite this meteoric growth in popularity, very few, if any, publications provide detailed information on sampling, sample preparation, analytical techniques and interpretive methodologies. Instead, most describe the final results and interpretations in depth without explaining how they were achieved. Similarly, very few papers describe any of the multitude of potential pitfalls associated with each step of a chemostratigraphy study, from the sampling of core/cuttings to interpreting data. Chemostratigraphy is almost never taught at undergraduate level, making it very difficult for students to acquire a sufficient level of knowledge on the subject. Though some companies offer formal training in chemostratigraphy, they do not, as a general rule, make these courses available to individuals working outwith the company.

I take a pragmatic long-term view that the only way the subject will develop as a whole will be for this information to be released to the wider geological community. Having taught chemostratigraphy to many young professionals, the following question often crops up—'where can I find a textbook detailing the methodologies and information I require in order to become a chemostratigrapher?' Up until now, it is my opinion that no such textbook exists. It is hoped that this book will inspire young professionals and others to take an interest in the subject.

Dhahran, Saudi Arabia Neil Craigie

Acknowledgements

I would first of all like to thank my employer, Saudi Aramco, for giving me permission to publish this work in the first instance and the following individuals working for the company who provided reviews of this work prior to submission to the publisher: Christian Scheibe, Nigel Hooker, Marco Vecoli, Abdullah Dhubeeb and Conrad Allen. As a chemostratigrapher, I found the comments of Christian Scheibe to be particularly useful. Strong words of thanks also go to Ian Francis and Alexis Vizcaino of Springer Nature. Without their support, I very much doubt it would have been possible to write this book.

I would like to thank several former work colleagues at RPS Ichron including Michael Leeds, Benjamin Lee and Daniel Atkin. The many hours we spent discussing the intricacies of reservoir correlation, and time we endured on oil rigs working on wellsite chemostratigraphy studies, were truly inspirational. Gratitude should go to former colleagues at Chemostrat Ltd., including Tim Pearce, Ken Ratcliffe and John Martin. Much of my knowledge in chemostratigraphy was gained from working with these individuals in the early 2000s. I completed my Ph.D. in chemostratigraphy at the University of Aberdeen in 1998 and would like to take this opportunity to thank my supervisors, Nigel Trewin and Malcolm Hole. I will never forget the advice and training they gave me from 1994 to 1998, and they are, at least in part, responsible for inspiring me to form a career in chemostratigraphy.

Last but by no means least, I would like to thank my wife Kathryn Craigie and my daughter Iona Craigie. They have had to get used to my absences during the many evenings and weekends spent writing this book, but have remained supportive throughout this process.

Contents

1 Introduction

Abstract

Chemostratigraphy may be defined as a correlation technique involving the application of inorganic geochemical data. It has become very popular as a hydrocarbon reservoir correlation tool over that last 20 years, mainly owing to improvements in analytical techniques, as it is now possible to analyze core, cuttings and field outcrop samples with a degree of efficiency that was not possible prior in the 1970's and 1980s. The advantages of chemostratigraphy are that it can be applied to sediments of any lithology, any age, found in any location and deposited in any environment. In addition to this, it may be utilized on core, sidewall core, cuttings and field outcrop samples, with an equal degree of ease. Perhaps the geatest benefit of chemostratigraphy, however, is that it offers a higher level of resolution than most other techniques. Using ICP (Inductively Coupled Plasma) and XRF (X-Ray Fluorescence) technologies, it is possible to acquire good quality data for around 55 and 42 elements respectively, in the range Na-U in the periodic table. This results in at least 42 potential variables that can be used for chemostratigraphic characterization and correlation. In fact, the number often exceeds 250 when elemental ratios are taken into account. The high levels of resolution are also explained by the low limits of detection of analytical instruments. Using modern XRF spectrometers it is possible to measure the abundances of most elements in concentrations of 1 ppm or less and, given that 1wt % is approximately equivalent to 10,000 ppm, it is obvious that minor changes in mineralogy and geochemistry can be recorded. The ICP technique offers even better levels of resolution/detection, with some trace elements and REE recorded at levels of less than 1 ppb. Consequently, very subtle changes in the mineralogy, such as variations in the distribution of specific heavy minerals and other accessory minerals occurring in the region of 0.1–2%, can be modeled using whole rock geochemical data. In spite of the aforementioned benefits of using chemostratigraphy as a reservoir correlation tool, numerous 'pitfalls' exist at every stage of a chemostratigraphy project. The study is likely to fail, for instance, if the sampling strategy is inadequate, or if cuttings samples have not been washed prior to analysis. An inability of the chemostratigrapher to

N. Craigie, *Principles of Elemental Chemostratigraphy*, Advances in Oil and Gas Exploration & Production, https://doi.org/10.1007/978-3-319-71216-1_1

recognize poor quality data or analytical drift are also potential reasons for failure. In addition to this, a number of challenges are associated with the interpretation of inorganic geochemical data and the proposition of correlation schemes. It is hoped that the information outlined in the following chapters will serve as a 'step-by-step' guide to chemostratigraphy and inspire the next generation of chemostratigraphers to further develop the technique.

1.1 Development of Elemental Chemostratigraphy

The definition of chemostratigraphy is stated as "the study of the chemical variations within sedimentary sequences to determine stratigraphic relationships" (https://en.wikipedia.org/wiki/Chemostratigraphy). However, this definition in the Wikipedia website may be slightly misleading as the technique can also be used as a stratigraphic aid in igneous rocks (Ramkumar 2015). Nevertheless, given that nearly every chemostratigraphy study is performed on sediments or sedimentary rocks, the application of the technique in other rock types is not covered in the present book.

The subject of chemostratigraphy can be broadly subdivided into isotope and elemental chemostratigraphy (Ramkumar 2015; Sial et al. 2015), but the present publication focuses on the latter. The applications of isotopes in chemostratigraphy would warrant the publication of a separate book but, for detailed information on this, the reader is referred to the excellent works of Veizer et al. (1980), Lindsay and Brasier (2002), Calver et al. (2004), Kaufman et al. (2007), Allen and Etienne (2008), Saltzman and Thomas (2012), Sial et al. (2015) and others. In the same way that biostratigraphy was 'morphed' into the subfields of palynology, micropaleontology and nanopaleontology, it is felt that chemostratigraphy has developed to such an extent that it should be viewed in terms of two distinct sub-categories: elemental and isotope chemostratigraphy. Much of the information presented in the following chapters relates to the application of elemental chemostratigraphy to correlate hydrocarbon-bearing reservoirs, but its use is certainly not restricted to the petroleum sector as methodologies are equally applicable to non-reservoir intervals, including field outcrop studies.

Chemostratigraphy was pioneered in the 1970s but few papers were published until the 1980s (e.g. Kishida and Riccio 1980; Zhengyong et al. 1985; Vuorinen et al. 1986) and early 1990s (e.g. Pearce and Jarvis 1991; MacLean and Barret 1993). In a recent search of chemostratigraphy publications in www.sciencedirect.com (detailing works published by Elsevier), the author found only 178 articles from before 1997 (of which most were written after 1992) and less than 340 articles per year were published from 1998 to 2003. A total of only 53, 31 and 40 articles were published in 2005, 2006 and 2007 respectively, whist the number increased to the range 63–84 from 2007–2012. This number exceeded 100 per annum in the period 2013–2016. Similar increases in the number of chemostratigraphy related articles have been recorded on other websites such as www.GeoscienceWorld.org, and www.Springerlink.com (Ramkumar 2015). Note that these figures are based on articles published on both elemental and isotope chemostratigraphy but, a cursory examination of titles and abstracts revealed the number of articles written in both subjects have increased exponentially over the years. Provision of a full list of papers published on elemental chemostratigraphy is beyond the scope of the present study, but some of the more noteworthy contributions are Pearce and Jarvis (1991), Pearce et al. (1999, 2005), Ratcliffe et al. (2006), Davies et al. (2013), Sano et al. (2013), Holmes et al. (2015), Madhavaraju et al. (2015),

Ramkumar (2015), Craigie (2015a, b), Craigie et al. (2016a, b), and Craigie and Polo (2017).

Other signs of the increase in popularity of chemostratigraphy are obvious to persons involved in the petroleum sector. Prior to 1992, nearly all chemostratigraphy work was completed by only two companies—Geochem Group Ltd, and Kingston Geological Services Ltd. From 1993 to 1999, the companies Chemostrat Ltd and IRES Ltd became the leading providers of this service. It is noteworthy, however, that these four companies were based entirely in the UK and, though they completed some international studies, much of their work was focused on projects in the North Sea region. Since that time, the number of companies involved in chemostratigraphy have increased sharply and now include Chemostrat Ltd, RPS Ichron, CGG, Halliburton, Schlumberger, Baker Hughes, IRES, StratoChem and others. Most of these companies complete studies throughout the globe and some have five or more regional offices. It is also apparent that at least some companies have witnessed a sharp increase in the number of staff that they employ in this field. For example, Chemostrat Ltd only employed 4 persons when the author joined the company in the year 2000, but employed around 100 by 2015 (Dr Tim Pearce, Director of Chemostrat Ltd—personal communication). Many oil companies also see the benefit of chemostratigraphy and employ specialists in this field.

The reasons for this increase may be manifold but arguably the main 'driver' involved improvements in analytical technology. ICP (Inductively Coupled Plasma) and XRF (X-ray Fluorescence) instruments are now capable of acquiring data for more than 40 elements in a very short timeframe. Furthermore, levels of analytical accuracy (closeness of result to a known value), precision (i.e. repeatability of results) and detection limits (typically 1 ppm by XRF and 1 ppb by ICP) have improved over the years. Related to this is the development of benchtop XRF spectrometers that became small and robust enough to be utilized at wellsite where data could be acquired in 'near real time'. This service was first introduced by the Chemostrat: Halliburton business alliance in the early 2000s but was then followed by IRES Ltd, RPS Ichron, Schlumberger and others. Another factor contributing to the popularity of chemostratigraphy may relate to improvements in software applications since the early 1990s. Prior to this time it would have taken much longer to achieve relatively simple tasks such as plotting profiles for individual elements, or creating binary diagrams, and some of this work may have had to be done by hand. With the advent of industry-standard applications such as ODM (Oilfield Data Manager), IC, Petrel and others, however, these tasks became much easier to accomplish. Ramkumar (2015) cites the economic liberalization in many countries following the end of the Cold War in the late 1980s to early 1990s as a possible explanation for the increase in usage of chemostratigraphy. According to Ramkumar (2015) "the end of the Cold War has heightened the energy demand that in turn necessitated hydrocarbon exploration in areas hitherto remained unexplored and/or less explored. The petroleum sector has clearly benefited from the development of chemostratigraphy as a reservoir correlation tool, enabling oil companies to produce improved exploration and development strategies."

In spite of the increase in popularity, chemostratigraphy is hardly ever taught at undergraduate level in universities and colleges and surprisingly little information has been published on methodologies relating to sampling, sample preparation/analysis and the interpretation of data. For instance, most papers illustrate details the 'final' chemostratigraphic scheme proposed by the author(s) but very few explain why certain key element/ratios were used instead of others, how the mineralogical affinities of elements were established, how the samples were selected and then prepared prior to analysis, or how data quality was measured. There is an almost non-existence of courses provided to the wider geological community on chemostratigraphy by oil service companies and academic institutions alike.

The author believes that the subject of chemostratigraphy can only grow and develop through the sharing of information. Perhaps the most appropriate analogies are in the fields of sedimentology and biostratigraphy, where the publication of textbooks has resulted in the flow of knowledge between academic institutions and companies alike, leading to the inspiration of many students to form careers in these subjects. In contrast to both sedimentology and biostratigraphy, very few geologists graduate from universities as specialists in chemostratigraphy. The route that most chemostratigraphers have taken were to join oil service companies as BSc, MSc or PHD graduates, and then to either learn the subject through the process of 'trial and error' or be provided with more formal training.

Many papers have been written on chemostratigraphy, yet the author only found one book dedicated entirely to the study of the science—a publication by Ramkumar (2015). This commences with a brief introduction to the technique of chemostratigraphy (Ramkumar 2015), followed by a series of articles written about on the applications of both elemental and isotope chemostratigraphy in specific field outcrop localities throughout the globe (e.g. Madhavaraju et al. 2015; Morath et al. 2015; Sial et al. 2015). One unfortunate aspect of this book is that these articles are written about different case studies and provide very little information on methodologies. The present book is intended to take the form of a comprehensive users guide to elemental chemostratigraphy, detailing all stages of a chemostratigraphy study from sampling to proposing a chemostratigraphic correlation scheme. It is believed to be the first of its kind to provide such an in depth step-by-step guide to chemostratigraphy, and it is hoped that this will inspire more graduates to form a career and the subject and to develop it as a science in the future.

1.2 Applications and Pitfalls

The obvious advantages of elemental chemostratigraphy are that it can be utilized in sediments of any lithology, any age, found in any location and deposited in any environment. This does not hold true for many other the disciplines. Biostratigraphy, for instance, often works best on mudrocks and siltstones deposited in suboxic-anoxic paleoenvironments. Conventional heavy mineral analysis is a commonly used technique to correlate reservoirs, but is normally only applied on the sand fraction.

A further advantage of the technique is that it can be applied to field outcrop, core, sidewall core and ditch cuttings samples with an equal degree of success. Some other techniques, such as petrography and heavy mineral analysis nearly always work far better on core than on cuttings. Perhaps the greatest advantage of chemostratigraphy, however, is that it offers a higher level of resolution than other techniques. Using ICP and XRF technologies, it is possible to acquire good quality data for around 55 and 42 elements respectively, in the range Na-U in the periodic table. This means that 42 or more variables could be used for the purpose of chemostratigraphic characterization and correlation. In reality the number is far greater than that and exceeds 300 variables when elemental ratios are taken into account. This differs from the more traditional XRD (X-Ray Diffraction) and petrography techniques where the number of variables are normally restricted to less than 15. Similarly, many schemes are based on trends in wireline logs but these do not normally number more than 12 in most wells.

Chemostratigraphy also hold the advantage that incredibly high levels of resolution are achieved by analytical instruments. Using modern XRF spectrometers it is possible to measure the abundances of most elements in concentrations of 1 ppm or less and, given that 1% is equivalent to 10,000 ppm, it is obvious that very subtle changes in mineralogy and geochemistry can be recorded. The ICP technique offers even better levels of resolution/detection, with some trace elements and REE (Rare Earth Elements) recorded at levels of less than 1 ppb. Consequently, very subtle changes in the mineralogy, such as variations in the distribution of specific heavy minerals and other accessory minerals in the region of 0.1–2% can be modeled using geochemical data. By contrast, when applying a standard 'point count' of 300 per slide, it is normally only possible to use petrographic data to recognize changes of 10% or more in the concentration of these minerals. For these reasons, elemental chemostratigraphy offers unique levels of versatility and resolution which cannot be achieved using other techniques.

It is unfortunate that the popularity of the technique and general lack of information on methodologies has resulted in many examples of malpractice, which the author believes has resulted in the private criticism of the technique vented by a small number of individuals working for oil/service companies, and by the public criticism of the technique on the part of academics (e.g. North et al. 2005; Hurst and Morton 2014). The failure of many studies can be traced to inadequate sampling and sampling preparation techniques. For example, a simple failure to wash cuttings to remove drilling additives, the incorrect 'picking' of ditch cutting cuttings to ensure good 'representivity', or failure to grind core/cuttings samples properly can result in the acquisition of poor quality data that cannot be utilized for subsequent chemostratigraphic purposes. Other failures occur during data generation and may relate to the inability of interpreters to recognize good and poor quality data. For example, analytical drift may be identified by analyzing standard reference on a frequent basis, but may never be recognized if these samples are not run in the first instance. Without good quality data, it is almost impossible to produce a chemostratigraphic scheme with any confidence. Sampling, sample preparation procedures, analytical techniques and the quality control of data are covered in Chap. 2 of this book.

In addition of the acquisition of data, there are many challenges in the interpretation of geochemical data. Many inexperienced chemostratigraphers produce schemes without gaining an understanding of the mineralogical affinities of the elements they are using for chemostratigraphic purposes. This should be achieved by comparing geochemical data with mineralogical data derived from XRD, petrographic and/or heavy mineral analysis. It is also important to use statistical and graphical techniques for this purpose. The techniques used to establish element: mineral links and the controls on geochemistry and mineralogy are described in detail in Chap. 3. This section also provides more general information on the geochemistry of minerals that are most likely to be encountered in clastic and carbonate sedimentary rocks.

A common fault made by many inexperienced chemostratigraphers is to interpret profiles plotted for elements and ratios in the same way they would for wireline log data. A pronounced upward increase in the GR (Gamma Ray) profile, for instance, may signify a decrease in grain size and be used for inter-well correlation purposes. Applying the same philosophy to geochemical data can be problematic, however, where a trend may be based on a small number of samples yielding local increases in elements that only exist in the immediate vicinity of the given well and cannot be used for correlation. Failure to take into account variations in grain size, carbonate dilution are other factors may also result in erroneous correlations. The techniques used to interpret data and the many pitfalls relating to the interpretation of data are discussed in Chap. 4.

Chapter 5 of this book explains how chemostratigraphy can be applied to the correlation of carbonate, clastic and unconventional (source rock) reservoirs. A more recent development in chemostratigraphy has been to use the technique as part of a multidisciplinary approach to reservoir correlation. The use of chemostratigraphy, in conjunction with reservoir sedimentology and biostratigraphy has obvious advantages, namely it is possible to produce schemes that are more detailed and confident. Most chemostratigraphic schemes are based on the recognition of changes in provenance, reflected by changes in the chemistry and abundance of detrital minerals. These changes in provenance are normally time-equivalent on a field scale between closely spaced wells (i.e. where wells are less than 10 miles apart), but this does not always hold true on a larger subregional-regional scale, where the same chemostratigraphic units can be deposited in different stratigraphic intervals at different times. This diachroneity can be modeled where biostratigraphic, sedimentological and seismic data are used in conjunction with chemostratigraphy. This inevitably means that the chemostratigrapher should be a good all round geologist with a basic knowledge of reservoir sedimentology, biostratigraphy, seismic interpretation, borehole image techniques and others. The importance of employing such an integrated approach to reservoir correlation is illustrated by some of the case studies presented in this chapter. Chemostratigraphy is increasing used in unconventional (source rock) reservoirs where the objectives are often twofold—to provide a chemostratigraphic scheme, and to use the inorganic geochemical data to model variations in unconventional reservoir properties such as TOC (Total Organic Carbon), clay abundance and brittleness. This subject is also covered in Chap. 5.

Chapter 6 discusses the applications of chemostratigraphy at wellsite. This has been in operation since the early 2000's but very few papers have been published on the subject and most take the form of short conference abstracts. Furthermore, very little, if any, information has been published on the unique challenges faced by the wellsite chemostratigrapher. Though most of the sampling, sample preparation, analytical and data interpretation procedures are applicable to both wellsite and conventional/laboratory based studies, the former presents additional challenges, many of which relate to the necessity for the wellsite chemostratigrapher to work at speed and often under considerable pressure. Adaptation of these procedures is often necessary, as is an ability to communicate with numerous members of staff working on the rig during the duration of drilling. For example, it is important for the chemostratigrapher to converse with the mud engineer to find out the type and quantity of drilling additives to be used and whether this will change during the drilling of the given well. The failure of at least some wellsite chemostratigraphy studies are traced back to a simple lack of communication on the part of the wellsite chemostratigrapher(s). Most of the information presented in the chapter has not been published previously and is based entirely on the authors experience of assisting in the development of the first wellsite chemostratigraphy services, and of working at wellsite during the drilling of numerous wells from the years 2004 to 2012.

1.3 Standardization of Terminology

One frustrating aspect of chemostratigraphy as a subject (both elemental and isotope chemostratigraphy) is that it has been developed in slightly different ways by separate individuals and companies, with terms never having been formally defined or universally accepted. Ramkumar (2015) describes a chemozone as "the unique rock record defined by chemostratigraphic indices and recognizable through unique geochemical signature(s) which in turn helps distinction of a designated rock record from other rock records and also correlation with applicable analogs at appropriate/applicable spatiotemporal scale." The author agrees with this definition and uses it as a general term to describe any unit or zone that has a unique geochemical signature in a given study section. Other authors use different terminologies. For example Montero-Serrano et al. (2010) use the term "chemofacies" which is understood to be synonymous with "chemozone". However, the word "chemofacies" may be slightly confusing as it may infer that chemozones reflect changes in depositional environment, which is often not the case. Sano et al. (2013), use the terms 'packages' and 'units'.

There is not particular problem with using this terminology, except that some confusion may occur as these are also used in sequence stratigraphy studies. The author prefers to use the following terms in decreasing hierarchical order:

Zones

Subzones

Divisions

Subdivisions

Rarely is it necessary to produce more than 4 orders of hierarchy, but the terms 'units', 'subunits' and 'beds' could be introduced if it is. For further information on the use and misuse of terminology relating to chemostratigraphy projects, the reader is referred to the publication of Ramkumar (2015). That fact that many of the terms mentioned by this author have not been universally accepted/applied, and are used in different context by different workers, led the author of the present book to provide no further discussion on this subject. Ultimately, the terms used in chemostratigraphy studies should be of much less importance than adoption of the correct methodologies to the interpretation of inorganic geochemical data.

1.4 The Future of Chemostratigraphy

As with the development of chemostratigraphy from the 1970's to present, it is highly probably that future advancements will be driven by improvements in analytical technology. ICP, XRF and other analytical instruments are becoming smaller, easier to use, more robust with time, and significant improvements have also been made to the speed and quality of data acquisition. Further enhancement may take the form of much smaller instruments available at lower price tags, thus encouraging more companies, academic institutions and individuals alike to take an interest in the technique. It is also believed, however, that the development of chemostratigraphy as a science can only happen if there is a greater willingness on the part of these groups to share information, including innovative interpretation techniques. It is hoped that the following chapters will go some way to improving the availability of information on the subject of elemental chemostratigraphy and may inspire the next generation of chemostratigraphers.

References

Allen, P.A., & Etienne, J.L. (2008). Sedimentary challenge to Snowball Earth. *Nature Geoscience, 1*(12): 817–825.

Calver, C.R., Black, L.P., Everard, J.L., & Seymour, D.B. (2004). U-Pb zircon age constraints in late Neoproterozoic glaciation in Tasmania. *Geology, 32*, 892–896.

Craigie, N. W. (2015a). Applications of chemostratigraphy in cretaceous sediments encountered in the North Central Rub' al-Khali Basin, Saudi Arabia. *Journal of African Earth Sciences, 104*, 27–42.

Craigie, N. W. (2015b). Applications of chemostratigraphy in middle Jurassic unconventional reservoirs in eastern Saudi Arabia. *GeoArabia, 20*(2), 79–110.

Craigie, N. W., Breuer, P., & Khidir, A. (2016a). Chemostratigraphy and biostratigraphy of Devonian, Carboniferous and Permian sediments encountered in eastern Saudi Arabia: An integrated approach to reservoir correlation. *Marine and Petroleum Geology, 72*, 156–178.

Craigie, N. W., Rees, A., MacPherson, K., & Berman, S. (2016b). Chemostratigraphy of the Ordovician Sarah Formation, North West Saudi Arabia: An integrated approach to reservoir correlation. *Marine and Petroleum Geology, 77*, 1056–1080.

Craigie, N. W., & Polo, C. A. (2017). *Applications of chemostratigraphy and sedimentology in a complex reservoir: A case study from the Permo-Carboniferous Unayzah group*. Central Saudi Arabia: Marine and Petroleum Geology (in prep.).

Davies, E. J., Ratcliffe, K. T., Montgomery, P., Pomar, L., Ellwood, B. B., & Wray, D. S. (2013). *Magnetic susceptibility (X stratigraphy) and chemostratography applied to an isolated carbonate platform reef complex; Llucmajor Platform, Mallorca*. SEPM Special Publication dedicated to the deposits, architecture and controls of carbonate margin, slope and basin systems.

Holmes, N., Atkin, D., Mahdi, S., & Ayress, M. (2015). Integrated biostratigraphy and chemical stratigraphy in the development of a reservoir-scale stratigraphic framework for the Sea Lion Field area, North Falkland Basin. *Petroleum Geoscience, 21*, 171–182.

Hurst, A., & Morton, A. (2014). Provenance models: The role of sandstone mineral-chemical stratigraphy. In R. A. Scott, H. R. Smyth, A. C. Morton, & N. Richardson (Eds.) (Vol. 386, pp. 7–26). Geological Society Special Publications.

Kaufman, A.J., Sial, A.N., & Ferreira, V.P. (2007) Preface to Special Issue of Chemical Geology on Precambrian Chemostratigraphy in honor of the late William T. Holser. *Chemical Geology, 237*(1-2): 1–4

Kishida, A., & Riccio, L. (1980). Chemostratigraphy of lava sequences from the Rio Itapicuru Greenstone Belt, Bahia State, Brazil. *Precambrian Research, 1*(2), 161–178.

Lindsay, J. F., & Brasier, M. D. (2000). A carbon isotope reference curve for ca. 1700–1575 Ma, McArthur and Mount Isa Basins, Northern Australia. *Precambrian Research, 99*(3), 271–308.

Maclean, W. H., & Barrett, T. J. (1993). Lithogeochemical techniques using immobile elements. *Journal of Geochemical Exploration, 48,* 109–133.

Madhavaraju, J., Hussain, S. M., Ugeswan, J., Nagarajan, R., Ramasamy, S., & Mahalakshmi, P. (2015). Paleo-redox conditions of the Albian-Danian carbonate rocks of the Cauvery Basin, south India: Implications for chemostratigraphy. In Ramkumar M (Ed.), *Chemostratigraphy—Concepts, techniques and application* (pp. 247–271). Amsterdam: Elsevier.

Montero-Serrano, J. C., Palarea-Albaladejo, J., Martin-Fernandez, J. A., Martinez-Santana, M., & Guitierrez-Martin, J. V. (2010). Sedimentary chemofacies characterization by means of multivariate analysis. *Sedimentary Geology, 228,* 218–228.

Morath, P., Carvert, L., & White, T. (2015). A chemostratigraphic model for the development of parasequences and its application to sequence stratigraphy and paleoceanography, Cretaceous Western Interior Basin, USA. In M. Ramkumar (Ed.), *Chemostratigraphy—Concepts, techniques, and applications* (Chap. 9, pp. 215–246). Amsterdam: Elsevier.

North, C. P., Hole, M. J., & Jones, D. G. (2005). Geochemical correlation in deltaic successions: A reality check. *Geological Society of America Bulletin, 117*(5/6), 620–632.

Pearce, T. J., & Jarvis, I. (1991). Applications of geochemical data to modeling sediment dispersal patterns in distal turbidites: Late quaternary of the Madeira abyssal plain. *Journal of Sedimentary Petrology, 62,* 1112–1129.

Pearce, T. L., Besley, B. M., & Wray, D. S. (1999). Chemostratigraphy: A method to improve interwell correlation in barren sequences—A case study using onshore Duckmantian/Stephanian sequences (West Midlands, UK). *Sedimentary Geology, 124,* 197–220.

Pearce, T. J., Wray, D. S., Ratcliffe, K. T., Wright, D. K., & Moscariella, A. (2005). Chemostratigraphy of the upper carboniferous schooner formation, southern North Sea. In J. D. Colinson, D. J. Evans, D. W. Holiday, & N. S. Jones (Eds.), *Carboniferous hydrocarbon geology: The southern North Sea and surrounding onshore areas* (Vol. 7, pp. 147–164). Yorkshire Geological Society, Occasional Publication Series.

Ramkumar, M. (2015). Toward standardization of terminologies and recognition of chemostratigraphy as a formal stratigraphic method. In M. Ramkumar (Ed.), *Chemostratigraphy—Concepts, techniques, and applications* (Chap. 1, pp. 1–22). Amsterdam: Elsevier.

Ratcliffe, K. T., Martin, J., Pearce, T. J., Hughes, A. D., Lawton, D. E., Wray, D. S., et al. (2006). A regional chemostratigraphically-defined correlation framework for the Late Triassic TAG-I Formation in Blocks 402 and 405a, Algeria. *Petroleum Geoscience, 12,* 3–12.

Sano, J. L., Ratcliffe K. T., & Spain, D. (2013). Chemostratigraphy of the Haynesville Shale. In U. Hammes & J. Gale (Eds.), Geology of the Haynesville gas shale in East Texas and West Louisiana, USA. *AAPG Memoir 105*, 137–154.

Saltzman, M.R., & Thomas, E. (2012). Carbon isotope stratigraphy. In F. M. Gradstein, J. G. Ogg, M. Schmitz, & G. Ogg (Eds.), *The Geological Time Scale*. Elsevier http://dx.doi.org/10.1016/B978-0-444-59425-9.00011-1

Sial, A. N., Gaucher, C., Ferreira, V. P., Pereira, N. S., Cezario, W. S., Chiglino, L., & Lima, H. M. (2015). Isotope and elemental chemostratigraphy. In M. Ramkumar (Ed.), *Chemostratigraphy—Concepts, techniques, and applications* (Chap. 2, pp. 23–64). Amsterdam: Elsevier.

Veizer, J., Holser, W.T., & Wilgus, C.K., 1980. Correlation of $^{13}C/^{12}C$ and $^{34}S/^{32}S$ secular variations. *Geochimica et Cosmochimica Acta,* V. *44*, pp. 579–588.

Vuorinen, A., Alhonen, P., & Suksi, J. (1986). Palaeolimnological and limnogeochemical features in the sedimentary record of the polluted Lake Lippajarvi in Southern Finland. *Environmental Pollution Series A, Ecological and Biological, 41,* 323–362.

Zhengyong, Q., Xiuen, Y., & Mingmei, J. (1985). Chemostratigraphic correlation of the middle and upper Proterozoic between the Yanshan and Shennongjia Basin. *Precambrian Research, V. 29*(1–3), 55–91.

Sampling, Sample Preparation and Analytical Techniques

2

Abstract

Very little information has been published on sampling and sample preparation procedures, yet these are paramount to the success of chemostratigraphy as, without good quality data, it is not possible to produce robust chemostratigraphic schemes. This is especially true of studies involving the analysis of cuttings samples, which have to be washed, sieved, meticulously 'picked' and ground prior to analysis. Core and field outcrop samples are simply described and then ground. The principal analytical techniques used to acquire inorganic geochemical data are ICP-OES (Inductively Coupled Plasma-Optical Emission Spectrometry), ICP-MS (Inductively Coupled Plasma-Mass Spectrometry) and XRF (X-ray Fluorescence). For the best quality data acquired for the largest number of elements, it is recommended that a combination of ICP-OES and ICP-MS are used. The XRF technique produces data for a lesser number of elements, but can be useful where rapid analysis of samples is required. Irrespective of the analytical technique used to acquire data, however, it is important to check the results for accuracy (closeness of result to 'known' values of particular elements), precision (repeatability of results) and detection limits, before the data can be utilized for chemostratigraphic purposes.

2.1 Introduction

The following chapter commences with information on sampling strategies and sample preparation techniques, followed by an outline of the principal analytical techniques used in chemostratigraphy studies. A discussion on data quality assessment is included towards the end of the chapter. Much of the information on sampling and sample preparation has not been published in any detail, hence the lack of referenced works on these subjects. Some publications include information on sample preparation and analytical procedures (e.g. Pearce et al. 2005; Craigie 2015a; Craigie 2015b; Madhavaraju 2015), but this is often summarized and lacking in detail. In spite of this, sample preparation and analytical procedures are considered by the

N. Craigie, *Principles of Elemental Chemostratigraphy*, Advances in Oil and Gas Exploration & Production, https://doi.org/10.1007/978-3-319-71216-1_2

author to be the most important aspects of any chemostratigraphy study. An erroneous interpretation can normally be rectified by a more detailed examination of the data, and by following guidelines relating to the interpretation of the datasets (see Chapters 3–5 for more details). If the data quality is poor in the first instance, however, it may be impossible to rectify the problem and produce a satisfactory interpretation.

2.2 Sampling Practices

The quality of data and resolution of the chemostratigraphic scheme may, to a large extent, be influenced by sampling practices. The quantity of samples may be dictated by budget, project objectives, the vertical extent of study sections, time deliverables and/or the availability of materials. As a general rule, core and cuttings samples should be taken at 3–5 and 10 ft intervals respectively, but this does not hold true for every study. Where very high levels of resolution are required over study sections of 100 ft or less, core samples should be taken at 1–3 ft intervals, though the frequency of sampling with regard to cuttings will always depend on how regularly these were taken during the drilling of the well. Conversely, where lower levels of resolution are required (e.g. over extended study sections of 1,500 ft), it may be acceptable to collect one sample every 20–30 ft in both cored and uncored intervals. In general, there are less restrictions placed on the collection of field outcrop samples, where the sampling strategy will be largely dictated by financial considerations and the timeframe for data acquisition. Nevertheless, a generalized rule of taking one sample per 3–5 ft is recommended in the vast majority of studies.

A common misconception persists that data quality is always better in core than in cuttings samples. The advantages of sampling core are that it is possible to achieve more precise depth control and to dictate the sampling strategy in accordance with the specific objectives of the study. This is particularly useful over relatively short study sections where the number of available cuttings samples could be limited. Furthermore, contamination of core materials is much less likely than for cuttings, which may contain significant amounts of drilling additives and 'cavings'. It is also easier to establish element: mineral links in cored intervals, where the same samples can be subjected to geochemical and mineralogical analysis. In spite of these obvious advantages, there can be some drawbacks of sampling core materials as the data is obtained from specific depth 'points' and often shows very 'spikey' trends when plotted in profile form, reflecting short term variations in source/provenance or depositional environment. On occasions, it is very difficult to model larger scale trends in these parameters unless a '3 point moving average' is applied to the dataset, with further smoothing sometimes required by 'de-spiking' the data (i.e. artificially reducing anomalously high or low values of particular elements/ratios or eliminating these results from the dataset). It is possible that 2-point and 3-point 'moving averages' could be applied to the data but, in the experience of the author, a 3-point 'moving average' works better.

In contrast to core, the geochemical profiles plotted for cuttings samples are generally more 'smooth', reflecting longer term fluctuations in source/provenance or depositional environment. Consequently, it is often easier to recognize trends and place chemostratigraphic boundaries using data acquired for cuttings samples rather than core. To illustrate this, Fig. 2.1 shows profiles plotted for the Zr/Cr, Zr/Ti and K/Na ratios in a well penetrating Paleozoic clastic sediments in Saudi Arabia. Note how variable the trends are in the cored section towards the top of the study interval compared with the smooth ones of the underlying cuttings.

Once the number of samples and sampling frequency has been established, considerable thought should be placed on determining the exact depth at which to take samples. For example, although it may have been decided to take samples every 3 ft over sections of core, it would probably be a mistake to sample at 3 ft intervals precisely. In most studies, the following rules should be considered:

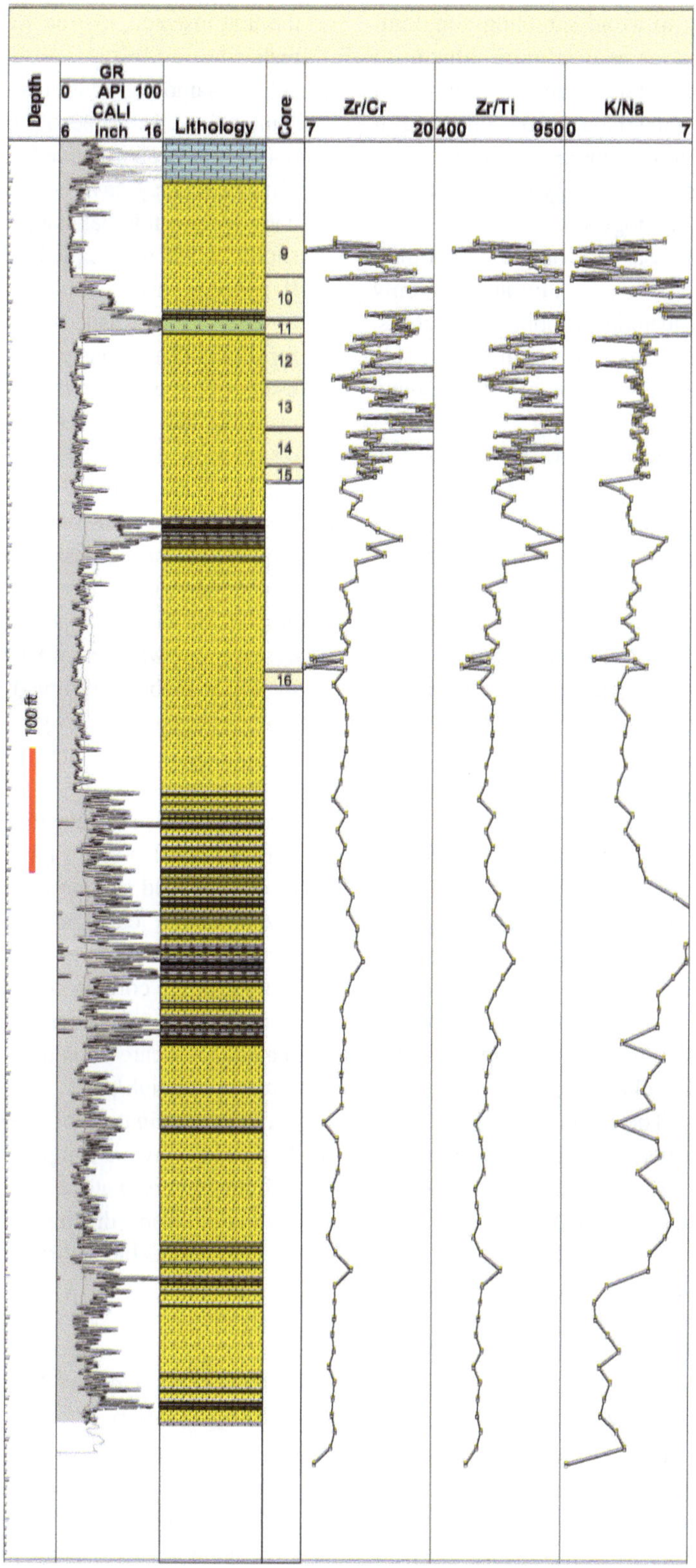

Fig. 2.1 Profiles plotted for Zr/Cr, Zr/Ti and K/Na to illustrate that data acquired from core samples often produce much more 'spikey' trends than that of cuttings samples in the uncored sections of the same well. It is often very difficult to recognize chemozone boundaries where trends are so variable. All depths are log depths in feet

(a) If possible, try to avoid sampling conglomerates and breccias as it is more difficult to obtain a 'representative' sample, in terms of grain size and mineralogy, from this type of material. The most appropriate advice is to sample the core immediately below or above beds of these lithologies.
(b) Where it is necessary to sample conglomerates and breccias, it is advisable to only sample matrix material as opposed to clasts, as the latter may be highly variable throughout the bed in terms of lithology and mineralogy.
(c) As far as possible, avoid taking samples in sections of core that contain 'rubble' as depth control may be poor.
(d) Try to increase the density of samples across significant stratigraphic boundaries such as formation boundaries and unconformities in order to characterise these in detail.
(e) Try to sample materials that are representative of the study section as a whole. For example, if the interval is dominated by medium grained sands, try to ensure that the vast majority of samples are taken from material of this grain size/lithology throughout the study section.
(f) As a general rule, it is advisable to avoid taking samples from unrepresentative materials. For example, if two or three c.10 cm sideritic beds exist in a study section of 400 ft, these should not be sampled as they are rare and unlikely to be correlative between wells. There can be exceptions to this rule (e.g. when sampling shales that may be less than 30 cm thick but represent regionally correlative flooding surfaces), but it is normally recommended to avoid sampling unrepresentative materials.
(g) If there are any doubts relating to the core material (e.g. has the core box been labelled properly), this material should not be sampled.
(h) Always make detailed sample descriptions. Record the lithology, grain size, colour and any sedimentary structures (e.g. bioturbation, cross lamination). This information may not be used in the study but, if it isn't recorded in the first instance, its true value will never be realized.
(i) When sampling cuttings, particularly in 'vintage' wells, some of the original labels may have been removed from the original samples or may have been wiped off sample bags. Where this has happened, the material should never be sampled.
(j) If possible, take samples from 'unwashed', as opposed to 'washed' cuttings. The reason for this is that a coarse sieve may have been used by the technician/mudlogger when washing the latter set and this may have resulted in the loss of 'fines', with the resulting sample being unrepresentative.
(k) To ensure adequate representivity, it is recommended that approximately 8 g or more of core material is taken in each sample. The amount of cuttings material will very much depend on whether washed or unwashed materials are being sampled. It recommended that a minimum of 5–8 g of washed and 15–25 g of unwashed material is taken for each cuttings sample.
(l) In order to ensure adequate 'representivity', it is recommended that a minimum of 2–3 cm^3 of core or field sample material is taken.
(m) Avoid taking core samples in sections of core where ink or paint has been added to the outside of the core on completion of drilling. Sometimes it is unavoidable to sample this material but care should always be taken to remove the ink/paint prior to analysis to avoid contamination of samples.
(n) If possible, avoid sampling from the outer 0.5 cm of core material as this may be contaminated from drilling additives, particularly in permeable sedimentary rocks.

2.3 Initial Stages of Sample Preparation

After the material is sampled, it will be processed before being analysed. For core samples, this is relatively simple as the sample only needs to be ground, as opposed to the washing and picking processes that are applied to cuttings prior to grinding. The grinding process may either be accomplished using a hand held pestle and mortar or by employing a mechanical grinder (Figs. 2.2 and 2.3). Grinding equipment may be fabricated from a variety of material, the most common being agate, tungsten carbide, steel and zirconium. However, for inorganic geochemical studies, agate is favoured. Zirconium grinders would result in considerable contamination of Zr in each sample (one of the most important elements used in chemostratigraphic purposes owing to the association between this element and detrital zircon), while steel grinders would result in contamination of Fe and probably a number of trace elements. Tungsten carbide grinders are often utilised in organic geochemistry studies, but are not advised when analysing samples for their inorganic geochemistry content, owing to contamination of certain trace elements including Co, Cu, W and others. Utilising grinders fabricated from agate may cause minor contamination of Si but, as this element exists in concentrations of greater than 15% in most samples, minor Si contamination in the region of 10 ppm or less is unlikely to have a significant effect on data quality. For this reason, it is advisable to use an agate grinder to micronize both core and cuttings samples (Fig. 2.4).

Although only 0.25 g is required for final ICP (Inductively Coupled Plasma) analysis and 3–5 g for XRF (X-ray Fluorescence), it is recommended that a minimum of 5 g of core material is ground to ensure adequate representivity. One common pitfall in the processing of both core and cuttings samples is not to grind the samples to a fine enough grade for analysis. Ideally each sample should be ground to a very fine powder comprising fragments of 10 μm in size or less. A very crude test to determine if a sample is of this grade involves smearing a small amount of micronized material on the back of a hand with a forefinger. If any granular material is felt, the sample needs to be reground. The consequences of not grinding samples to the correct grade are serious. In XRF analysis, the X-ray beam may contact individual grains, rather than a homogenous ground 'mass', resulting in unrepresentative

Fig. 2.2 Retsch planetary ball mill PM 400 used to micronize rock samples (after www.retsch.com)

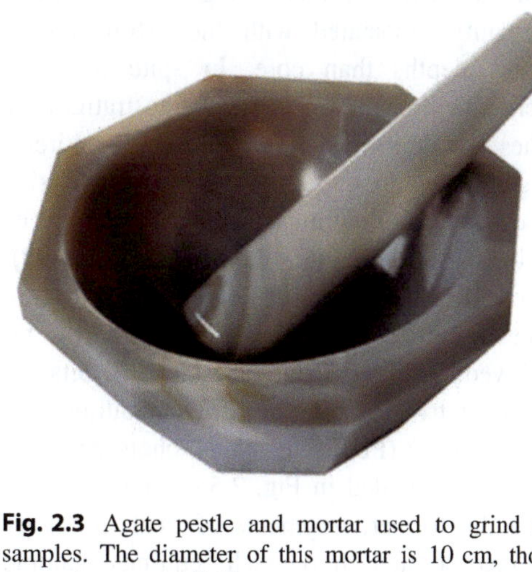

Fig. 2.3 Agate pestle and mortar used to grind rock samples. The diameter of this mortar is 10 cm, though many sizes are available

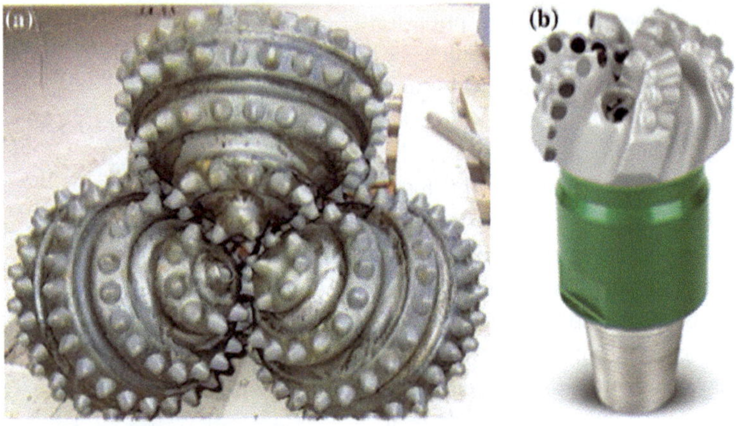

Fig. 2.4 **a** Photograph of a traditional roller cone drill bit, **b** photograph of a more modern PDC drill bit (after http://horizontaltech.com/images/products/categrory-photos/drill-bits-pdc.jpg)

results and the over- or under-estimation of certain elemental abundancies. In ICP analysis, the larger grains may not be fully digested prior to analysis, and this can also result in spurious data being generated from unrepresentative materials.

During the drilling process, well drill cuttings samples are created by the chipping and/or shearing motion of the drill bit. These are then sampled at the drilling rig from the shale shakers by a mudlogger, who does a variety of analysis on subsamples before archiving the samples for future analysis. Based on calculated lag time from drilling to separation at the shale shakers and the mud pump rates, it is possible to calculate depth. However, there is a greater amount of uncertainty associated with the estimation of cuttings depths than core. In spite of these uncertainties, many modern chemostratigraphy studies are performed mostly, or even entirely, using cuttings (e.g. Craigie 2015a). The size shape and type of cuttings depends on the type of lithologies drilled, rate of penetration (ROP), hydrostatic pressure and mud weight, though the greatest influence is the type of drill bit employed. The two main types of drill bits are traditional tri-cone bits and polycrystalline diamond compact (PDC) cutter bits, photographs of which are provided in Fig. 2.5a, b respectively. Using the former, fragments of 'real' lithology are incorporated into each cuttings sample (Fig. 2.5a). The fragments produced by PDC bits are often much smaller and, in many instances, these take the form of reconstituted fragments. A combination of the shear forces ahead of the PDC cutter and fluid hydrostatic pressure results in collapse of the rock and pore network fabric. The resulting residue is then immediately re-cemented into stacked structures reminiscent of vermiculite (Fig. 2.5b). The reconstitutes fragments are considered 'in situ', but are difficult to use for SEM, petrography or digital image analysis as original fabrics/grains are normally destroyed, with the fragments comprising a mixture of sheared formation rock and entrained mud-weighting material (Usman and Meehan 2016). Even determining the dominant lithology of a reconstituted fragment may be difficult, though this may be achieved by breaking the sample up to look at individual grains and by carefully examining wireline log characters. A clue of lithology can also be gleaned from the colour of the sample as lighter colours are normally typical of sandstone or limestone lithologies, whilst darker ones indicate the presence of mudrocks. Caution should, however, be applied when using colour alone to determine lithology.

The process of preparing cuttings samples is similar to that of core but this material needs to be washed and 'picked' prior to being ground. The samples should first be placed in a beaker

Fig. 2.5 **a** Photograph of 'real' llithology fragments derived from drilling of sandstone and mudrock lithologies using a traditional cone bit. The scale bar is 1.6 mm, **b** photograph of reconstituted cuttings fragments derived from a modern PDC drill bit (after Usman and Meehan, 2016)

containing liquid detergent and water added before being mixed with a stirring rod. The sample is then transferred to a 10–30 μm sieve where it is water washed. This material is then placed in a small tray on a 'hotplate' to dry. The temperature of the hotplate is normally set in the range 150–300 °C, though it is recommended that a temperature of less than 250 °C is used to dry organic-rich samples. Samples that have already been washed by the mudlogger at well-site, or after completion of the drilling of the well, may not have to be rewashed but should be carefully examined to determine if they contain any drilling additive contaminants. If there are any doubts over the cleanliness of cuttings samples, they should be rewashed using the aforementioned method.

The next step in the preparation of cuttings samples involves passing the dried sample through two sieves—one with a mesh size of 3 mm to remove the largest fragments, many of which are in the form of 'caved' material, the other comprising a c.10 micron mesh in order to remove the smallest (dust) fragments. A magnet is then passed over the sample to remove any magnetic metallic fragments, normally consisting of broken fragments of the drill bit. The remaining material should be carefully 'picked' to ensure representivity. The lithology at any given depth may be determined by considering sample descriptions and then comparing these with composite and wireline logs. Where ICP-OES and ICP-MS instruments are used to acquire data, it is necessary to pick around 0.25 g for analysis, equating to approximately 200 fragments. In samples of mixed lithology it is essential to only pick material of one dominant lithology in a given sample. For example, it the logs and samples indicate that 70% of the 10 ft section in which a cuttings sample was taken comprises mudrock, only fragments of this lithology should be picked. Each sample should be carefully described in terms of lithology, grain size and colour prior to being ground using an agate grinder. Where fragments of 'real lithology' were drilled, it should be a fairly simple task to select the dominant lithology of a given sample. Where a PDC drill bit was used and reconstituted fragments are present, these should be picked (in preference to 'real' lithology fragments), as they are generally considered to be 'in situ'. Care should, nevertheless, still be taken to select fragments of one dominant lithology where possible. It is noteworthy that most reconsistituted fragments are less than 3 mm in size, but there are occasions where they exceed this and may even occur in the 1–3 cm range. Under these circumstances, it may be necessary to select fragments for analyses that exceed 3 mm in size.

Where the XRF analytical technique is employed, it will be necessary to grind 3–5 g of cuttings material. In most studies it is too impractical and time consuming to 'pick' this volume of individual representative fragments, so the process of 'negative picking' should be employed. This involves spending around 10–15 min picking out non-representative fragments, caved materials and foreign fragments. Please note, however, that the amount of time spent 'picking' the sample varies considerable and depends on the size of fragments, proportion of each lithology and the percentage of foreign materials in the sample. It is very unlikely that all of this material can be removed from the samples, but a basic description should be included, detailing the percentage and type of non-representative fragments recorded in each sample.

2.4 Final Stages of Sample Preparation—ICP Analysis

Once the sample is micronized it may either be placed in a sample cup in powdered form or pressed to create a pellet before being placed in an XRF analyser. For ICP analysis, it is necessary to liquefy the sample. This may be achieved using the aqua regia, acid digestion or alkali fusion techniques, the details and relative merits of which are discussed in the following paragraphs.

2.4.1 Aqua Regia Leaches

Partial (aqua regia leaches) digestions are generally not used for geological studies but are employed for environmental ones. The technique is described by Jarvis (1991) and Jarvis and Jarvis (1992). The process involves leaching samples with a refluxed HCl–HNO_3 mixture at 95 °C for several hours. This is followed by dilution and then analysis. The advantages of this technique are that it is relatively inexpensive and almost 100% of base metals are recovered (including Cu, Cd, Co, Mn, Ni, Pb), with minimal loss of volatile elements. The main disadvantages are that minerals hosting more stable high field strength elements, such as Ti, Zr, Cr, Nb, Ta and Y remain undissolved. Given the importance of acquiring data for these elements in nearly every chemostratigraphy study, this technique is not recommended. In environmental studies, however, this is used on a regular basis where it may not be necessary, or even desirable, to achieve complete digestion of study materials.

2.4.2 Acid Digestion

Acid digestion is the second most commonly used sample preparation technique in ICP studies of rock materials. This is described by Jarvis (1991) and Jarvis and Jarvis (1992). According to Jarvis and Jarvis (1992) "samples are typically attacked with HNO_3–HF–$HClO_4$ mixtures which are then evaporated to incipient dryness. The residues are digested in a further aliquot of the acid mixture and a second evaporation undertaken, followed by final dissolution of samples in 1 M HNO_3." This technique is favoured by many geologists and geochemists and enables near complete digestion of most elements in the range Na-U in the periodic table. Unfortunately the following drawbacks are noted:

(a) Si is volatised as silicon tetrafluoride during the sample preparation and some other elements, including B and Se will also be lost.
(b) Choride ions from remnant $HClO_4$ may result in interference problems with the determination of As and V, and to a lesser degree Cr, Fe, Ga, Ge, Se, Ti and Zn. Potential interferences should be carefully assessed before the data is validated.
(c) Repeated attacks are often required with HNO_3–HF–$HClO_4$ mixtures in order achieve complete digestion of many refractory heavy minerals such as zircon, rutile, ilmenite, staurolite, garnet, tourmaline and others. Even after such attacks, it may prove very difficult, if not impossible, to achieve complete digestion of these minerals. This is considered the principal disadvantage of utilizing acid digestion as a sample preparation technique, and why the author is of the opinion that it is generally not recommended for routine chemostratigraphy studies.

2.4.3 Alkali Fusion

The traditional way of completing alkali fusion preparations is described by Jarvis and Jarvis (1992) and Totland et al. (1992). The powdered sample is first mixed with lithium metaborate ($LiBO_2$) at 1050 °C and then dissolved in 0.5 M HNO_3. Jarvis and Jarvis (1992) point out that some workers prefer to use sodium peroxide (Na_2O_2) or sodium carbonate (Na_2CO_3) fluxes where it is necessary to acquire data for B. As a more rapid and efficient way of preparing samples,

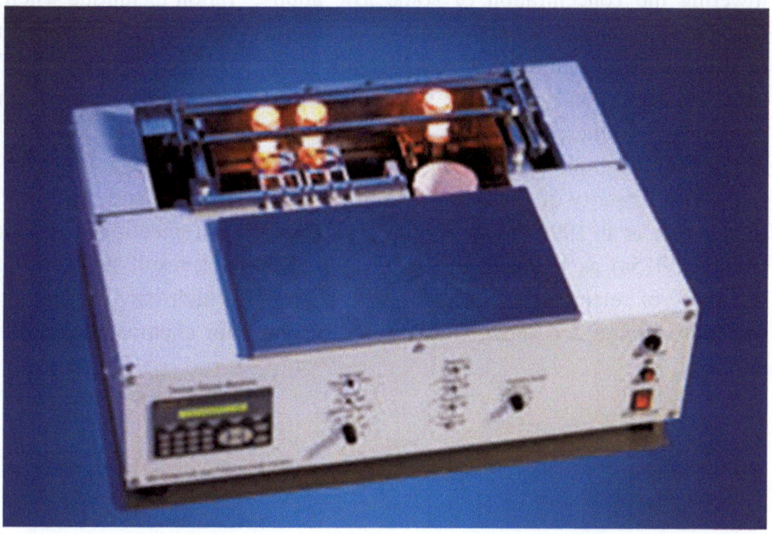

Fig. 2.6 Photograph of Fluxana automatic fusionmachine (from: http://portableas.com/index.php/products/sample-prep-fusion)

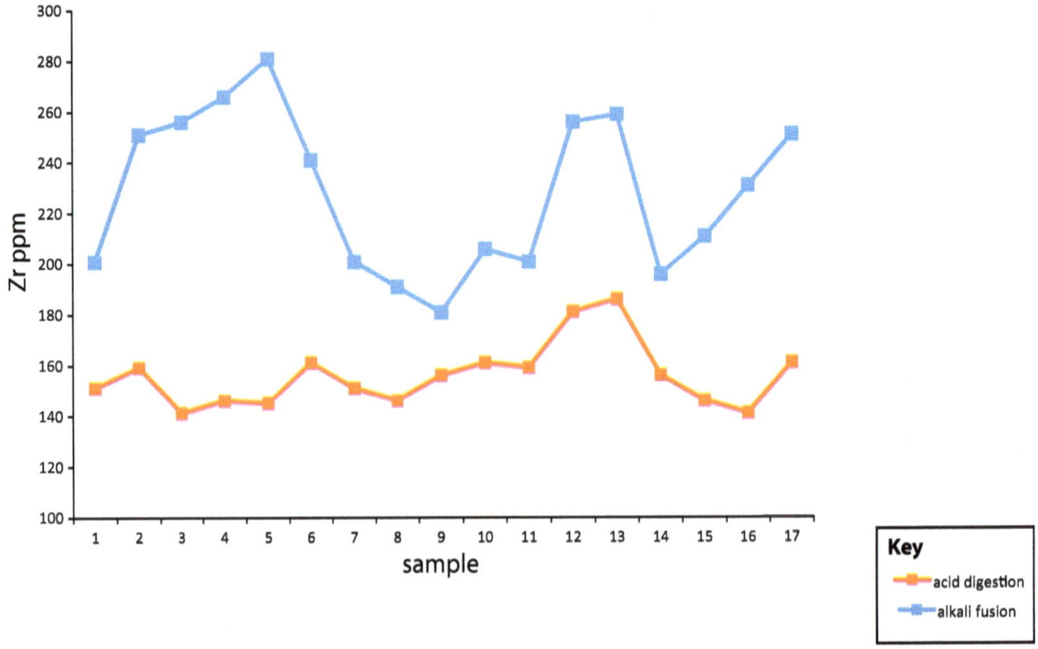

Fig. 2.7 Graph showing the comparison between values of Zr (ppm) when the same samples were subjected to acid digestion and alkali fusion preparations prior to ICP analysis. In both instances, the data for Zr were acquired by ICP-OES. Clearly, recovery of Zr was much higher using the alkali fusion technique and there is very little resemblance in the curves displayed in this graph

an automatic fussion machine can be used in preference to the furnace to prepare the melt prior to dissolvation in acid (Fig. 2.6). This is a much quicker and more efficient way to prepare samples, though both techniques are acceptable.

The main disadvantage of using alkali fusion is that increased dilution factors may result in lowering the concentration of some low abundance trace elements below the limit of detection. In addition to this, the technique is much more time consuming and expensive than acid digestion. In spite of these drawbacks, however, this is the preferred sample preparation technique used in the vast majority of chemostratigraphy studies (e.g. Pearce et al. 1999, 2005; Wright et al. 2010; Craigie 2015a) as it guarantees a much higher recovery of elements associated with heavy minerals. Figure 2.7 shows the results of the element Zr by ICP-OES analysis perfomed on 17 samples. The most obvious feature of this graph is that recovery of Zr is much higher when using the alkali fusion technique. In addition to this, there is very little comparison between the trends of the two curves, providing further confirmation that acid digestion is an inferior sample preparation technique when attempting to acquire data for stable high field strength elements.

The processes of washing, picking and grinding samples (in addition to liquefying samples for ICP analysis and preparing discs/pellets for XRF) can be very time consuming and tedious, but failure to follow the aforementioned practices will, more often than not, result in the acquisition of poor quality data, which may be difficult or even impossible to use for chemostratigraphic purposes.

2.5 Commonly Used Analytical Techniques

The following paragraphs describe the most commonly used instruments for the acquisition of elemental data. These include ICP-OES (Inductively Coupled Plasma-Optical Emission Spectrometry), ICP-MS (Inductively Coupled Plasma-Mass Spectrometry) and XRF (X-ray Fluorescence). The purpose of this section is not to describe the instruments in terms of their underlying physics and engineering in any detail, but to offer a brief overview of how each one operates and the relative benefits and drawbacks of employing these technologies to acquire inorganic geochemical data. Section 2.5.1 essentially summarises the paper of Jarvis and Jarvis (1992) and, for more detailed information on ICP technology, the reader is encouraged to research this excellent work and references therein. Advancements in ICP technology have taken place since this paper was written over 20 years ago but, in the opinion of the author, few publications provide such a user friendly account of ICP for the non-engineer/physicist.

2.5.1 ICP Technology

A key characteristic of ICP-OES and ICP-MS is that both involve the generation of a plasma which may be defined as any luminous volume of partially ionised gas. In spectroscopy, however, the term is generally restricted to electrically excited discharges. Figures 2.8 and 2.9 show basic diagrammatic illustrations of

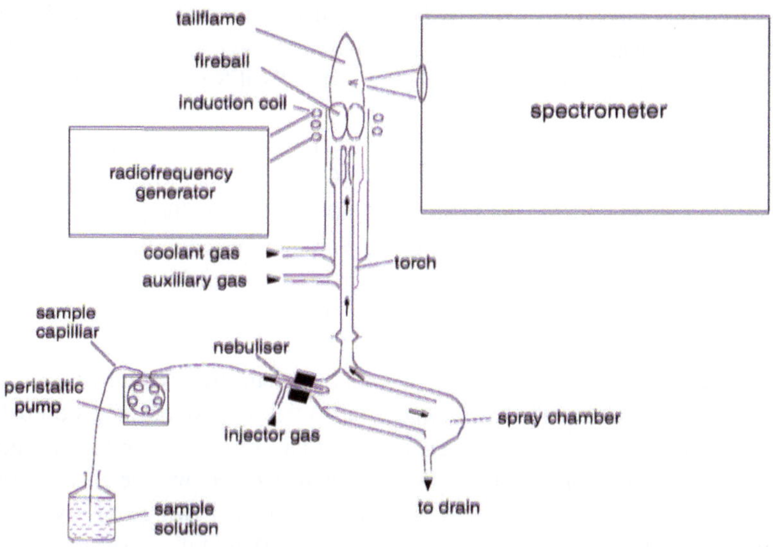

Fig. 2.8 Diagrammatic illustration of ICP-OES (after Jarvis and Jarvis 1992). See text for further explanation

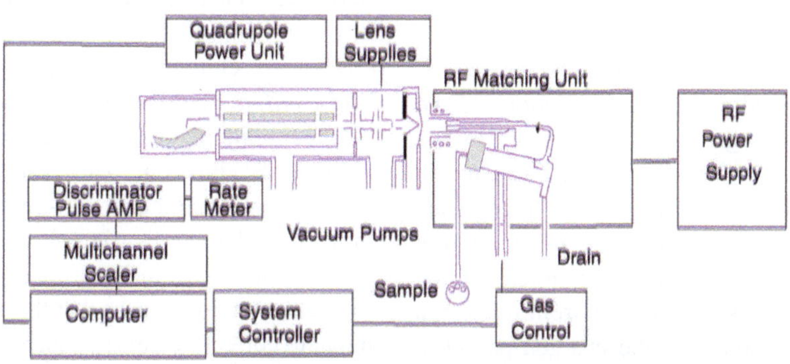

Fig. 2.9 Diagrammatic illustration of ICP-MS (modified, after Jarvis and Jarvis 1992). See text for further explanation.

ICP-OES and ICP-MS instruments respectively. Radiofrequency (RF) magnetic fields induced by a copper coil wound around the top of a glass torch, are responsible for creating the plasma in ICP spectrometry. The liquefied sample is nebulised to form a fine aerosol, before being transported to the centre of the instrument where it undergoes desolvation (removal of the solvent from the liquid sample), vaporisation to molecular level and dissociation into atoms. Some of the atoms are ionised. The excitement of both atoms and ions in the plasma occurs as they revert to their ground states in the tail flame and light (photons) is emitted. Variation in light are measured using ICP-OES, whilst ICP-MS involves extraction of ions from the plasma into a mass spectrometer for analysis.

2.5.1.1 ICP-OES

In ICP-OES instruments a nebuliser shatters the liquid into fine droplets using an Ar gas stream (Fig. 2.8). On entering the spray chamber, larger fragments (>10 μm) are removed, only enabling the finest spray to pass into the plasma. The larger droplets are incompletely dissociated in the plasma and contribute noise to the analyte signal, hence the reason they are removed in the spray chamber. Baffles within the spray chamber result in condensing of the larger particles, which then flow to drain. The torches of most ICP-OES instruments comprise three concentric glass tubes which are designed to support a stable Ar plasma at its top while allowing a sample to be injected into its core. Coiled around the upper part of the plasma torch, is a water-cooled copper tube which is connected to a RF generator. This creates an oscillating RF magnetic field within the Ar flowing through the torch. A high-voltage spark is used to seed the Ar stream passing through the torch with electrons. These electrons are then subjected to intense oscillations of the radiofrequency magnetic fields generated by the induction coil and collide with Ar atoms causing ionisation. The magnetic fields force each charged particle formed in the gas stream to flow in a closed annular path within the torch. Ohmic heating of the gas stream results from resistance to the induced motion of the charged particles. Such transfer of energy by inductive coupling (avoiding the use of electrodes), causing the formation of a high-temperature plasma, is a definitive feature of both ICP-OES and ICP-MS instruments.

Ar is the preferred plasma gas for the following reasons:

(a) being inert it generally does not react chemically with the samples;
(b) ionisation of elements is achieved as it has a first ionisation energy of 15.75 eV;
(c) it has optical transparency;
(d) its low thermal conductivity causes heat to be retained in the plasma fireball, enabling stable operation of moderate power inputs.

Nitrogen has been utilized for the same purpose, but the problem with using this gas is that it produces increased spectral and isobaric interferences.

In ICP-OES most spectrometers are designed to measure the full range of wavelengths from the upper part of the vacuum ultraviolet (170 nm) to the lower limit of visible light (780 nm). Quantification of elemental abundances are related to the separation of a complex emission spectrum into its component wavelengths, with sufficient sensitivity and resolution to precisely measure light intensities at the characteristic wavelength of each analyte. Central to the acquisition of good quality data is to select the correct line for each element. Complete coincidence of wavelets, or partial overlaps (identified by a 'wing' on a wavelet) should be avoided. Thankfully, the effects of both can be avoided, or at least minimized, by selecting an alternative analytical line. Furthermore, by using narrower slits or higher spectral orders, it may be possible to eliminate/minimize the effects of wing overlap. Another complication, known as continuum interference, results from enhanced background emission. This can be rectified, however, by applying an off-peak background correction. The fact that interferences from concomitants normally increase in a linear manner with concentration, means that they are relatively simple to correct. In general, such corrections are very effective where interference factors are low (<10% of the analyte signal) but the precision and accuracy of determination may be compromised where larger corrections have to be applied. There are a large number of potential emission lines for most elements. For example, over 100 exist for Fe alone. In spite of this, it should be a relatively simple task for the analyst to choose an analyte by selecting one of the c.12 most sensitive lines sufficiently free of interferences. The analyst should consider the intensity of the peak relative to any potential interferences before selecting it. Corrections may not result in any improvement in accuracy and may be unnecessary where the intensity of the peak is large enough. The specific lines chosen will depend on the composition of the samples, together with the resolution and sensitivity of the specific ICP-OES instrument used for analysis.

As a general rule, ICP-OES instruments are employed to acquire data for the major elements Al, Si, Ti, P, Ca, Mg, Mn, K, Na and Fe, though certain high abundance trace elements (e.g. Zr, Cr, Ba, Cu) can also be measured using this technique. Concentrations of low abundance trace elements and REE are much more difficult to measure and these are normally analyzed by ICP-MS.

2.5.1.2 ICP-MS

The basic principals of ICP-MS analysis are very similar to that of ICP-OES. Sample preparation procedures are exactly the same-the liquid sample passes through a nebulizer before entering a spray chamber where the larger particles are removed. An Ar gas plasma forms in the torch as it does in ICP-OES analysis, the main difference being that the torchers are normally positioned vertically in ICP-OES instruments and horizontally in ICP-MS. The latter technique involves the physical extraction of ions from the plasma into a mass spectrometer and measurement using an ion detector. This process is rapid, with a spectrum for the entire mass range of ^{7}Li to ^{238}U achieved in less than 2 min.

Figure 2.9 shows a basic diagrammatic illustration of an ICP-MS instrument. Ions are extracted via a sampling cone aperture, with a 'skimmer' or second aperture mounted behind this. Immediately behind the skimmer is an electrostatic lens system, the purpose of which is to focus the ions which pass into the mass spectrometer, where a quadrupole system acts as a mass filter. Detection of ions is achieved using electron multiplier detectors. Excellent sensitivities are achieved for almost all elements in the periodic table as it is possible to count individual ions and there are very low background signals. Perhaps one of the greatest drawbacks of ICP-MS, however, is that owing to counting losses in the electronics, the response of these detectors becomes non-linear at very high count rates. For this reason, ICP-MS instruments are normally only used to acquire data for relatively low concentration trace elements and REE.

Interferences may cause erroneous results and these fall into three broad categories, the first being polyatomic ions. These form in the plasma and are caused by a reaction between the most abundant ions in the plasma, namely Ar, H, O and N. In addition to this, the acid used to dissolve the powdered sample during preparation causes problems. For example, interferences exist with determination of V and As when using Cl-based acids are used, or in samples containing naturally high levels of Cl. For this reason HNO_3 is the preferred acid to use during the acid digestion or alkali fusion procedures. The second source of interference involves doubly charged ions. The majority of ions forming in the spectra are singly charged (M^+), so the scale of measurement of mass (M/Z or amu) is that for a singly charged ion of each isotope. However, if an element forms a proportion of doubly charged ions (M^{2+}), then the signal from them is seen at a mass to charge ratio of two (i.e. at half the mass of the parent ion). Where the second ionization energy of the element is lower than the first ionization energy of Ar (i.e. the plasma support gas), some doubly charged ions can form. Fortunately, this is rarely a problem as ICP-MS systems can be optimized to give low levels of M^{2+} ions in the region of 0.2–0.5% of the parent ion. A third cause of interference may occur where there is incomplete association in the plasma and/or recombination in the tail flame, resulting in the occurrence of oxide species.

These species exist 16, 17 or 18 amu above the parent ion and may cause an interference. As an example of this effect, the seven naturally occurring isotopes of Ba occur adjacent to the LREE's, so barium oxide species are a potential source in interference on some LREE's.

Despite the list of potential interferences, however, few actually cause a problem in routine ICP-MS analysis. In most instances, alternative isotopes are selected which are free from any interference effects.

In summary, ICP-OES is used to acquire data for major and high abundance trace elements, whilst ICP-MS instruments measure the abundance of trace elements and REE. By using a combination of the two techniques, it is possible to acquire the highest quality data (in terms of accuracy, precision and detection limits) for the largest array of elements (typically 50–55) in the range Na-U in the periodic table. For this reason, combined ICP-OES and ICP-MS is favored in most chemostratigraphy projects. The main drawbacks are as follows:

(a) Expensive as separate ICP-OES and ICP-MS instruments are required.
(b) The instruments are inherently unstable and should only operate under strict laboratory conditions free of any vibrations, where the electricity current in constant and temperatures are strictly controlled.
(c) Machines are highly complex and require skilled operators.
(d) Lengthy sample preparation procedures means that sample throughput is slow in comparison with XRF.
(e) The ICP instruments are large and cumbersome, with a requirement for full wet lab facilities.

Given the physical space needed to accommodate ICP instruments and related laboratory equipment, together with the financial commitment and technical expertise requirements, many chemostratigraphers outsource this work to third parties. When choosing a laboratory to undertake this work, the following questions need to be asked:

(a) Does the lab provide both ICP-OES and ICP-MS analysis? Many provide one but few have both.
(b) What are the manufacturers of the instruments and when were they purchased? Please note that the quality of data may be compromised if instruments are more than 15 years old.
(c) How regular is the servicing and what experience does the engineer have in servicing the instruments? In general it is recommended that ICP instruments are serviced every 12 months by an experienced engineer.
(d) If one of the ICP instruments were to malfunction, what impact would this have on the delivery of results? What arrangement does the third party have with the manufacturer to facilitate the repair of the instrument in a timely fashion? Does the lab have replacement instruments to run samples in the event of such a malfunction?
(e) How experienced is the lab technician in the analysis of rock samples? It is recommended that labs are only used where the technician(s) has vast experience of analysing rock samples. There are many ICP laboratories, but most only have experience in the analysis of water and/or environmental materials.
(f) Ask for a complete list of elements that can be analysed and approximate detection limits for each one.
(g) Confirm that the alkali fusion technique will be employed. If so, then ask the technician to describe this procedure in full. If the alkali fusion procedure is not used, or the technician is unable/unwilling to describe it, then choose to analyse your samples in another laboratory.
(h) What standard reference materials are analysed to check on instrumental drift, accuracy and precision? Ask to view the result of repeated analysis of such reference standards.
(i) What is the sample throughput and roughly how many samples are expected to analysed per week and per month?
(j) What system(s) do the lab operate to ensure the efficient tracking of samples? Are ISO

standards of data management, or equivalent, adhered to? Will unsampled materials be returned?
(k) What clients have the lab worked for and would it be possible to share some results?
(l) Is there more than one analyst/technician available to analyse samples? The reason this question should be asked is that an individual may have to take time off work owing to sickness, vacations etc. There should always be more than one person capable of/available to analyse samples.
(m) What is the cost of analysis? Cost is a major consideration when outsourcing most ICP analytical work, but please also be advised that the cheapest option is not always the best. When choosing a laboratory to analyse samples, points a-l should be assessed in conjunction with financial considerations.

Note that many of the aforementioned questions relate to logistical issues such as servicing contracts and staff availability, yet these are often the factors that 'make or break' a project and are overlooked when choosing a laboratory to analyse samples.

2.5.2 X-ray Fluorescence

The discovery of X-rays took place in the late 19th century, though it wasn't until 1948 that the first commercial X-ray Fluorescence spectrometer was built (Beckhoff et al. 2006). Since that time XRF has become the instrument of choice for many geochemists and chemostratigraphers over the last four decades owing to the robustness of instruments and relative ease of use. Furthermore, XRF machines are generally less expensive than ICP and preparation procedures are much simpler, enabling the rapid analysis of samples. The following paragraphs provide a somewhat brief introduction to the theory of XRF and the relative merits of the most commonly utilized XRF instruments. For more detailed information on XRF, the reader is referred to the work of Beckhoff et al. (2006) and references therein. Most of the information presented in the following paragraphs was gleaned from this publication and from the website http://learnxrf.com/category/xrf-principles/.

2.5.2.1 XRF Theory

Ejection of an electron from its atomic orbital can occur by the absorption of a light wave (photon) of sufficient energy, though the energy of the photon (hv) must be greater than that with which the electron is bound to the nucleus of the atom. After ejection of an inner orbital electron, an electron from a higher energy level orbital will be transferred to the lower energy level orbital. A photon may be emitted from the atom during this transition and this fluorescent light is called the characteristic X-ray of the element. There is equality between the energy of the emitted photon and the difference in energies between the two orbitals occupied by the electron making the transition. In each element, the energy difference between two specific orbital shells is always the same. Consequently, the photon emitted when an electron moves between these two levels will always have the same energy. The identity of a given element can, therefore, be obtained by determining the energy (wavelength) of the X-ray light (photon) emitted by that element.

The number of photons per unit time (i.e. the count rate or peak intensity) for a particular energy (wavelength) of fluorescent light emitted by an element is related to the amount of that analyte in the sample. Based on peak intensities, it is then possible to quantify the concentration of specific elements.

The two principal types of XRF instrument used in chemostratigraphy studies are Energy Dispersive X-ray Fluorescence (ED-XRF) and Wavelength Dispersive X-ray Fluorescence (WD-XRF), the details of which are discussed in Sects. 2.5.2.2 and 2.5.2.3 respectively.

2.5.2.2 Energy Dispersive X-ray Fluorescence (ED-XRF)

The basic principal of ED-XRF is that X-rays are 'fired' onto the surface of the sample, with X-rays emitted from that sample detected by an X-ray detector (Fig. 2.10). The detector and detector electrons resolve spectral peaks due to

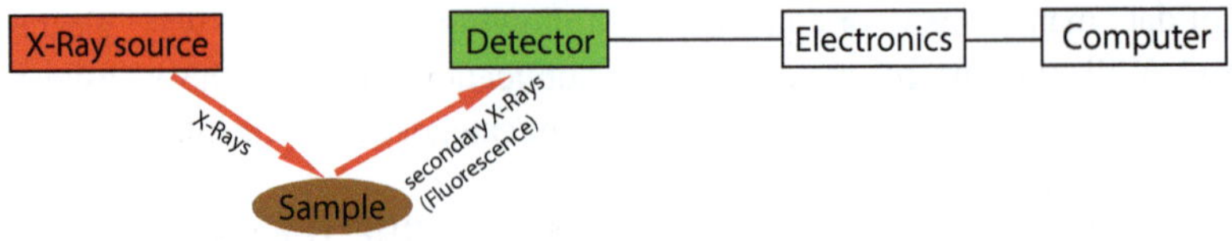

Fig. 2.10 Basic principle of EDXRF analysis

different energy X-rays. According to the website http://learnxrf.com/category/xrf-principles/, "It wasn't until the 1960s and early 1970s that electronics had developed to the point that high-resolution detectors, like lithium drifted silicon, Si(Li), could be made and installed in commercial devices. Computers were also a necessity for the success of ED-XRF even if they were often as large as the instrument itself."

The two most important features of any XRF instrument are the X-ray source and detector, with the former generally taking the form of a 50–60 kV, 50–300 W X-ray tube. Detectors are designed to produce electrical pulses that vary with the energy of the incident X-rays. Different types of detectors are used by each manufacturer but include Peltier cooled Si(Li), proportional counters, Peltier cooled PIN diode detectors and sodium iodide (NaI) detectors.

In addition to the source and detectors, many ED-XRF instruments employ X-ray tube filters to absorb some energies of source X-rays more than others, so that counts are reduced in the region of interest, while producing a peak that is well suited to exciting the elements of interest. As an alternative to filters, secondary targets may be used, where the secondary target material is excited by the primary X-rays from the X-ray tube, emitting secondary X-rays that are characteristic of the elemental composition of the target. Secondary targets require roughly 100 times more primary X-ray intensity but generally yield lower background and better excitation than the filter option. A more recent development in XRF technology is the application of polarizing targets as secondary targets. The advantage of polarizing XRF is that X-rays are partially polarized when they are scattered off a surface, thus increasing the signal to noise ratio. In order to further minimize the scatter, and hence the background at the detector, the target and sample are placed on orthogonal axis. Many ED-XRF instrument can be used to acquire data for core, cuttings and field samples but, in recent years, most chemostratigraphers have found polarizing XRF instruments, such as the Rigaku NEX CG (Fig. 2.11), or the Ametek Xepos to be particularly useful. These machines are capable of acquiring data for trace elements in concentrations of around 1 ppm or less, and generally produce data of high precision and accuracy, provided they are properly calibrated for the specific analyzed matrix.

Over the last 20 years, manufactures have responded to consumer demands to make ED-XRF instruments smaller, more robust and

Fig. 2.11 Photograph of the Rigaku NEX-CG polarization EDXRF instrument (*Courtesy* of Rigaku)

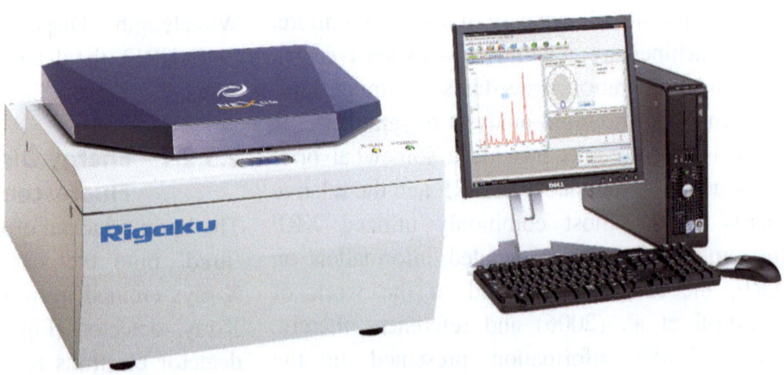

easier to use. Many are small enough to be transported to field locations (including wellsite), enabling the rapid acquisition of data. Instruments such as the Oxford Instruments Twin Ex, Rigaku NEX CG or the Ametek Xepos weigh less than 80 kg and are small enough to operate in confined spaces, which is often the case at wellsite. The applications of wellsite chemostratigraphy using data generated by EDXRF are discussed in more detail in Chap. 6.

In addition to this, some companies (e.g. Innov-X, Thermoscientific, Bruker) have developed hand held ED-XRF instruments. Weighing around 2 kg or less, these instruments can be used for fieldwork studies, in core laboratories or at wellsite. They are battery powered and normally less that 5 W (compared with other ED-XRF machine which are c.50 W), making them safe to handle. The instruments operate with silicon drift detectors. As they have large surface areas and better resolutions than their alternatives (i.e. Si-PIN and CdTe detectors), this means that they can differentiate between the broader X-ray peaks as well as detect and measure lower energy X-rays (Longoni et al. 1998; Struder et al. 1998; Young et al. 2016).

Although there are some advantages to using handheld EDXRF instruments, such as ease of use, low cost of equipment, speed of analysis, limited sample preparation and ease of transportation, they are less commonly used in routine chemostratigraphy projects owing to the lower beam energies under which they operate. This results in lower energies of the secondary returning secondary X-rays, making it more difficult to detect and measure elements, particularly lighter ones such as Na and Mg (Young et al. 2016). Depending on the sample, matrix and calibration method, it may not be possible to acquire data for low concentration trace elements (e.g. Ta and Cs where concentrations are often less than 5 ppm in sedimentary rocks) and REE. Furthermore, analytical precision (i.e. repeatability of results) is often compromised owing to such low energy sources (Young et al. 2016).

In a detailed test of handheld ED XRF instruments, Young et al. (2016) found differences in data quality, depending on whether a smooth (sawed) or rough (natural) surface had been analyzed in many instances, with the former producing the best results. For example, one sample produced 54 and 43% Si for smooth and rough surfaces respectively.

In summary, handheld XRF instruments are considered useful for quick-look evaluations or in studies where there are no requirements to generate data for a large number of trace elements/REE, and where there is no need for low detection limits (i.e. <10 ppm) or high levels of precision, accuracy and resolution. For example, it may be possible to use this technique to recognize differences in bulk lithology (e.g. sandstone vs. shale) and mineralogy without utilizing the data for produce a high resolution chemostratigraphic scheme, based on more subtle change in accessory heavy minerals. Handheld EDXRF instruments may be further developed in the future but, at the time of writing, they are not recommended for acquiring data in the vast majority of chemostratigraphy projects.

2.5.2.3 Wavelength Dispersive X-ray Fluorescence (WD-XRF)

WD-XRF instruments (Fig. 2.12) have been in used in a number of geochemical studies and the earliest XRF spectrometers were WD-XRF instruments, as the method works without high resolution solid state detectors. Instead of this, they utilize diffractive optics to provide high spectral resolution. Owing to the fact that the overall efficiency of WD-XRF instruments is often low, X-ray tubes are normally rated at 1–4 kW, though some lower powered systems operate at 50–200 W (http://learnxrf.com/category/xrf-principles/). The main difference between WD-XRF and EDXRF is that in the former, the fluorescent X-rays emitted by the sample are directed into a diffraction grating monochromator (Fig. 2.13). Further improvement to the WD X-ray system is achieved by the use of collimators to limit the angular spread of X-rays. The detector is not relied on for the systems resolution, so it can be a proportional counter or another low-resolution counter capable

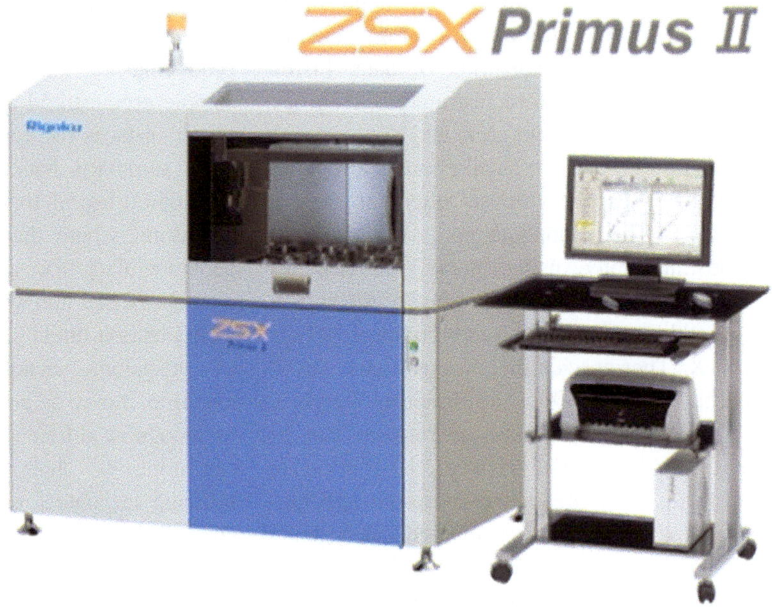

Fig. 2.12 Photograph of the Rigaku ZSX Primus II WDXRF instrument (*Courtesy* of Rigaku)

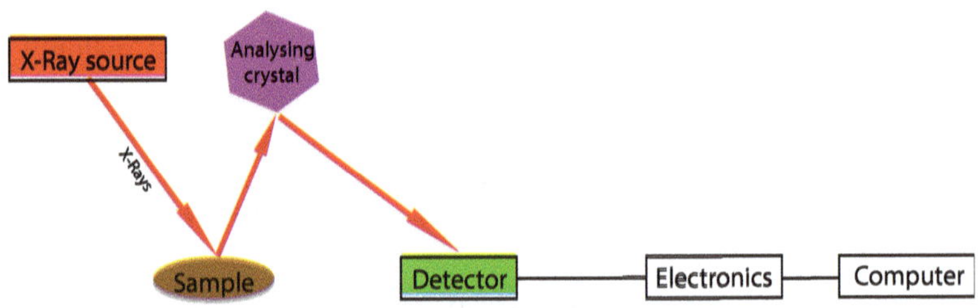

Fig. 2.13 Basic principle of WDXRF analysis

of detecting a million or more counts per second (http://learnxrf.com/category/xrf-principles/).

WD-XRF is ideal for analyzing single elements as all of the components can be set to form a fixed single WD X-ray channel. A number of fixed single channels exist in a 'simultaneous WD X-ray analyzer' and these are usually formed in a circle around the sample with the X-ray tube facing upward in the middle. Some WD-XRF instruments have a 'sequential WD X-ray analyzer' which uses a goniometer to allow the angle to be changed, enabling one element after another to be measured in sequence. In addition to this, combined sequential/simultaneous instruments are also available (http://learnxrf.com/category/xrf-principles/). In general, the number of analyzed elements and data quality (in terms of accuracy, precision and detection limits) is superior by WDXRF than EDXRF, though the instruments are normally much more expensive and, being larger and less 'robust', are generally only suitable for conventional laboratory-based (as opposed to field-based) chemostratigraphy studies.

2.5.2.4 Core Scanning EDXRF

A relatively recent development in XRF technology is core scanning energy-dispersive XRF instruments. These enable core to be analysed rapidly at the millimetre to submillimetre scale and can process cores up to 1.8 m in length. According to Kelloway et al. (2014) the volume of data generated by such systems is significantly higher than that from traditional bulk sampling. For example, analysis of a 1.8 m core every 100 μm would potentially result in data for

18,000 separate analysis evenly spaced along the core. This may be particularly useful where it is necessary to model changes in geochemistry/mineralogy on a small scale. In unconventional shale reservoirs, for instance, it could be possible to recognise relatively 'short lived' increases in base level resulting in the accumulation of organic matter within beds of less than 1 cm in thickness. Such subtle changes in geochemistry may be very difficult to identify using sampling intervals associated with conventional XRF analysis.

In spite of the obvious benefits of core scanning EDXRF, however, there are some challenges. Kelloway et al. (2014) point out that rough surfaces can be a problem. Air between the sample surface and the X-ray detector results in attenuation of fluorescing X-rays, and changing distances between the detector and sample may cause a false impression that the elemental concentration is changing along the core. Ideally the core should be cut along its length to produce a flat surface but this can be difficult in rock types prone to crumbling. This is often true of coals which typically contain brittle organic material interlayered with harder mineral-rich laminae. To alleviate this problem, some automated core scanners are set up in such a way that a constant distance is maintained between the detector and sample surface during measurement.

Perhaps one of the most difficult issues to resolve is where data are acquired from cores containing coarse sand, conglomerates/breccias (especially if they are clast supported) and highly variable grain sizes. In conventional XRF analysis, the core samples are ground to a fine dust prior to analysis, ensuring that a homogeneous 'mass' is analyzed. Where coarse grains or even clasts are present, it is possible that the resulting geochemistry from particular samples will not reflect the 'true' geochemistry of that rock. For example, the geochemical signatures yielded by granite clasts and matrix sandstone may be very different, resulting in data that is very difficult to interpret. In conventional XRF studies, this problem would be solved by only analysing the matrix and grinding the sample prior to analysis.

A further drawback relates to difficulties in converting element intensities measured by EDXRF core-scanners to element concentrations. For this reason, results are often presented in the form of count rates (counts per unit per unit area), or as ratios of counts, count rates, or intensities of elements (Flood et al. 2016; Weltje and Tjallingii 2008).

Owing to the aforementioned difficulties it is perhaps unsurprising that few chemostratigraphy project involve the use of ED-XRF core-scanners but where they are employed, log ratio calibration equations should be applied to quantify elemental composition. This involves reduced major axis regression of log-ratio transformed geochemical data to provide a robust calibration scheme. The details of how this is achieved are discussed by Weltje and Tjallingii (2008), Weltje et al. (2015) and Flood et al. (2016).

It is unlikely that core scanning ED-XRF instruments would ever replace conventional XRF instruments, even if good quality data can be obtained using log ratio calibration equations, as analysis is entirely restricted to core. Most chemostratigraphy projects include the analysis of cuttings or a combination of core and cuttings. In spite of this, ED-XRF core scanners may be useful where the objectives are to recognize particularly thin laminae of under 1 cm, in sections where the core is relatively intact, and where the lithologies contain consistent and fine grain sizes (ideally in silt or clay grade material).

2.5.2.5 Calibration Methods for XRF Analysis and Other Important Factors

Irrespective of the particular make and model of XRF spectrometer used in a given study, or whether EDXRF or WDXRF are employed, one of the principal factors governing the quality of data relates to the choice of calibration method, and in particular, to the choice of standard reference materials used to build that method. XRF instruments can be employed in multiple applications and can be utilized to analyse both liquid and solid samples, so different calibration methods should be used for different matrices. For example, a method developed for steel samples containing high levels of Fe may be useless if it were applied to quartz arenites with Fe contents lower than 0.5%. In most chemostratigraphy studies, separate methods are developed for clastic and carbonate samples, as the latter contain much higher concentrations of Ca, Mg, Sr and other elements. When preparing a method for clastic sediments, it is recommended that around 50 samples are selected to cover a range of compositions expected. For example, values of Zr in sandstones and shales range generally from 60 to 500 ppm, so it is necessary to select an approximately equal number of samples that contain levels of this element in the 60–100 ppm, 100–200 ppm, 200–300 ppm, 300–400 ppm and 400–500 ppm ranges. Developing a method where nearly all of the standards yield Zr values of the 400–500 ppm range would probably result in erroneous results when analyzing samples with lower values of this element. Many XRF instruments are purchased with an 'industry-standard' calibration designed for the analysis of a wide range of materials, but such methods are rarely useful for the analysis of geological materials and have to be refined by the addition of standard reference materials.

Published reference standards (e.g. from US Geological Survey) can be used for calibration purposes. Unfortunately, these samples are often very well characterized for major elements but not so for trace elements and REE. Consequently, it is normally necessary to use published standards in addition to internal non-published reference standards that have been analyzed by ICP and/or XRF, provided the data is considered to be of good quality for each sample. Analyst's often find it most challenging to develop calibrations for carbonate sediments as published standard reference materials normally only exist for pure limestone with very high concentrations of Ca. As many studies on carbonates involve the analysis of argillaceous, arenaceous and dolomitic limestones, as well as 'pure' limestones, it is necessary to develop a method using samples of calcareous mudrock, argillaceous limestone, dolomitic limestone, dolostone and arenaceous limestone to reflect this level of variability.

A potential problem with calibrating XRF instruments may relate to the fact that many standard reference materials are acquired from igneous rocks (Michael Dix, Weatherford geologist—personal communication). The elements contained in these rocks are often found in minerals of a completely different density than those found in clastic and carbonate sediments. For example much of the Fe occurs in olivines and pyroxenes in basalts but may occur in less dense minerals such as chlorite in clastic sediments. This theory has not been fully tested and much more research has to be done on the influence of mineral density on calibrations. At the present time, however, it is recommended that XRF calibrations are not based entirely on standards from igneous rocks.

The ease of which XRF may be calibrated very much depends on the software. Some are very straightforward, whilst others require an in-depth knowledge of the software and should only be calibrated by skilled engineers. One positive aspect of XRF calibrations for chemostratigraphy projects is that only two methods are normally required (clastic and carbonate method) and, once developed, should be applicable for all chemostratigraphy projects in the future, with minor refinements required on an infrequent basis. The only complications may occur when major parts of the XRF instrument have to be replaced such as the X-ray tube or detector. Under these circumstances, it may be necessary to refine the method to ensure that similar values are generated for the same samples analyzed before and after these repairs.

Another potential complication relates to the type of collimator installed in the XRF instrument. Many XRF spectrometers are fitted with a Zr collimator as standard but, unless this is replaced with one fabricated from a different material (usually Au or Pd), the resulting concentrations of Zr will be overestimated. Table 2.1 shows the results of XRF analysis performed on 5 samples using a Zr and Au collimators, with the former yielding 'unrealistically' high values for this elements in the region of 4000 ppm when the 'shale average' for Zr is less than 500 ppm.

Table 2.1 Comaprison beween data analysed by EDXRF using a Au collimator (run 1) and one fabricated from Zr (run 2)

Depth	Run	Ba	Cr	Cu	Sc	Sr	Y	Nb	Th	U	Zr
Sample 1	Run 1	211.10	14.42	16.22	1.95	157.56	17.47	12.47	7.58	1.39	175.89
Sample 2	Run 1	218.43	28.52	11.71	3.82	41.53	7.64	8.99	5.73	0.75	137.01
Sample 1	Run 2	210.20	15.11	15.11	1.95	156.23	17.55	12.00	7.58	0.98	4023.34
Sample 2	Run 2	225.88	28.22	12.33	8.91	43.22	6.89	8.45	5.62	0.68	3980.22

Note that the results are similar except for Zr, where values are much higher in run 2

When outsourcing XRF analysis to an external laboratory the following factors should be considered:

(a) Make, manufacturer and type of instrument(s). Is the instrument capable of generating data for at least 35 elements in the range Na-U, with limits of detection for most trace elements in the region of 1 ppm? If this is not the case, then the particular XRF should be deemed unsuitable for chemostratigraphy studies.
(b) Is the collimator fabricated from Au or Pd, as opposed to Zr?
(c) How have the calibration methods been constructed and what standard reference material were used for this purpose? Always ask to view the composition of each reference standard. Are the methods suitable for the type of samples that you will be analysing?
(d) What experience do the operators have with the analysis of rock samples?
(e) Ask to view data generated for clastic and carbonate samples in recent studies? Do these numbers make sense for the analysed matrix? For example, a Ca value of 60% would not be expected when analysing a sample of quartz arenite.
(f) Ask the laboratory to demonstrate levels of analytical accuracy and precision. Always ask to view data generated for the same standard reference materials analysed at different times.
(g) Has the detector of X-Ray tube been replaced recently? If so, has this had any effect on data quality?
(h) What is the sample throughput? Don't be impressed by claims that samples can be analysed in less than 5 min. This may be the case but, in order to acquire the best quality results, it is normally necessary to have an analytical time of at least 15 min per sample. Extended analytical times of more than 30 min per sample are unnecessary in most projects and may result in unnecessary time wasting in the project.
(i) How quickly will the lab be able to generate results? The analytical time per sample may be in the region of 15 min but, if there is a backlog of samples and/or a lack of laboratory staff, this will have a negative impact on the date of delivery.
(j) Will all of the samples be analysed on the same XRF instrument or will different instruments be used? To avoid any potential complications, the same instrument should be used to analyse each sample in a given project, though it is accepted that this is not always possible (e.g. when an instrument fails).
(k) Cost of analysis is rarely, if ever, mentioned in academic papers and textbooks, but this should be significantly lower than ICP analysis, owing to the lower cost of XRF instruments, lower servicing/repair costs and less detailed laboratory procedures. As with ICP analysis, the cost of analysis should be considered along with the aforementioned factors when choosing a laboratory to analyse samples by XRF.

2.5.3 Comparison Between ICP, EDXRF and WDXRF

In order to acquire the best quality results for the largest number of elements, the favored option for nearly all chemostratigraphers would be combined ICP-OES and ICP-MS analysis. However, this is generally more time consuming and expensive than XRF analysis. The choice of analytical instrument to employ in a given chemostratigraphy project depends on a number of factors which are summarized as follows:

(a) Available budget. Combined ICP-OES and ICP-MS is normally much more expensive than XRF analysis, so less samples can be analysed for the same cost.
(b) Project timings. XRF data can be acquired much more rapidly than that of ICP but timeframes for the delivery of results will vary from project to project. As a very general rule, a full dataset for 100 samples should be acquired within 5 days using ICP-OES and ICP-MS. For XRF, that number ranges from 200 to 500. Where there is a requirement for rapid analysis, XRF is invariably the preferred option.
(c) Amount of available sample material. Combined ICP-OES and ICP-MS can be completed on 0.25 g or less, but at least 3 g is needed for XRF. Under most circumstances, this does not present any problems as it is usually possible to acquire more than 10 g of core, cuttings or field samples in any given study. In some projects, however, particularly those undertaken on historic wells drilled prior to 1990, there may be less than 1 g of material left for analysis. Similarly, there may be restrictions on the amount of core that can be extracted from archived material. Under these circumstances, ICP analysis may be favoured.
(d) Analytical requirements. Where the best quality of data is required for the largest number of elements, it is advisable to employ ICP-OES and ICP-MS. For example, if one of the 'key' elements used to place chemostratigraphic boundaries in a scheme developed for existing wells is Ta, then the ICP option would be advised as this element normally exists in very low concentrations that do not exceed 3 ppm. Conversely, if the scheme is dependent on variations in Zr, Cr and Th (these elements normally exceed 8 ppm), then either ICP or XRF could be utilized.
(e) Is the framework likely to be used for wellsite applications in the future? If the intention is to build a framework that could be employed for this purpose (e.g. for the geosteering in reservoirs), then the initial framework developed for existing wells should be based on XRF data, as ICP instruments are deemed unsuitable for such applications. More detailed information on the applications of chemostratigraphy at wellsite are discussed in Chap. 6.

An approximately equal number of chemostratigraphy studies are performed using ICP and XRF data. Where the latter is the chosen analytical option, some thought should be put into the choice of XRF instrument to employ and, in particular, whether WD-XRF or ED-XRF is preferred. As a very general rule, WD-XRF is capable of generating data for more elements and achieving higher levels of analytical precision and accuracy. The disadvantages associated with WD-XRF instruments are that they are normally more unstable than their ED-XRF counterparts and are more prone to malfunction. Furthermore, being much larger than most modern ED-XRF instruments, they are much more difficult to transport to field locations such as the offshore drilling installations. Many modern EDXRF spectrometers have the capability of generating data of similar quality as that of WDXRF (in terms of detection limits, accuracy and precision), but with the added advantage of being small and robust enough to use in the field. They are nearly always easier to operate than WD-XRF analyzers and are less likely to malfunction. Polarization ED-XRF instruments such as the Ametek Xepos or Rigaku CG XRF are particularly impressive in terms of the array of elements that they are capable of generating data for, and the levels of analytical accuracy and precision achieved. A further advantage of ED-XRF analysis is that it is normally less expensive than that of WD-XRF.

In summary, chemostratigraphy projects may be completed using data acquired by either ICP or XRF, with the choice of analytical option largely dependent on available budgets, project time-frames and the specific elements that are needed.

2.6 Other Analytical Techniques

ICP-OES, ICP-MS and/or XRF are used in almost all chemostratigraphy projects but there are occasional exceptions. The following paragraphs provide a brief summary of some of the techniques, other than ICP-OES, ICP-MS and XRF analysis, that have been used to generate data for chemostratigraphy studies.

2.6.1 Atomic Absorption Spectrophotometry (AAS)

The AAS technique is described in detail by Price (1972) and summarized by Rollinson (1993). The basis of AAS is that atoms of an element can absorb radiation. This happens during atomization of the element when the wavelength of light absorbed is specific to each element. The basic components of an AAS instrument are an atomizing device, a light source and a detector. As a result of atomic absorption, a lowering of response occurs in the detector during the atomization of a sample in a beam of light. This can be calibrated and is sensitive at the ppm level. The sample is initially prepared as a solution using the same acid digestion or alkali fusion procedures used in ICP-OES and ICP-MS analysis. This is then aspirated via a nebulizer and atomized in an acetylene-air or acetylene-nitrous oxide flame. For more detailed information on this technique the reader is referred to the work of Price (1972). It is possible to use AAS to acquire data for all of the major elements and most trace elements in the range Na-U in the periodic table. One significant advantage of AAS over XRF is that light elements such as Be and Li can be analysed but have atomic numbers too low to analyse by XRF. Another benefit is that the AAS instruments are normally much less expensive that ICP and have low running costs (Rollinson 1993). In spite of these positive features, however, the main disadvantage of using AAS is that the sample is element—specific, meaning that only one element can usually be analyzed at a time, although this issue has been partly resolved by fitting instruments with multiple-turret lamp holders. Even though it is possible to analyse 3–4 elements or more at a time, the AAS instrument cannot compete with more rapid methods of analysis, namely ICP-OES, ICP-MS and XRF. It would simply take too long to acquire data for the 40–55 elements per sample required in the majority of chemostratigraphy projects. For that reason, AAS is rarely used for this purpose.

2.6.2 Spark Source Mass Spectrometry (SSMS)

SSMS is rarely used in chemostratigraphy studies but can acquire data for around 40 trace elements simultaneously, with detection limits of 1–10 ppb, and has been described in detail by Taylor and Gordon (1977), Kronberg et al. (1988), Jochum and Hofman (1989) and Rollinson (1993). The samples are mixed with spiked graphite before being briquetted into rod-shaped electrodes, with a vacuum discharge being generated between two sample electrodes. Based on mass detected on photoplates situated in the focal plane of the mass spectrometer, it is possible to identify individual elements. Analysis of the mass spectra and ion intensities are determined from line blackenings on the photoplate.

2.6.3 Neutron Activation Analysis (INAA and RNAA)

Neutron activation analysis is mainly used to produce data for trace elements and REE. The two types of neutron activation analysis are Instrumental Neutron Activation Analysis (INAA) and Radiochemical Neutron Activation Analysis (RNAA). INAA utilizes a powdered rock or mineral sample, while RNAA involves the chemical separation of selected elements. In INAA c.100 mg of powdered sample is placed in a neutron flux in a neutron reactor along with standards. The sample and standards are then subjected to irradiation for a period of up to 30 h. The neutron flux produces new, short lived radioactive elements which yield gamma radiations. Different gamma radiations are emitted for each element, with the intensities of the radiations being proportional to the amounts of the isotopes present. Isotopes with different half-lives are measured by gamma ray spectrometry (i.e. counting) performed at set intervals (hours, days and weeks). The next step involves making corrections for overlapping lines in the spectrum, with concentrations determined by comparison with the standards analyzed at the same time. According to Rollinson (1993), the method is very sensitive for some high field strength elements, the platinum group elements and REE.

Chemical separation of elements, occurring after irradiation but before counting, is advised where elemental values are below 2 ppm. This approach increases sensitivity and is known as RNAA. For more detailed information on INAA and RNAA, the reader is referred to the work of Muecke (1980).

In general, INAA and RNAA are considered too time consuming to be used in routine chemostratigraphy studies.

2.6.4 Laser-Induced Breakdown Spectroscopy (LIBS)

LIBS utilizes laser pulses to ablate a target and the emission spectrum of the resulting plasma is collected and analyzed. Prior to the introduction of high quality benchtop XRF instruments in the early 2000s, Halliburton used LIBS instruments to acquire data in near real time at wellsite in a number of studies in the North Sea, North Africa, South America and elsewhere. More recently, a LIBS based instrument has even been used to acquire geochemical data from the Shaler outcrop in Gale crater, Mars (Anderson et al. 2015). In theory it is possible to acquire data for as many elements as XRF and there is an added advantage that it is now possible to detect light elements such as H, Li, B, C and N that are not routinely analyzed by XRF or ICP (Anderson et al. 2011). As with XRF, the quality of data determined by LIBS is largely determined by the calibration method. The strength of an element's emission lines in the LIBS plasma is dependent on matrix effects (including self-absorption, the degree of ionization in the plasma, the degree of laser-to-sample coupling, and the abundance of other elements (Anderson et al. 2011; Clegg et al. 2009). Initially, quantification of elements was achieved using univariate calibration based on the area under emission lines for the element of interest. More recent work has demonstrated that multivariate techniques such as partial least squares are better at accounting for matrix effects and therefore yield more accurate results (Clegg et al. 2009; Tucker et al. 2010; Anderson et al. 2011). It is also possible to use neural networks as an alternative method of multivariate calibration for LIBS (Anderson et al. 2011).

In general, the data quality, in terms of number of analyzed elements, detection limits, accuracy and precision are considered to be superior by XRF and ICP, hence the reason that LIBS instruments are rarely used in chemostratigraphy studies. Another problem with LIBS relates to grain size and whether the samples have been ground far enough. Repeated analysis are undertaken on a sample, be it a pressed pellet or the flat/sawed surface of a hand specimen, with the results 'averaged' to give the 'overall' geochemical composition of the sample. However, if the samples have not been ground far enough, or if there are inconsistencies in the grain size of unground hand specimens, this can lead to erroneous results.

In recent years, several benchtop and even hand-held LIBS instruments have been developed and some LIBS instruments are capable of acquiring data for light elements such as Li and B. It is unfortunate that most do not deliver such good quality data for heavy elements, most of which are required in chemostratigraphy projects (Dr Christian Scheibe, Saudi Aramco chemostratigrapher—personal communication).

2.6.5 Electron Microprobe Analysis

The basic principles of electron microprobe analysis are similar to XRF, the difference being that the sample is excited by a beam of electrons as opposed to an X-ray beam. Secondary X-rays are analyzed according to their wavelength, with the peak area measured relative to a standard, and concentrations related to intensities, making appropriate corrections for that matrix. The two principal types of electron microprobe analysis are energy-dispersive and wavelength-dispersive analysis, which utilize energy versus intensity and wavelength versus intensity spectra respectively. Of the two instruments, Energy dispersive electron microprobe analysis is more rapid but less precise than the wavelength method (Rollinson 1993).

Electron microprobe analysis is mainly used to acquire data for major elements and, by employing a beam of 1–2 μm in diameter, it is possible to analyse individual mineral grains in thin section. Despite this advantage, the electron microprobe instrument is rare used in chemostratigraphy projects as data for more elements and of a higher quality are acquired by XRF and ICP. Furthermore, the small beam size may represent a disadvantage for whole rock analysis as the small 'field' of analysis may be unrepresentative of the sample as a whole.

2.6.6 Downhole Geochemical Logging Tools

In addition to the aforementioned analytical techniques, geochemical data can be acquired in boreholes via wireline logging tools. The spectral gamma ray (SGR) provided results for U, Th and K, whilst more specialist tools such as Litho Scanner (www.slb.com/services/characterization/petrophysics/wireline/lito_scanner.aspx) and Gem™ Elemental Analysis Tool (www.halliburton.com/en-US/ps/wireline-perforating/wireline-and-perforating/op) provide data for other elements. Unfortunately, these specialist tools generally only provide data for major elements and a small number of trace elements, so they are rarely used for chemostratigraphic purposes. Instead, they are normally utilized to predict changes in bulk lithology and mineralogy. It is hoped that, as these tools are developed in the future, it will be possible to acquire data for more trace elements and that they may be used to identify chemostratigraphic boundaries.

2.7 Assessment of Data Quality

Irrespective of the choice of analytical technique employed in a given study, it is important to conduct checks on data quality before embarking on interpretation.

Data quality is dependent on the following three factors:

- Detection limits
- Precision
- Accuracy

Detection limits are typically around 0.01 ppm or less for ICP analysis, 0.1–2 ppm for WDXRF and 1–10 ppm for EDXRF (limits of detection may be greater than 10 ppm for hand held XRF analysers and EDXRF spectrometers with X-ray tubes of less than 40 kV), though precise levels of detection varies according to specific elements and the type of XRF/ICP instrument employed. Where limits of detection fall outside these ranges, this should be of concern. It should also raise concern where there is a change in the levels of detection recorded by the same instrument over time. If there are any doubts relating to detection limits, it is recommended that the instrument is serviced/repaired.

Table 2.2a Data produced for repeated analysis of a standard reference material

(a)

	Al	Si	Ti	Fe	Mn	Mg	Ca	Na	K	P	Y	Cr	Zr
Run 1	18.40	51.20	0.96	9.62	0.0222	2.35	5.46	1.30	3.17	0.1289	21.5	152.0	130.5
Run 2	18.90	51.20	0.95	9.86	0.0201	2.35	5.41	1.29	3.17	0.1289	20.8	152.7	131.3
Run 3	18.11	51.20	0.95	9.49	0.0226	2.37	5.44	1.31	3.17	0.1287	22.2	153.0	131.3
Run 4	15.40	42.10	0.94	4.40	0.0000	5.10	3.80	2.55	1.24	0.1100	10.4	249.2	125.6
Run 5	18.10	55.20	1.55	9.30	0.1000	2.40	3.20	1.45	2.40	1.2000	22.1	12.1	179.8
Run 6	18.50	60.20	0.11	10.22	0.0000	8.20	2.40	0.55	3.10	0.2220	5.5	122.3	127.7

The data quality is considered good for runs 1–3 but runs 4–6 show poor levels of precision (i.e. repeatability) for all elements

Note that concentrations of major elements are quoted as wt% oxide values, while the trace elements of Y, Cr and Zr are quoted as ppm

Precision refers to repeatability of results. It is recommended that the same three samples (preferably standard reference materials) are analysed at least once, but preferably two or three times, per day to test for precision and any instrumental drift. Corrections may be made to the data if there are minor fluctuations in precision. However, more serious fluctuations may require the instrument to be serviced/repaired. Examples of data showing good and poor levels of precision are presented in Table 2.2a. This table shows that a standard reference sample was run three times on day 1, with very similar results produced for the elements and, hence, good levels of precision. By contrast, poor levels of precision were produced when the same test was conducted on day 2 (runs 4–6), with very different values of each element yielded on the three runs. As a general rule, values should be within 8% or less of each other when samples are run repeatedly.

Accuracy refers to the closeness of a result to the 'known' values of particular elements and is assessed by analysing standard reference materials. Table 2.2b illustrates examples of good and poor levels of accuracy. This shows that levels of accuracy are good for most elements, with the exception Cr which displays much higher values than expected. As with precision, the aim is to achieve results within around 8% or less of the known values of particular standard reference materials. Unfortunately, this is often more difficult to achieve by XRF owing to coelution of spectral peaks and the setup of the particular method/calibration employed. In many ways, it is more important to achieve good analytical precision than accuracy, particularly in XRF analysis. If results are consistently higher or lower than the 'known values' of elements, it is perfectly acceptable to use the particular instrument in chemostratigraphy studies, provided precision is good.

2.8 Quality Control Checks

A number of checks should be made before the data is considered suitable to interpret. These are summarised as follows:

1. Analytical precision and accuracy. This is determined by analysis of standard reference materials (see Sect. 2.7 for discussion)
2. Are the limits of detection acceptable for each element (see Sect. 2.7 for discussion)
3. Do the results obtained from major elements roughly match the bulk lithology recorded in hand specimen and by wireline log? If not then check if the core samples have been labelled correctly and there have been no problems relating to sample preparation or analysis. For example if a quartz arenite is analysed, low Si/Al values in the region of 2 would be unexpected.

Table 2.2b Data produced for repeated analysis of a standard reference material. The data quality is considered good for most elements except for Zr which displays poor levels of accuracy (values consistently higher than known value) Note that concentrations of major elements are quoted as wt% oxide values, while the trace elements of Y, Cr and Zr are quoted as ppm

(b)

	Al	Si	Ti	Fe	Mn	Mg	Ca	Na	K	P	Y	Cr	Zr
run 1	9.22	37.88	2.88	12.48	0.25	12.66	13.69	3.20	1.27	1.19	28.6	188.2	245.3
run 2	9.24	37.89	2.68	12.58	0.20	12.61	13.66	3.18	1.25	1.20	28.1	189.3	245.6
run 3	9.88	37.18	2.65	12.41	0.19	12.50	13.63	3.14	1.26	1.19	28.1	187.0	241.8
run 4	9.22	37.66	2.24	12.46	0.19	12.63	13.67	3.16	1.28	1.18	28.0	187.3	245.4
run 5	10.42	38.02	2.59	12.45	0.22	12.67	13.63	3.18	1.28	1.18	28.5	188.9	249.3
Expected Value	**10.30**	**36.50**	**2.78**	**12.47**	**0.20**	**12.50**	**13.60**	**3.18**	**1.29**	**1.18**	**28.00**	**125.0**	**248.0**
	good	good	good	good	good	good	good	good	good	good	good	poor	good

4. Are any values of major elements and trace elements 'unrealistically' high. If so, then checks should be made on sample processing and analysis. For example, the element Zr is normally in the range 100–400 ppm in clastic sediments, so a value of 5000 ppm should be viewed with suspicion.
5. Are values of Ba anomalously high? This element is nearly always concentrated in drilling additives and, though it is highest in cuttings samples, values should never exceed 20,000 ppm. Where values consistently exceed this figure, the samples should be re-washed to ensure the removal of drilling additives.
6. Do the sample depths provided in the spreadsheet match those on a sampling sheet? This may seem a pedantic issue, but many projects have been ruined by relatively minor issues such as the labelling or mislabelling of samples or mix-up of well names. It is always advisable to address these matters before releasing the data for interpretation or publication.
7. Have a sufficient number of samples been analysed in each study section? If not, then it is advisable to undertake additional sampling and analysis before embarking on interpretation.

2.9 Summary and Conclusions

Sampling, processing and analysis are the most time consuming parts of any chemostratigraphy study but are also the most important. For financial reasons or to save time, analysts may feel compelled to cut corners by paying less attention to these factors. It is noteworthy, however, that the principal reasons for the failure of chemostratigraphy studies relate to incorrect sample preparation and analytical procedures being followed. If the steps outlined in this chapter are adhered to, then data quality should be acceptable. In such circumstances, errors may have been made with respect to the interpretation of the data but, if the data cannot be relied on in the first instance, the study will almost certainly be a failure.

References

Anderson, R., Bridges, J. C., Williams, A., Edgar, L., Ollila, A., Williams, J., et al. (2015). ChemCam results from the Shaler outcrop in Gale crater, Mars. *Icarus, 249,* 2–21.

Anderson, R. B., Morris, R., Clegg, S. M., Bell, J., Wiens, R. C., Humphries, S. D., et al. (2011). The influence of multivariate analysis methods and target grain size on the accuracy of remote quantitative chemical analysis of rocks using laser induced breakdown spectroscopy. *Icarus, 215,* 608–627.

Beckhoff, B., Kanngieber, B., Langhohh, N., Wedell, R., & Wolff, H. (2006). *Handbook of practical X-ray fluorescence analysis*. New York: Springer.

Clegg, S., Sklute, E., Dyar, M., Barefield, J., & Wiens, R. (2009). Multivariate analysis of remote laser-induced breakdown spectroscopy spectra using partial least squares, principal component analysis, and related techniques. *Spectrochimica Acta Part B: Atomic Spectroscopy, 64,* 79–88.

Craigie, N. W. (2015a). Applications of chemostratigraphy in Middle Jurassic unconventional reservoirs in eastern Saudi Arabia. *GeoArabia, 20*(2), 79–110.

Craigie, N. W. (2015b). Chemostratigraphy of the Silurian Qusaiba Member, Easatern Saudi Arabia. *Journal of African Earth Sciences, 113,* 12–34.

Flood, R. P., Bloemsa, M. R., Weltje, G. J., Barr, I. D., O'Rourke, S. M., Turner, J. N., et al. (2016). Compositional data analysis of Holocene sediments from the west Bengal Sundarbans, India: Geochemical proxies fro grain-size variability in a delta environment. *Applied Geochemistry, 75,* 222–235.

Jarvis, I. (1991). Sample preparation for ICP-MS. In K. E. Jarvis, A. L. Gray, & S. Houk (Eds.), *Handbook of inductively coupled plasma-mass spectrometry* (pp. 172–224). Blackie, Glasgow.

Jarvis, I., & Jarvis, K. E. (1992). Plasma spectrometry in the earth sciences: techniques, applications and future trends. *Chemical Geology, 95,* 1–33.

Jochum, K. P., & Hofman, A. W. (1989). Fingerprinting geological material using SSMS—comment. *Chemical Geology, 75,* 249–251.

Kelloway, S., Ward, C. R., Marjo, C. E., Wainwright, I. E., & Cohen, D. R. (2014). Quantitative chemical profiling of coal using core-scanning X-ray fluorescence techniques. *International Journal of Coal Geology, 128–129,* 55–67.

Kronberg, B. I., Murray, F. H., Daddar, R., & Brown, J. R. (1988). Fingerprinting geological materials using SSMS. *Chemical Geology, 68,* 351–359.

Longoni, A., Fiorini, C., Leutenegger, P., Sciuti, S., Fronterotta, G., Struder, L., et al. (1998). A portable XRF spectrometer for non-destructive analysis in archaeometry. *Nuclear Instruments and Methods in Physics Research Section A: Accelerators, Spectrometers, Detectors and Associated Equipment, 409,* 407–409.

Madhavaraju, J. (2015). Geochemistry of late cretaceous sedimentary rocks of the Cauvery Basin, South India: constraints on paleoweathering, provenance, and end cretaceous environments. In M. Ramkumar (Ed.), *Chemostratigraphy: Concepts, techniques and applications*. Amsterdam: Elsevier.

Muecke, G.K. (1980). *Neutron activation analysis in the geosciences*. Mineralogical Association of Canada Short Course Handbook.

Pearce, T. J., Besley, B. M., Wray, D. S., & Wright, D. K. (1999). Chemostratigraphy: a method to improve interwell correlation in barren sequences—a case study using inshore Duckmantian/Stephanian sequences (West Midlands, UK). *Sedimentary Geology, 124,* 197–220.

Pearce, T. J., Wray, D. S., Ratcliffe, K. T., Wright, D. K., & Moscariello, A. (2005). Chemostratigraphy of the upper carboniferous schooner formation, southern North Sea. In J. D. Collinson, D. J. Evans, D. W. Holiday, N. S. Jones (Eds.), *Carboniferous Hydrocarbon geology: The southern North Sea and surrounding inshore areas* (Vol. 7, pp. 147–164). Yorkshire Geological Society, Occasional Publication Series.

Price, W. J. (1972). *Analytical atomic absorption spectrometry*. London: Heyden.

Rollinson, H. (1993). Using geochemical data: Evaluation, presentation, interpretation.

Struder, L., Lechner, P., & Leutenegger, P. (1998). Silicon drift detector—The key to new experiments. *Naturwissenschaften, 85,* 539–543.

Taylor, S. R., & Gordon, M. P. (1977). Geochemical applications of spark-source mass spectrography, III. Element sensitivity precision and accuracy. *Geochimica et Cosmochimica Acta, 41,* 1375–1380.

Totland, M., Jarvis, I., & Jarvis, K. E. (1992). An assessment of dissolution techniques for the analysis of geological samples by plasma spectrometry. *Chemical Geology, 95*(1–2), 35–62.

Tucker, J. M., Dyar, M. D., Schaefer, M. W., Clegg, S. M., & Wiens, R. C. (2010). Optimization of laser-induced breakdown spectroscopy for rapid geochemical analysis. *Chemical Geology, 277,* 137–148.

Usman, A. and Meehan, D.N., 2016. Unconventional Oil and Gas Resources: Exploitation and Development. CRC Press, 860pp.

Weltje, G. J., & Tjallingii, R. (2008). Calibration of XRF core scanners for quantitative geochemical logging of sediment cores; theory and application. *Earth and Planetary Science Letters, 274,* 423–438.

Weltje, G. J., Bloemsma, M. R., Tjallingii, R., Heslop, D., Röhl, U., & Croudace, I.W. (2015). Prediction of geochemical composition from XRF core scanner data: A new multivariate approach including automatic selection of calibration samples and quantification of uncertainties. In I. W. Croudace, & R.G. Rothwell

(Eds.), *Micro XRF studies of sediment cores: Applications of a non-destructive too for the environmental sciences* (pp. 507–534). Springer, Dordrecht, NL, (Developments in Paleoenvironmental Research, 17).

Wright, A. M., Spain, D., & Ratcliffe, K. T. (2010). *Application of inorganic whole-rock geochemistry of shale resource plays*. Paper presented at Canadian Unconventional Resources and international petroleum conference held in Calgary, Alberta.

Young, K. E., Evans, C. A., Hodges, K. V., Bleacher, J. E., & Graff, T. G. (2016). A review of the handheld X-ray fluorescence spectrometer as a tool for field geologic investigations on Earth and planetary surface exploration. *Applied Geochemistry, V, 72,* 77–87.

Website: http://learnxrf.com/category/xrf-principles/.

3 Geochemistry and Mineralogy

Abstract

Provided the data are considered to be of high quality (in terms of precision, accuracy, detection limits etc.), the mineralogical affinities of elements and controls on geochemistry and mineralogy should be understood before utilizing the inorganic geochemical data to propose chemostratigraphic schemes. Some elements are almost exclusive to particular minerals (e.g. Hf and Zr in zircons), but many are not. For example, Ti may be linked with a number of heavy minerals including rutile, anatase, sphene, titanomagnetite, magnetite and illmenite. Similarly, K is often concentrated in K feldspars, but may also be associated with clay minerals (e.g. illite) and micas. For this reason, it is necessary to gain an understanding of element:mineral links by comparing geochemical and mineralogical data acquired from the same samples, and by employing statistical and graphical techniques.

3.1 Introduction

On receiving high quality data, the temptation to commence the production of chemostratigraphic schemes is often immense, but this should not proceed without a comprehensive understanding of the mineralogical affinities of elements, and the controls on geochemistry and mineralogy in any given study. The most common element: mineral associations are discussed in the first part of this chapter, but it should never be over looked that most elements are contained in more than one mineral. The controls on geochemistry and mineralogy are different in each study and should be established by first comparing geochemical and mineralogical data acquired from the same samples. It is then necessary to employ statistical and graphical techniques for the same purpose. The latter part of the chapter outlines the various techniques that should be utilized in every chemostratigraphy study to gain an understanding of the mineralogical affinities of elements.

N. Craigie, *Principles of Elemental Chemostratigraphy*, Advances in Oil and Gas Exploration & Production, https://doi.org/10.1007/978-3-319-71216-1_3

3.2 Common Element Mineral Links and Controls on Geochemistry and Mineralogy

The most common element:mineral links are outlined in the following paragraphs, commencing with a discussion on heavy minerals. This is followed by sections on paleoredox/paleoenvironmental indicators, clay minerals and carbonate minerals. As many chemostratigraphy studies are conducted on cuttings, the chapter would not be complete without a brief mention of how the presence of drilling additives can affect the distribution of particular elements.

3.2.1 Stable Elements Associated with Detrital Heavy Minerals

The elements most likely to be almost exclusively associated with heavy minerals include Zr, Hf, Ti, Ta, Nb, Cr, Y, Th and HREE. These are considered high field strength elements, their distributions being largely unaffected by post depositional weathering/diagenesis. For this reason alone, they tend to be utilized in isolation, or in the form of ratios, to place chemostratigraphic boundaries. In addition to this, P, U and other trace and rate earth elements may also be concentrated in heavy minerals. The following paragraphs provide a description of the most common heavy minerals found in sedimentary rocks. The bulk compositions and formulae of heavy minerals were published in the decades prior to 1980, but very little was known about the incorporation of REE and trace elements in these minerals until single-grain microbeam methods, particularly electron microprobe analysis, became widely available from the 1980s (Mange and Morton 2007). More advanced technologies, such as the sensitive high resolution ion microprobe (SHRIMP) and laser ablation inductively coupled plasma mass spectrometry (LA-ICPMS) have been developed in more recent years, enabling trace elements and isotopic compositions to be analysed in single-mineral grains. The focus of most studies involving these techniques has been on the radiometric dating of detrital heavy mineral grains, particularly zircons and apatites, but important information has been gleaned on the trace element and REE compositions of zircon, monazite, apatite, chrome spinel, ilmenite, garnet and other heavy minerals (Belousova et al. 2002a; Morton and Yaxley 2007; Mange and Morton 2007). This information has undoubtedly assisted our understanding of the distribution of REE and trace elements when measured in bulk rock samples for chemostratigraphy studies.

The following paragraphs summarize details on most of the heavy minerals found in sedimentary rocks and information on their geochemical compositions. Please note that this is by no means exhaustive and the reader is referred to the works of Mange and Morton (2007), Scott et al. (2014) and other publications mentioned in these paragraphs for additional information.

3.2.1.1 Zircon

The heavy mineral zircon ($ZrSiO_4$) occurs in a wide variety of igneous and metamorphic rocks but it is most abundant in silicic igneous rocks such as granitoids. The elements Zr and Hf are almost exclusively linked to zircon but other elements associated with this mineral include U, Th, Y, Nb, Ta, Sc and REE (Mange and Morton 2007). The quantity of these trace elements depends on host rock composition (Heaman et al. 1990; Hoskin and Ireland 2000; Belousova et al. 2002a), so such variations have possible applications in provenance and correlation studies. As an example of this, Owen (1987) suggested that Hf contents in zircons could be used for provenance discrimination. More recently, Heaman et al. (1990) found that a variety of host rocks could be differentiated based on variations in Lu, Sc, Th/U, Lu/Sm and Hf in zircon. Belousova et al. (2002a) demonstrated an increase in the total REE content from ultramafic, through mafic, to silicate whole rock compositions. Scheibe (2002) also used Zr/Hf for provenance discrimination in North Sea sandstone reservoirs, assuming that Zr and Hf were exclusively concentrated in zircons. There are also variations in the distribution of particular REE in zircons. According to Mange and Morton (2007), zircons from carbonatites and kimberlites have relatively flat chondrite normalised REE patterns, whilst those from granites and pegmatites are depleted in LREE. It is known that metamorphic and igneous zircons generally have Th/U ratios in the range 0.001–0.1 and 0.2–1.5 respectively (Varva et al. 1999; Hartmann et al. 2000). In a comprehensive study on the provenance of Carboniferous sandstones in the Pennine Basin, UK, Hallsworth et al. (2000) found substantially lower U and Th levels in the westerly-derived Clifton Rock in comparison to sandstones shed from the south (Dalton Rock, Halesowen Formation) or from the north (Rough Rock, Ashover Grot and pre-Marsdenian sandstone).

In most studies zircon is considered the most stable heavy mineral but that does not necessarily hold true in zircons containing high values of U and Th. These zircons are more likely to become metamict (i.e. to have their crystal structure compromised), as they are exposed to a higher radiation dose through the emission of α particles (Holland and Gottfried 1955). Metamict zircons are much more likely to be mechanically unstable and subjected to dissolution during weathering compared with non-metamict grains (Balan et al. 2001). Hallsworth et al. (2000) proposed that repeated cycles of weathering and transportation resulted in loss of U and Th- enriched zircons in the aforementioned study of the Pennine Basin.

3.2.1.2 Monazite

The monoclinic mineral monazite [(Ce, La, Nd, Th)PO_4] is present in silicic igneous rocks and in regionally metamorphosed argillaceous sediments (Nash 1984; Chang et al. 1998), with compositional differences considered to be related to paragenesis. The most abundant elements in this mineral are Ce, La, Nd, Th and P, though U, Y and other REE may also exist in varying quantities. In a detailed study of monazites, Gonçalves et al. (2016) report La, Ce and Nd concentrations in the hundreds of thousands ppm, whilst Th values ranged from 68,252 to 102,874 ppm. The REE Pr, Sm and Gd yielded concentrations of tens of thousands ppm, whereas thousands of ppm (<10,000 ppm) were recoded of Gd, Tb and Dy. Lesser amounts of Eu, Ho, Er, Tm, Yb and Lu (i.e. generally <400 ppm) were measured, with values of U ranging from 972 to 3093 ppm.

Distinct compositional domains, many on the scale of 5–10 μm are often located in individual grains and represent successive generations of mineral growth, which record details of the geological history of the host rocks. Fleischer and Altschuler (1969) found the quantity of LREE in monazites occurring in igneous rocks to be highly dependent on the host rock composition, with the La/Nd ratios of granites and granite pegmatites being much lower than those of alkaline igneous rocks and carbonatites. Overstreet (1967) proposed that concentrations of Th in monazites increased with metamorphic grade in metapelites, from the lowest Th contents being associated with greenschist facies, increasing through amphibolite facies, with the highest values found in granulite facies. Jonasson et al. (1988) also presented evidence for an increase in Y with metamorphic grade. Making inferences about pressure/thermal history or metamorphic grade, based solely on the contents of Th and Y, however, should be avoided as these observations do not always hold true (Mange and Morton, 2007). For example, in a study of metapelites in the eastern Mojave Desert, USA, Kingsbury et al. (1993) found no association between abundance and metamorphic grade.

3.2.1.3 Xenotime

Xenotime (YPO_4) is less commonly observed than monazite but, according to Hetherington et al. (2008) this may be explained by the smaller grain size of this mineral, meaning that it is less likely to be identified. In addition to this, xenotime has a bulk rock composition that is not so stable, resulting in possible destruction of grains. For those reasons, this mineral is rarely recorded in routine heavy mineral studies and is rarely used to differentiate provenance in sedimentary rocks. In spite of this, xenotime often contains a significant proportion of high field strength elements and deserves some attention in this chapter. In igneous rocks, this mineral is commonly found in pegmatites and peraluminous granites. Xenotime also occurs in regionally metamorphosed politic and semipelitic rocks ranging from those of the chlorite to the cordierite +K-feldspar and cordierite+garnet zones (Pyle and Spear 1999), though it has also been recorded in some granulite-facies and migmatite zones (Heinrich et al. 1997; Spear and Pyle 2002).

The mineral is tetragonal with a zircon-type structure. It has an ideal formula of ATO_4, where A represents yttrium (Y^{3+}), and T a tetrahedrally coordinated atom of phosphorous (P). The element Y is hosted in an eight oxygen coordinated polyhedral characterized by two sets of four equivalent bond distances oriented orthogonal to each other, forming a dodecahedral cation site. REE may substitute for Y as they have similar ionic radii. This is particularly true with respect to the slightly smaller HREE (Gd to Lu) but other REE may also be accommodated in the A site (Heatherington et al. 2008).

Simple iso-structural substitution may occur in xenotime by exchange of the phosphorous ion PO_4^{3-}) in the T site. For example, Graeser et al. (1973) and Ondrejka et al. (2007) reported complete solid-solution with the Y-arsenate end-member chernovite ($YAsO_4$). The naturally occurring Y vanadate end-member wakefieldite (YVO_4) also exists, though this mineral is rare and the existence of a solid-solution between YVO_4 and xenotime (YPO_4) is poorly understood (Heatherington et al. 2008). Another end-member is Fergusonite ($YNbO_4$) which has been recognised in granites and pegmatites. According to Ondrej et al. (2007), however, the substitution of Nb into YPO_4 and $YAsO_4$ is limited. In spite of this, the ability of xenotime to contain both Nb and V should not be ignored.

As with the monoclinic monazite, complex substitutions in the A-site are also noted in xenotime, though they are fewer (Heatherington et al. 2008). The cheralite-monazite substitution forms a complete solid-solution, resulting in the incorporation of Th and Ca into monazite, though this is of less significance in xenotime. The Ca^{2+} ion is of a similar size to that of Y and HREE but is crystographically less favourable in xenotime, and Ca rarely exceeds 10% of the A site (Förster 2006). The elements Fe and Mn are other examples of divalent cations detected in xenotime crystals (Förster 2006; Ondrej et al. 2007). Heatherington et al. (2008) proposed a cheralite-like substitution with Th and U.

Thorite ($ThSiO_2$) is involved in a complex coupled substitution with xenotime, via the vector $[Y]_{-1}[PO4]_{-1}[Th][SiO_4]$ (Förster 2006). There is continuous compositional exchange between the two end members and the minerals are iso-structural. In theory, U^{4+} substitution in the A site should be more favourable than that of Th^{4+} because of its slightly smaller octahedral sites (Shannon and Prewitt 1970). Substitution of U is explained via an exchange with end- member coffinite ($USiO_4$) along the vector $[Y]_{-1}[PO4]_{-1}$ $[U][SiO_4]$. Smits (1989) proposed a continuous solid-solution between xenotime and coffinite, similar to that between xenotime and thorite, but this has not been recorded in natural samples (Förster 2006; Heatherington and Harlov 2008).

Twinned crystals of xenotime and zircon have been recorded in granites and pegmatites, as have intimate overgrowths of one another (Neumann 1985; Vallini et al. 2002; Wall et al. 2008). Intermediate xenotime-zircon compositions have also been identified suggesting the existence of a solid solution between the two minerals. This has not been proven, however, as there are gaps in the series (Bea 1996; Förster 2006; Johan and Johan 2004).

The elements S, F and Cl are minor constituents of xenotime but the mechanisms by which they substitute into REE phosphates are not particularly well understood (Heatherington et al. 2008)

3.2.1.4 Rutile, Anatase and Brookite

The three main polymorphs of TiO_2 are rutile anatase and brookite. In general, rutile is considered to be much more stable (Luiz da Silva 2017) and is more abundant than anatase and brookite in sedimentary rocks. Rutile (TiO_2) is an ultrastable heavy mineral found in a range of igneous and metamorphic rocks. In addition to being one of the principal host minerals for Ti, rutiles may also contain Ta, Nb, Cr, Fe and even Zr (Mange and Morton 2007). Zack et al (2002) used Nb/Ti and Cr/Ti ratios in rutiles to distinguish sedimentary provenance. Following this, Zack et al. (2004) demonstrated that it was possible to differentiate two separate sources of rutile (metapelites and metabasites) based on their concentrations of Nb and Cr. In addition to this, they found that Zr in metapelitic rutile, where it coexists with zircon and quartz, is extremely temperature dependent, indicating that the Zr content can be utilised to measure maximum metamorphic temperatures. This is considered to be the first single-grain geothermometer parameter to be used in heavy mineral provenance studies.

In a study of the Beryl Field, North Sea (UK sector), Preston et al. (1998) noted the common presence of Nb-rich and Nb-poor rutiles in the same assemblage, suggesting that these grains are independent of size and/or hydraulic controls.

Much less information is available on the trace element geochemistry of anatase and brookite, but it is proposed that these minerals are similar to rutile, containing appreciable amounts of Nb and Cr.

3.2.1.5 Titanite (Sphene)

Titanite (general formula $XYOZO_4$, ideally $CaTiOSiO_4$), or sphene as it is often named, crystallises from intermediate magmas that form diorites and granodiorites, during hydration reactions that convert clinopyroxene to amphibole or oxidation during post-magmatic re-equilibration (Frost et al. 2000). This suggests that titanite may only be found in eruptive rocks on rare occasions (Frost and Lindsey 1991). In metamorphic rocks, titanite has been recorded in calc-silicate rocks (greenschist to granulite facies) and marbles (greenschist to eclogite grade), in metabasites (upper greenschist and lower amphibolite facies), in low-grade pelitic rocks and in metamorphosed granitic rocks (orthogenesis) (Tiepolo et al. 2002). It is important to document the chemistry of this minerals as it is a major host of high field strength trace elements including Nb, Ta, Zr, Hf, Th, U and REE (Spandler et al. 2016 and references therein).

According to Tiepolo et al. (2002), the tetrahedron is fully occupied by Si in natural titanites, whereas the most important chemical species substituting at the octrahedral sites are Ti, Al, Ta, Sb, Nb and Fe^{3+}. Angel (1997) and Knoche et al. (1998) document complete substitution of Ti by Si at the octahedral site at high temperatures and pressures, whilst complete Al substitution has been noted at high pressure by Troitzsch et al. (1999). The [7]-fold-coordination site is dominated by Ca but Sr, Ba, U, Th and REE probably substitute for Ca at this site. The incorporation of trace elements and REE involve complex processes, but are mainly dependent of the composition of the melt (Watson 1976), pressure, temperature (Adam and Green 1984) and the crystal-chemical behaviour of the solid phase (Tiepolo et al. 2002). For more detailed information on trace and REE incorporation in titanite, including recorded concentrations of these elements, the reader is referred to the works of Tiepolo et al. (2002) and Spandler et al. (2016).

3.2.1.6 Ilmenite

A number of studies have been completed on the chemistry of ilmenite ($FeTiO_3$). Basu and Molinaroli (1991) found TiO_2 contents in the ranges of 50–60 and 40–50 wt% in ilmenites hosted by metamorphic and igneous rocks respectively. Similarly, Schneiderman (1995) also noted that metamorphic rocks had ilmenite grains with higher levels of TiO_2 than igneous ones. Less information is available on the trace element composition of ilmenite but She et al. (2015) report Cr, Ni, V, Mg, Nb, Ta and Sc in grains of this mineral collected from the Taihe layered intrusion, Emeishan large igneous province, SW China.

3.2.1.7 Magnetite and Titanomagnetite

Magnetite (Fe_3O_4) is very commonly identified in plutonic igneous rocks and, according to Nadoll et al. (2014), typically contains Mg, Al, Ti, V, Cr, Mn, Co, Ni, Zn and Ga in concentrations ranging from 10 to 1000 ppm. This is consistent with the findings of She et al. (2015) who also report Mn, Ga, Ta, Sc, Hf and Zr in grains of magnetite taken from the Taihe layered intrusion, Emeishan large igneous province, SW China. This mineral has an inverse spinel structure with the general stoichiometry AB_2O_4 (Bragg 1915; Fleet 1981), where A represents a divalent cation such as Mg, Fe^{2+}, Ni, Mn, Co or Zn, and B is represented by a trivalent cation such as Al, Fe^{3+}, Cr, V, Mn or Ga (Nadoll et al. 2014). With a 4+ charge, the element Ti can also occupy the B site when substitution is coupled with a divalent cation (Wechsler et al. 1984). Ferric (Fe^{3+}) and ferrous (Fe^{2+}) iron atoms randomly occupy octahedral sites in the magnetite structure, whilst tetrahedral sites are exclusively occupied by the smaller ferric iron atoms Fe^{3+}, $[Fe^{2+} Fe^{3+}]O_4$ (Nadoll et al. 2014).

At temperatures exceeding 600 °C, continuous solid solution exists between magnetite and ulvospinel (titanomagnetite$_{ss}$—Ti_xFe_3—xO_4) and their oxidation products (titanomaghemite). This process involves coupled substitution of Ti^{4+} for Fe^{3+} in the octrahedral sites and Fe^{2+} for Fe^{3+} in the tetrahedral sites (Buddington and Lindsley 1964; Nadoll et al. 2014). The trace element chemistry of titanomagnetite is likely to be very similar to that of magnetite, but is not so well documented.

3.2.1.8 Apatite

Apatite has the formula $Ca_5(PO_4)3(OH, F, Cl)$ and occurs as a common accessory mineral in almost all igneous and metamorphic rocks (McConnel 1973; Nash 1984; Chang et al. 1998). The trace elements Sr, Y, U, Th and Mn and REE are known to substitute for Ca in igneous apatite, usually in trace amounts (Nash 1984; Ayres and Watson 1993; Chang et al. 1998; Belousova et al. 2002b). According to Mange and Morton (2007), trace element abundances are largely controlled by the degree of host rock fractionation. For example, Sr becomes relatively depleted during fractionation, whilst Y, Mn and REE are relatively enriched. Belousova et al. (2002b) noted lower Th contents in highly fractionated rocks, possibly caused by crystallisation of monazite which removes Th and LREE from the melt. The same study showed higher quantities of U in apatites from granites and granite pegmatites than in dolerites.

The total REE abundances in apatites are largely influenced by the REE content of the host rock, the highest REE values being associated with apatites from alkaline rocks (Chang et al. 1998; Belousova et al. 2002b). Relative enrichments in REE are also controlled by the composition of the host rock, particularly the degree of fractionation. Apatites from highly fractionated rocks such as granite pegmatites, show strong relative LREE depletion in comparison to less fractionated mafic rocks which are relatively enriched in these elements. For this reason, various geochemical ratios involving LREE versus MREE or HREE can be used to distinguish igneous rock type host rocks. La/Nd ratios, for example, are lower in apatites derived from granites and granite pegmatites than from those from alkaline rocks (including pegmatites) (Mange and Morton 2007).

In many studies, apatites have negative Eu anomalies on chondrite-normalised REE plots, being particularly pronounced in more fractionated rocks (Mange and Morton 2007). This may be caused by concentrations of Eu^{2+} from the melt in feldspars (Budzinski and Tischendorf (1989). This led Belousova et al. (2002b) to relate the magnitude of the Eu anomaly to the composition of the host rock.

3.2.1.9 Chrome Spinel

Chrome spinel $[(Mg, Fe^{2+})(Cr, Al, Fe^{3+})_2O_4)$ is found in ultramafic to mafic rocks and is extremely sensitive to bulk rock composition and petrogenesis of the host rock (Mange and Morton 2007). The elements Cr, Mg and Al are the principal constituents of this mineral but behave differently during fractional crystallisation or partial melting. Al is strongly partitioned into the melt, whilst Cr and Mg are partitioned into the solid. Temperature exerts the principal control on the partitioning of Mg and Fe^{2+} between spinel, silicate melts and minerals (Irvine 1965, 1967; Dick and Bullen 1984; Hisada and Arai 1993). Of the high field strength elements, Cr is the main component of chrome spinel but Hisada et al. (2004) record TiO_2 concentrations in the region of 4.5 wt%. Further information on the use of chrome spinels to differentiate changes in provenance is discussed by Mange and Morton (2007). Most heavy mineral ion microprobe studies utilize variations in Cr, Mg, Fe and Ti to differentiate chrome spinels from different sediment sources.

3.2.1.10 Garnet

Of all the heavy minerals, garnet is most widely used for the discrimination of provenance through heavy mineral analysis, as it is one of the most abundant heavy minerals and is relatively 'stable' during weathering and burial diagenesis. The chemistry of garnet is controlled by the bulk composition of the host rock and the pressure and temperature of formation in both igneous and metamorphic rocks (Mange and Morton 2007). The majority of naturally occurring garnets have the general formula $(Mg, Fe^{2+}, Mn, Ca)_3(Al, Cr, Ti, Fe^{3+})_2Si_3O_{12}$, but Deer et al. (1992) records several possible substitutions. The two isomorphous series of garnets are as follows:

a. pyralspite garnets with the formula $[(Mg, Fe^{2+}, Mn)_3Al_2Si_3O_{12}$
b. ugrandite garnets with the formula $[Ca_3(Al, Cr, Ti, Fe^{3+})_2Si_3O_{12}]$

Complete and continuous variations are found within the individual series but there is no evidence for continuous variation between the two series (Mange and Morton 2007).

Morton and Mange (2007) describe four different types of garnet assemblages labelled "Type A, "Type B", Type C" and "Type D". Type A garnets yield high levels of Mg, but low Ca. These are generally attributed to high-grade (granulite facies) metasedimentary rocks or charnockites. With low concentrations of Mg and variable Ca, type B garnets are derived from amphibolite facies metasedimentary terrains. Type C garnets produce high Mg and Ca values and are supplied by mafic and ultramafic gneisses. Garnets from metabasic rocks have lower Mg contents than those derived from ultramafic rocks such as pyroxenites and perodotites. This led Mange and Morton (2007) to propose a subdivision of Type C garnets with subtypes Ci and Cii defined by Mg concentrations of less than and greater than 40% respectively. It is possible to use these subtypes to assess the relative contributions from mafic and ultramafic metamorphic sources. With elevated Fe^{3+} and –Ca (andradite-grossular) Type D garnets are often associated with metamorphosed calcareous sediments. However, other potential origins of these garnets include various reactions involving clinopyroxene, scapolite, plagioclase, wollastonite, calcite and other minerals (Dasgupta and Pal 2005).

A number of other types of garnet have been recognised in igneous and metamorphic rocks, such as yamatoite ($Mn_3V_2Si_3O_{12}$), kimzeyite ($Ca_3(Zr, Ti)_2(Al, Si)_3O_{12}$) and yttragarnet ($Y_3Al_2Al_3O_{12}$), but have not been recorded as detrital grains in sedimentary rocks (Mange and Morton 2007). Vanadium garnet goldmanite ($Ca_3V_2Si_3O_{12}$) is the only exception to this rule, having been identified in Palaeocene sandstones in the northern North Sea (Hallsworth et al. 1992).

In addition to containing variable concentrations of Ca, Mg, Fe, Ti, Al, Si and Mn, garnets are also known to be enriched in HREE, providing a further possibilities to differentiate garnets from different source areas (Mange and Morton 2007).

3.2.1.11 Pyroxene

Occurring in almost every type of igneous and metamorphic rocks, pyroxenes are considered the most important group of ferromagnesium rock-forming minerals. The two principal types of pyroxenes are:

(a) orthopyroxenes—these are orthorhombic with the formula $(Mg, Fe)SiO_3$.
(b) clinopyroxenes—these are monoclinic and are members of the four component system $CaMgSi_2O_6$–$CaFeSi_2O_6$–$Mg_3Si_2O_6$–$Fe_3Si_2O_6$.

The compositions of pyroxenes are dominated by Mg, Fe, Si and Ca but they are also known to contain relatively high concentrations of Ti, Ni and Cr. The reason they have not been studied in greater detail for the purpose of provenance determination is that they are relatively susceptible to post depositional weathering/diagenesis are often removed from sedimentary rocks. In some instances, however, they are preserved. For example, in a study of detrital clinopyroxenes from the River Nile and its delta, Schneiderman (1997) found that those derived from the Red Sea Hills produced higher values of Ca, but lower Cr and Ti, compared to the clinopyroxenes sourced from the Ethiopian Highlands.

3.2.1.12 Other Heavy Minerals

There are many other heavy minerals in addition to the ones already mentioned but they generally fit into one of the following categories:

(a) Heavy minerals that are very rarely identified in sedimentary rocks, such as olivines which are highly unstable.
(b) Heavy minerals that are present in most sedimentary rock but either contain very low concentrations of trace elements and REE, or very little is known about their trace element/REE chemistry (e.g., staurolite, chloritoid)
(c) Heavy minerals that contain 'marker' elements that cannot be routinely analysed by XRF or ICP. One very obvious example of a mineral in this category is tourmaline, which contains appreciable quantities of B. Unfortunately this element is too light to analyse by XRF, and cannot be analysed by ICP owing to the use of Lithium Metaborate in the alkali fusion sample preparation procedure.

3.2.1.13 The Importance of Thorium

Individual concentrations of Th, and Th/K ratios, have been used since the 1970s to predict the quantity and type of clay minerals in sandstones and, depressingly, these parameters are still used in the present day for this purpose by some mineralogists, geologists and petrophysicists. Serra et al. (1980) and Fertl (1983) used Th/K ratios to predict clay mineralogy and several workers noted crude relationships between Th and kaolinite (Plier and Adams 1962; Scott 1968, Tieh et al. 1980). However, Hurst (1990) and Hurst and Milodowski (1994) pointed out that the crystal structure of kaolinite could not accommodate Th and that this element is predominantly associated with heavy minerals, particularly monazite. In the study of Hurst and Milodowski (1994) twelve sandstones were subjected to XRD, XRF, Neutron Activation Analysis (NAA) and thermogravimetric analysis. In addition to this, the samples were examined by SEM using backscatter electron imaging modes with energy-dispersive X-ray microanalysis, electron microprobe analysis and autoradiography of natural alpha-particle emission. They found that almost all of the Th in each sample was concentrated in heavy minerals, though a very small proportion of Th was contained in the clay fraction, almost certainly hosted by clay and silt sized heavy mineral grains. Since this study, it is now almost universally accepted by the scientific community that Th is not associated with clay minerals.

In the vast majority of chemostratigraphy studies, it can be assumed that Th is concentrated in heavy minerals but there is one exception. In a study of Middle Devonian lacustrine sediments, Craigie (1998) found this element to be concentrated in biogenic phosphate by substitution of Ca^{2+} by Th^{4+} in the calcium phosphate lattice. Occurrences of such thoriferous biogenic phosphate are, however, extremely rare and, in this instance, mobilisation of Th from the Caledonian aged hinterland rocks probably took place under conditions of high or low pH combined with groundwaters enriched in carbonate, phosphate and/or sulphur. In general, the element Th is considered to be highly immobile under 'standard' aqueous conditions and is exclusively concentrated in detrital heavy minerals.

3.2.1.14 Stability of Heavy Minerals and the Effects of Weathering and Hydrodynamic Sorting

Heavy minerals are generally more stable than micas and feldspars, and at least as stable as quartz. In spite of this, the effects of post depositional weathering/diagenesis should never be ignored as some heavy minerals are much more resistant to physical/chemical attack than others (Hurst and Morton 2014). Zircon, rutile, tourmaline and monazite are very stable, with moderate levels of stability displayed by amphibole, epidote, garnet, staurolite, xenotime and titanite. The heavy minerals kyanite, andalusite and sillimanite are relatively unstable, whilst olivine and pyroxene are amongst the most unstable heavy minerals. Temperature is widely considered to exert the principal control on heavy mineral mobility, with temperatures exceeding 60 °C generally required for mobilisation. The exceptions include the most unstable heavy minerals such as olivines, pyroxenes and amphiboles which are known to become unstable at temperatures below this. Indeed, Walderhaug and Porten (2007) estimate that complete dissolution of calcic amphibole in sandstones from the Norwegian Sea took place at temperatures of around 40 °C. What is less well understood than the influence of temperature are the effects of host rock chemistry and the composition of groundwaters. Clearly, this is a subject that requires more attention in the future. Elements that are highly mobile, such as Ca, which is hosted by clinopyroxene, amphibole, epidote, titanite and other heavy minerals, are released during the dissolution process. Elements that have low mobility, however, are also known to be released. (Hurst and Morton 2014). For example, the release of Ti and Nb must accompany the dissolution of titanite grains. Any chemostratigraphic and heavy mineral study should take account of weathering/diagenesis when using mineralogical or geochemical data to model changes in provenance or for correlation purposes.

In addition to weathering, the effects of transportation and hydrodynamic sorting should not be ignored. For example, in a study of Neogene-Recent sediments in the Amur River and its delta in North Sakhalin Basin, Nicholson et al. (2014) attributed a positive correlation between the garnet/zircon ratio and distance from the river mouth to hydrodynamic sorting across the delta system. They noted that lighter garnet grains were transported further than denser zircons, meaning that this parameter could not be used to recognise changes in provenance in this region. The importance of heavy mineral stability, transportation and hydrodynmamic sorting in chemostratigraphy studies is discussed in more detail in Chap. 4.

3.2.1.15 Use of Feldspars and Other Minerals for Provenance Determination

As a general rule, other detrital minerals such a quartz, feldspar and mica are rarely used for discrimination of provenance and their distributions are not routinely modelled using geochemical data for chemostratigraphic purposes. Quartz is the most abundant mineral in clastic sedimentary rocks, but the distribution of this mineral is largely controlled by depositional environment, hydrodynamic sorting and post depositional diagenesis. Consequently, it would be very difficult to use Si abundances, the main proxy for quartz, to infer changes in provenance or for chemostratigraphic purposes. Micas and feldspars are also controlled by these factors and are particularly susceptible to the effects of weathering. In some studies, feldspars are utilized to differentiate provenance but only in conjunction with heavy mineral data, and where the effects of chemical and physical attack are minimal. One example of this is a study by Nascimento et al. (2015) of the provenance of sands derived from the Amazon and Madeira rivers. Some chemostratigraphy studies use variations in K and Na, where they are associated with K feldspar and plagioclase respectively, to place correlative boundaries but, as with the aforementioned heavy mineral study, K and Na are utilized in conjunction with heavy mineral bearing, high field strength elements. Furthermore Na and K should only be used in chemostratigraphic studies where petrographic data indicate that there has been no, or very little, destruction of feldspar grains during transportation and burial diagenesis.

Micas are sheet silicate minerals with the general formula:

$$X_2Y_{4-6}Z_8O_{20}(OH, F)_4$$

in which X is K, Na or Ca, or less commonly Ba, Rb or Cs; Y is Al, Mg, or Fe or less commonly Mn, Cr, Ti or Li; Z is mainly Si or Al, but may also include Fe^{3+} or Ti. The most common forms in clastic sediments are muscovite which is an aluminium potassium silicate and biotite, a potassium, magnesium, iron aluminium silicate. Muscovite is considered to be a significant host for the elements K and Rb, while biotite is known to contain elevated concentrations of Mg and Fe. Klemme et al. (2011) used LA-ICP-MS to identify concentrations of trace elements and REE in metamorphic derived white micas. A Ti value of 4720 ppm was recorded but this is not considered to represent a high value for this element which is predominantly linked with heavy minerals. With the exception of Ba, Rb and Li, all of the other trace and REE were recorded at levels lower than 70 ppm. The elements Nb, V, Cr, Mn, Zn and Ga were recorded in the 30–60 ppm range but, as with Ti, these elements are more likely to be hosted by other minerals and the presence of mica is unlikely to have a significant impact on the whole-rock abundance of these elements unless these minerals are particularly abundant. Other trace elements and REE occurred in concentration of less than 1 ppm. For these reasons, and the fact that micas are known to be chemically unstable, they are rare modelled geochemically to indicate changes in source/provenance or for chemostratigraphic purposes.

3.2.2 Discussion on the Mineralogical Affinities of Elements

In conclusion, the high field strength elements of Zr, Hf, Nb, Ta, Th, Ti, Y, Cr and HREE are considered to be the most useful ones to utilize in chemosratigraphy studies as they are highly immobile, being largely unaffected by post-depositional weathering/diagenesis and almost exclusively concentrated in detrital heavy minerals. One unfortunate aspect of whole rock analysis by ICP or XRF is that it is often very difficult to relate abundances of these elements to specific heavy minerals in the absence of heavy mineral analysis. For example, it may be impossible to determine whether Th is concentrated in zircons, monazites, apatites or all three minerals. The only significant exceptions to this rule are Zr and Hf which are almost exclusive to zircon. Consequently, close associations between Zr and other elements such as Th, U and HREE, revealed by PCA and binary diagrams, may indicate a link between these elements and zircon. The distributions of other heavy minerals are notoriously difficult to model using whole rock inorganic geochemical data though the following generalizations can be made:

Zr and Hf: zircon
Ti, Nb, Ta: rutile, anatase, titanite titanomagnetite, magnetite ilmenite
Cr: mainly chrome spinel, titanite, titamomagnetite, magnetite ilmenite
Y: mainly xenotime, monazite and apatite
Th and HREE: found in a wide variety of heavy minerals.

Caution should be applied, however, when using these generalized rules. For example, Y may be most closely associated with xenotime, monazite and apatite but, if these minerals are absent in the first instance, Y may be found in other heavy minerals (e.g. garnets). Even where heavy mineral data are available, the distributions of some heavy minerals, such as garnets, may be very difficult to model using geochemical data, as the composition of individual grains may be highly variable. In some studies, individual grains have been analysed by LA-ICP-MS but this is highly expensive and rarely performed on a routine basis. In other instances, the 'marker' elements for heavy minerals are not analysed. For example, tourmaline is considered one of the main sources of B, but this element is too light to analyse by most XRF instruments, and is not measured by ICP owing to contamination from lithium metaborate flux used in the alkali fusion sample preparation technique (see Chap. 2 for details).

The elements U and P are often used in chemostratigraphy studies, with the latter being associated with phosphatic heavy minerals such as apatites, monazites and xenotimes. U may be concentrated in a variety of heavy minerals but is abundant in zircons, monazites, xenotimes and apatites. Caution should be applied when using these elements for chemostratigraphic purposes, however, as they may be influenced by changes in paleoenvironment. For example, P may be concentrated in biogenic phosphate, the distribution of which may be controlled by redox conditions. Similarly, U is often abundant in organic matter and should not be used to model changes in source/provenance where this is the case. When using U and P for chemostratigraphic purposes it is very important to integrate sedimentological data derived from core/field logging. For example, Triassic aged fluvial 'red bed' sediments from the northern North Sea were deposited under oxic conditions where very little, if any, organic matter was preserved. In a study of these sediments, P and U would almost certainly be concentrated in heavy minerals. Conversely, the same elements would probably be linked to organic matter in a study of black coloured organic rich shales encountered in the Upper Jurassic Kimmeridge Clay Formation in the same region.

3.2.3 Elements Influenced by Changes in Paleoredox

Next to changes in provenance, one of the most important parameters that chemostratigraphers try to model, with a view to correlation, are variations in paleoredox. The following paragraphs describe the elements that are most likely to be influenced by this factor and the mechanisms by which each one is adsorbed. Determining changes in paleoredox typically means distinguishing between oxic, suboxic and anoxic environments. Oxic conditions occur where O_2 concentrations in bottom waters exceed 2 ml O_2/1 H_2O. Levels of O_2 range from 0.2 to 2 ml O_2/1 H_2O in suboxic environments and are below 0.2 ml O_2/1 H_2O during anoxic conditions (Tyson and Pearson 1991). Sulphidic or nonsulphidic conditions may persist during anoxia, with sulphidic conditions also known as euxinic when hydrogen sulphide exists within the water column. Euxinic conditions are normally found in semi enclosed basins, such as the Black Sea and Cariaco Trench. A byproduct of sulfate-reducing bacteria is H_2S and the presence and absence of free H_2S in the water column is used to distinguish suphidic (euxinic) from non-sulphidic environments.

Under oxic conditions, organic matter is broken down by methanogenic bacteria via an oxidative-reductive disproportionation of carbon (Tribovillard et al. 2006). Oxygen depleted conditions, and ultimately anoxia, may occur at the sediment-water interface or within sediments, when the oxygen demand exceeds the supply. Anoxia may also develop in the water column of stagnant or confined water masses where there is a lack of circulation, or in places where intense organic matter degradation consumes O_2 faster than it is replenished, even in open-marine conditions.

High levels of Mo, Cu, Co, Ni, Zn, Cr, U and V are often associated with anoxic paleoenvironments and it may be possible to identify sediments deposited under these conditions where values of these elements exceed their 'average crustal rock' or 'average shale' concentrations (Wedepohl 1971, 1991; McLennan 2001). However, caution should be applied when using this technique for paleoenvironmental modelling as these elements may also be concentrated in clay, carbonate and heavy minerals. For example, U is commonly associated with zircons, monazites and apatites, so high levels of this element may simply reflect an increase in the proportion of U-bearing heavy minerals, rather than the onset of anoxia. Similarly, Cr is often linked with chrome spinels, titanomagnetites and other heavy minerals. Carbonate minerals are often characterised by elevated concentrations of Zn, whilst clay minerals are known to contain variable quantities of Mo, Cu, Co, Ni, Zn and V. When identifying anoxic sediments or sedimentary rocks the following factors should be considered:

(a) Color of mudrocks should be recorded. Dark grey, black or dark brown colored shales are normally deposited under anoxic conditions, whilst light grey, green and reddish colors are generally linked with oxic environments.
(b) The presence of pyrite is normally indicative of anoxia. However, this should be treated with caution as pyrite can also be precipitated hydrothermally.
(c) One of the best ways to identify anoxic sediments is to consider the associations between two or more elements using eigenvector analysis, correlation coefficient analysis and crossplots. For example, where the distribution of U is similar to that of Zr and Hf, this element would almost certainly be concentrated in detrital zircons. Conversely, an association of U with Mo, Cu, Ni and Co should be more likely to indicate anoxic environments. Enrichments in U, V, Cu, Co, Mo, Zn and Ni generally typify oxygen depleted environments. The element Cr may also be used for this purpose but care should be taken when interpreting Cr anomalies as this element is relatively stable and is much less susceptible to changes in paleoredox.
(d) The presence of fossils in sediments may be used to identify anoxic sediments, particularly where soft tissues are preserved. For example, in a study of Middle Devonian lacustrine sediments in N.E. Scotland, Craigie (1998) used the presence of fossil fish in dark grey to black mudrocks to identify deep water anoxic sediments.
(e) Normalisation of trace elements to Al is often a much more effective technique to identify enrichments associated with the development of anoxia, since sediments often contain high proportions of particular minerals that dilute trace element abundances of the sample. This is particularly true of biogenic dilutants such as calcium carbonate and opal. Al-normalised data is used in many chemostratigraphy studies but caution should be applied as high Mo/Al, Cu/Al and Ni/Al ratios, for example, cannot always be linked to anoxic sediments. High values of Al-normalised data may simply indicate a depletion of clay minerals related to environmental conditions or diagenetic factors unrelated to redox. Similarly, low Al-normalised ratios may reflect an increase in the proportion of detrital or authigenic clay minerals, again unrelated to redox.
(f) High TOC levels of greater than 2% normally typify anoxic paleoenvironments. Where TOC data are unavailable, values of this parameter can be estimated using the Schmoker ratio calculated from the density log (Craigie 2015 and references therein). However, anoxic environments are not universally defined by high TOC as the supply of organic matter from hinterland areas is often at least, if not more important, than preservation. In a study of Jurassic sediments in Saudi Arabia, Craigie (2015) identified beds with low TOC that were deposited under anoxic conditions.

In summary, it is necessary to use a multidisciplinary approach to identify sediments deposited in anoxic paleoevironments, rather than relying solely on geochemical data. In spite of this, concentrations of U, Mo, Co, Cu, Zn, Ni, V and Cr have been used on countless occasions to identify changes in paleoredox and this has aided chemostratigraphic correlation, identification of TOC enriched sediments/potential hydrocarbon source rocks, and the modelling of paleoenvironments. The following paragraphs provide details of how these elements and others can be used to model changes in redox.

3.2.3.1 Vanadium

V exists in the quasi-conservative form of vanadate oxyanions (HVO_4^{2-} and $H_2VO_4^{-}$) in oxic waters. In pelagic and hemipelagic sediments, this element is tightly coupled with the redox cycle of Mn (Hastings et al. 1996). Vanadate is known to absorb onto both Mn- and Fe oxyhydroxides (Wehrly and Stumm 1989) and possibly kaolinite (Breit and Wanty 1991). V^{5+} is reduced to V^{4+} under mildly reducing conditions and forms vanadyl ions (VO^{2-}), related hydroxyl species VO(OH)-3 and insoluable hydroxides $VO(OH)_2$ (Tribovillard et al. 2006). In more strongly reducing (i.e. euxinic) environments, the presence of free H_2S produced by bacterial sulfate reduction causes V to be further reduced to V^{3+}, which can be absorbed by geoporphyrins or be precipitated as the solid oxide V_2O_3 or hydroxide $V(OH)_3$ phase (Breit and Wanty 1991; Wanty and Goldhaber 1992; Tribovillard et al. 2006). According to Calvert and Pederson (1993), Algeo and Maynard (2004), and Tribovillard et al. (2006), this two-stage reduction process of V may lead to the formation of separate V carrier phases of contrasting solubilities under non-sulfidic anoxic versus euxinic conditions. Substitution of V^{3+} for Al occurs in the octrahedral sites of authigenic clay minerals or in recrystallizing clay minerals (Breit and Wanty 1991).

3.2.3.2 Uranium

Arguably, more attention has been given to U than any other element with regard to the recognition of sediments enriched in organic matter and deposited in anoxic paleoenvironments. This may be partly explained by the fact that this is one of the few elements that can be analysed downhole by spectral gamma wireline logging tools and can be measured in the field by hand held spectral gamma counters, without the requirement for more elaborate analysis by XRF or ICP. U has two stares, U^{4+} and U^{6+}, but U^{6+} is much more easily mobilised that the relatively stable U^{4+}. For this reason, most of the dissolved U in oxic waters is in the form of U^{6+}. In anoxic environments U^{6+} is reduced to U^{4+} which then normally precipitates as uraninte (UO_2) (Tribovillard et al. 2006). Less commonly, U substitutes for Ca in biogenic phosphate, also under anoxic conditions (Craigie 1998). The enrichment of U occurs within the sediment, not the water column, so the oxygen penetration depth and the sedimentation rate may be significant. A slower sedimentation rate, for instance, may enable more time for the diffusion of uranyl ions from the water column into the sediment (Crusius and Thompson 2000). Several authors, including Klinkhammer and Palmer (1991), and Tribovillard et al. (2006) report the removal of U from the water column to be accelerated by the formation of organometallic lignands in humic acids. The absorption of U on organic substrates is discussed in detail by Wignall and Maynard (1993) and Algeo and Maynard (2004). In theory U uptake can occur without bacterial activity. However, the reduction process would be very slow without bacterial activity, so there is normally a direct link between the intensity of sulfate reduction activity and the amount of reactive organic matter (Tribovillard et al. 2006).

Tribovillard et al. (2006) discuss the possibility of U remobilization affecting the abundances of this element in sediments. This can occur if oxygen penetrates to a depth where authigenic U has accumulated, for example by an increase in bottom water oxygen abundance, or a decrease in organic matter flux. Under these conditions at least some, if not all, of the U may be lost from the sediments to the overlying water column.

Along with Mo, U is thought to be one of the most useful elements in the identification of anoxic paleoenvironments. In spite of this, some caution should still be applied when using U for this purpose as U can also exist in organic matter under oxic conditions. In a detailed study of organic rich soils in an alpine region of Switzerland, Regenspurg et al. (2010) found U to be mainly bound to soil organic matter rather than to minerals and was mainly in the U^{6+} state. Retention in soil by absorption to soil minerals, or precipitation as U^{6+} minerals such as uranyl hydroxide, or calcium uranyl phosphate (e.g. uranophane), may occur at neutral to alkaline pH conditions. Regenspurg et al. (2010) also point out that uranyl complexes readily with organic molecules such as acetate, oxalate or humic acid. This may provide an explanation for high U concentrations reported in humic rich environments such as peats and bogs. Consequently, high levels of U may simply be related to the presence of organic matter rather than the existence of anoxic environments. Another potential complication when interpreting U data is that the element is abundant in heavy minerals, particularly monazites and zircons. Before using abundances of U to infer changes in depositional conditions or levels of organic matter, the mineralogical affinities of the element should be clearly understood, through statistical/graphical evaluations of the datasets and careful comparison with available mineralogical data.

3.2.3.3 Nickel, Copper and Zinc

The process of Ni, Cu and Zn incorporation into sediments under anoxic conditions is described by Tribovillard et al. (2006) and the following paragraph is a summary of this work. These elements, along with Co, Mo, V and Cr, are absorbed onto Fe-Mn-oxyhydroxides and then released upon reductive dissolution of these oxyhydroxides at or below the water-sediment interface. These trace elements are then available for new reaction and to form in new minerals. In reducing environments, Cu, Ni and Zn are known to form in pyrite and, as a consequence of this, are often used as 'key' indicators of anoxia.

Ni behaves as a micronutrient in oxic marine environments and may exist as a soluble Ni carbonate ($NiCO_3$) or within humic or fulvic acids (Calvert and Pedersen 1993; Algeo and Maynard 2004; Tribovillard et al. 2006). Accelerated scavenging of Ni may take place by complexation with organic matter. This may also result in Ni enrichments in the sediment (Piper and Perkins 2004; Nameroff et al. 2004; Naimo et al. 2005). Release of Ni from organometallic complexes to pore waters may occur during the decay or organic matter. In moderately reducing environments, Ni may be released from the sediments to the overlying waters, owing to the absence of sulphides and Mn oxides. Under (sulfate-) reducing conditions, however, Ni is likely to be absorbed as the insoluble NiS into pyrite (solid solutions), in spite of the slow kinetics of this process (Tribovillard et al. 2006). On rare occasions, the Ni complexed with organic matter may also be incorporated into tetrapyrrole complexes and can be preserved as Ni-geoporphyrins under anoxic/euxinic conditions (Grosjean et al. 2004; Tribovillard et al. 2006).

The element Cu mainly occurs as organometallic lignands and, to a lesser degree, $CuCl^+$ ions in solution in oxic marine environments. Complexation of Cu with organic matter is common, as is adsorption onto particulate Fe-Mn oxyhydroxides, both processes accelerating scavenging and subsequent sediment enrichment (Tribovillard et al. 2006). The release of Cu to pore waters and the overlying water column may result from the dissolution of Fe-Mn-oxyhydroxides during the decay of organic matter. In reducing environments, particularly bacterial sulfate reducing environments, Cu^{2+} is reduced to Cu^{1+} and may be incorporated in solid solution in pyrite. Alternatively, Cu^{1+} may form its own sulfide phases, such as CuS and CuS_2 (Morse and Luther 1999; Tribovillard et al. 2006).

In oxic marine environments, Zn acts as a micronutrient, being present as soluble Zn^{2+} cations or $ZnCl^+$. However, the element is mostly present as complexes with humic or fulvic acids (Calvert and Pedersen 1993). Adsorption of Zn onto particulate Fe-Mn oxydydroxides may also occur (Fernex et al. 1992). Release of Zn to pore waters and the overlying water column may take place upon organic matter decay. The element may be incorporated as ZnS as a solid solution phase in pyrite or, to a lesser degree, it may form its own sulphides, these processes occurring under anoxic conditions (Tribovillard et al. 2006).

Although these three elements have been used to define ancient flooding surfaces deposited in anoxic paleoenvironments (e.g. Craigie 1998, 2015), caution should be applied when using them for this purpose as they are often concentrated in carbonate and/or clay minerals. In addition to this, Ni and Cu can exist in high concentrations in detrital heavy minerals such as pyroxenes. Consequently, elevated values of these elements may not always be linked to anoxia. The element Mo is also abundant in pyrite and, in general, is much more useful in defining these environments as it is less likely to be absorbed by clay or carbonate minerals (Tribovillard et al. 2006). The incorporation of Mo into sediments deposited in anoxic environments is discussed in detail in the following paragraphs.

3.2.3.4 Molybdenum

Being mainly in the form of molybdate (MoO_4^{2-}), Mo is not concentrated by ordinary plankton and is not readily adsorbed by natural particles, including the surfaces of clay minerals (Goldberg et al. 1998). The element is, however, easily captured by Mn-oxyhydroxides at the sediment surface (Tribovillard et al. 2006 and references therein). Liberation of adsorbed Mo to pore waters occurs during subsequent reduction (Crusius et al. 1996). This process can concentrate Mo close to the sediment water interface, when the oxic-anoxic interface is encountered, though it does not explain the process of Mo fixation within the sediment (Tribovillard et al. 2006). The fixation of Mo is thought to occur during reduction, with the element being scavenged directly from the water column or captured from pore waters supplied diffusively with Mo from the water column (Calvert and Pedersen 1993; Crusius et al. 1996; Zheng et al. 2000). Other workers favour a direct role played by sulphide (Helz et al. 1996; Vorlicek et al. 2004). In such models, fixation in the presence of dissolved sulphide does not simply result from MoS_2 or MoS_3, but instead mineralization occurs via organic thiomolybdates and inorganic Fe–Mo–S cluster complexes, possibly taking the form of solid-solution components in Fe-sulphides. It is possible that H_2S/HS- transforms Mo from a conservative element to particle-reactive species in marine conditions (Helz et al. 1996). Replacement of oxygen atoms in MoO_4^{2-} by soft lignands, such as S donors are likely, and an important step in this inorganic pathway is the transformation of MoO_4^{2-} to thiomolybdates which are particle-reactive and thus prone to scavenging (Erickson and Helz (2000). For this reaction to occur, there is a necessity for the persistence of sulfidic conditions (Erikcson and Helz 2000). Bonds are formed between Mo and metal rich (notably Fe) particles, sulfur-rich organic molecules (Helz et al. 1996; Tribovillard et al. 2006) and iron sulphide (Vorlicek et al. 2004) once the thiomolybdate switch has occurred. Vorlicek et al. (2004) described the formation of Fe–Mo–S cluster compounds which are retained on pyrite surfaces, arguing that a reduction step is favourable, if not necessary, in that case.

The importance of organic matter in the fixation of Mo deserves some consideration, with one role of organic matter being a carrier for Fe and other trace metals. A second role is more direct involving O–S group attached to acromolecular detritus being directly inserted into MoO_4^{2-}, forming covalent bonds between Mo and the macromolecule (Tribovillard et al. 2006). Thiols, including humic-bound thiol groups, can also activate Mo (Helz et al. 1996). The work of Adelson et al. (2001) suggested that organic thiomolybdates are formed by replacement of oxygen in the first coordination sphere by S, resulting in covalent S bonds between Mo and sulfidized macromolecules.

In summary, it is highly probable that most Mo is concentrated in pyrite in sediments deposited in anoxic environments, though the possibility of some being associated with organic matter should not be ignored. Whether the precise mineralogical affinities of Mo and the mechanism(s) by which it is concentrated in sediments can be established for a given study or not, the association between this element and anoxic paleoenvrionments have been observed in many studies throughout the globe (e.g. Craigie 1998, 2015; Meyers et al. 2005; Wright et al. 2010; Algeo and Tribovillard 2009). One exception to this, however, has been noted in restricted depositional environments, where increasing restriction results in lower deepwater aqueous Mo concentrations, resulting from Mo removal to the sediment in excess of resupply. This causes lower Mo/TOC ratios with decreasing aqueous Mo abundances (Algeo 2004). Such effects have only been observed in restricted environments, but their importance in the recognition of ancient anoxic paleoenvironments probably requires more attention in the future.

Tribovillard et al. (2006) also note that that both Mo and V can exist in nitrogenase, an enzyme used by nitrogen-fixing bacteria. They also recognise, however, that they are likely to be a relatively low contributor to the total abundance of Mo in sedimentary rock.

3.2.3.5 Cadnium

The element Cd is not normally analysed during 'standard' chemostratigraphy studies using the XRF or ICP techniques. In spite of this, the element requires some consideration, given its strong association with anoxic sediments. Unlike Cu, Ni and Zn, Cd only has one state (Cd^{2+}) and has a nutrient-like behaviour in the water column and sediment, implying a relatively short residence time (Morford and Emerson 1999; Tribovillard et al. 2006). Delivery of Cd to the sediment occurs via and association with organic matter. During organic matter decay, Cd is released to the pore waters/water column. Under reducing conditions, the element is thought to form a separate insoluble sulphide phase (CdS) in the presence of H_2S, rather than coprecipitating with FeS (Tribovillard et al. 2006).

3.2.3.6 Manganese

Concentrations of Mn are normally highest close to oxic-anoxic boundaries and in suboxic environments (Craigie 1998, 2015; Calvert and Pedersen 1993; Wirth et al. 2013). Intense leaching of Mn, induced by reducing water conditions, results in increased concentrations of dissolved Mn^{2+} in bottom waters. Conversely, rapid Mn^{2+} oxidation can occur via autocyclic reactions and/or through microbial metabolism after mixing with oxygenated waters (Wirth et al. 2013). Mn-(oxyhydr)oxide and Mn-carbonate crusts can eventually form as a result of such intense Mn^{2+} oxidation and have been recorded by several workers, including Bellanca et al. (1999) and Sabatina et al. (2011). Wirth et al. (2013) conclude that Mn (oxyhydr)oxides and Mn-carbonates have therefore been brought in association with generally anoxic depositional environments that are episodically perturbed by oxygenation events. These may take the form of eustatic sea-level changes, changes in primary productivity or inflows of oxygenated waters into marine or lacustrine basins (Calvert and Pedersen 1996; Wirth et al. 2013). The formation of Mn-(oxyhydr)oxides in the water column are triggered by the injection of O_2-rich waters. Such Mn-(oxyhydr)oxides may be transformed to Mn-carbonates in the presence of reducing anoxic sediments (Calvert and Pedersen 1996). Where conditions are conducive to the preservation on these Mn minerals, such as rapid burial before Mn minerals are reduced to Mn^{2+} and released to the water column, sedimentary Mn enrichments can thus be taken as indicators of sporadic water-column ventilation during otherwise anoxic conditions in the past.

In spite of the importance of this element as a paleoredox indicator, Mn may also be concentrated in clay and carbonate minerals, the formation of which are often unrelated to paleoenvironment. For this reason, it is important to consider mineralogical, sedimentological and/or paleontological data, as well as the association of Mn concentrations with other elements, before utilizing elevations in this element to infer the existence of suboxic conditions.

3.2.3.7 Chromium

The Cr^{6+} species normally occurs in oxygenated seawater, with subordinate Cr^{3+} (Calvert and Pedersen 1993). In anoxic environments Cr^{6+} is reduced to Cr^{3+}, forming aquahydroxyl cations which can complex with humic/fulvic acids or be absorbed to Fe- and Mn-oxyhydroxides (Algeo and Maynard 2004). Although some studies show a clear association between Cr and the development of anoxia (e.g. Craigie 1998), many do not. Tribovillard et al. (2006) discussed the reasons for this. One possible explanation may be that this element has structural and electronic incompatibilities with pyrite and Cr^{3+} uptake by authigenic Fe sulphides is very limited. It also does not form an insoluble sulphide. A consequence of organic matter remineralization by sulfate-reducing bacteria is that Cr is not readily trapped within the sediments as a sulphide, and may be lost to the overlying water column by diffusive/advective transport during sediment compaction. A further, and arguably more important, reason for the difficulties in linking Cr concentrations with paleoredox, is that this element is often concentrated in detrital heavy minerals such as chrome spinel (see earlier section of this chapter for detailed explanation). It is entirely plausible that increases in Cr may be related to changes in provenance and/or hydronynamic sorting rather than paleoredox. It is highly recommended that Cr is used in conjunction with other redox sensitive elements and parameters, if it is to be employed as an indicator of anoxic paleoenvironments.

3.2.3.8 Cobalt

The element Co exists as the dissolved cation Co^{2+} or is complexed with humic/fulvic acids (Saito et al. 2002; Whitfield 2002; Achterberg et al. 2003). The formation of insoluble CoS occurs under anoxic conditions, which can be absorbed by authigenic Fe-sulfides (Huerta-Diaz and Morse 1992). However, Tribovillard et al. (2006) point out that CoS incorporation into authigenic sulphides is limited by very slow kinetics. As with Cr, Co is often concentrated in detrital minerals, providing a further reason why this element is often unused as a proxy for the development of anoxia.

3.2.3.9 Phosphorous and Phosphorites

Phosphorous is essential to all forms of life and being linked to the supply of organic matter, the element may be used to model changes in organic productivity and depositional environment. Craigie (1998), for instance, linked elevated P values to the preservation of fossil fish bones and scales preserved in deep water lacustrine conditions, encountered in Middle Devonian sediments in NE Scotland. Unfortunately, the link between P, organic productivity and depositional environment is often more complicated, and is summarized in the following paragraphs.

The phytoplankton necromass that reaches the sediment-water interface provides the principal source of P to the sediments, though fish scales and bones also play a role. During bacterial degradation of organic matter in the sediment, under oxic, suboxic or anoxic conditions, P is released as PO_4^{3-}. Most P is released to the water column with only around 1% of organic P trapped in the sediments (Benitez-Nelson 2000). Under certain conditions, however, P released to pore waters can reach high concentrations and be precipitated as authigenic phosphatic minerals, most commonly as carbonate fluoroapatite, aka francolite (Tribovillard et al. 2006). According to Trappe (1998), Piper and Perkins (2004) and Tribovillard et al. (2006), francolite may precipitate rapidly by replacing a short-lived and poorly crystallized precursor, or slowly by replacing calcite. Obviously, the concentration of P has to be high enough in the first instance to enable the precipitation of phosphatic minerals. This may result from a high supply of organic matter as was the case in the aforementioned study by Craigie (1998). However, high organic matter productivity is not the only prerequisite to the formation of these minerals. Tribovillard et al. (2006) point out that P enrichment can be effectuated through the redox cycling of iron in low-productivity areas of oceans, where P is absorbed onto iron-oxide coatings. They also mention the involvement of Mn in P retention and an alternative method of P uptake by bacterially precipitated polyphosphates.

In addition to P forming in frankolite and other phosphatic minerals, Kraal et al. (2017) found that calcium carbonate can play an important role in P retention in anoxic and sulphidic sediments.

They also noted that calcium carbonate may be more important for P retention in carbonate rich sediments with low HPO_4^{2-} concentrations and thus limiting P mineral authigenesis. It is significant, however, that P bearing carbonates are relatively unstable and it is possible that P could be released to the water column during carbonate dissolvation, resulting from changes in pH over time. A further complication is that P may exist in the form of detrital phosphatic heavy minerals such as apatite. Consequently, increases in the levels of this element may simply be related to hydrodynamic sorting of changes in source/provenance. In summary, P is not always a good paleoenvironmental indicator.

Craigie (1998) found an obvious association between P and the preservation of fossil fish in deep water lacustrine environments. High levels of Y, U, Th, Sr and REE are concentrated in this material. In this study, elevated concentrations of these elements and P were used to identify anoxic paleoenvironments, developed during 'wet' periods of the Middle Devonian (in the Orcadian lacustrine basin, NE Scotland), resulting from maximum lake expansion. High concentrations of U in biogenic phosphate often indicate deposition under anoxic conditions, with U^{4+} substituting for Ca^{2+} in the francolite structure in such environments. Whilst this reaction is well documented, Jarvis et al. (1994) and Trappe (1998) argue that U can also be adsorded by francolite during oxic conditions. As with P, it is not possible to use U, or the concentration of other elements associated by biogenic phosphate, to make inferences on paleoproductivity of paleoredox in the absence of supporting geochemical, sedimentological and/or paleontological data.

3.2.3.10 The Importance of U and Mo

One conclusion from the aforementioned review of literature on the employment of trace elements to the identification of anoxic paleoenvironments is that U and Mo are the most useful elements for this purpose, but the distributions of these elements do not always mirror each other. Based on their own research and extensive literature reviews, Algeo and Tribovillard (2009) and Tribovillard et al. (2012) point out that absorption of U in marine sediments commences at the Fe^{2+}:Fe^{3+} redox boundary (i.e. suboxic conditions) at a much earlier stage than the uptake of authigenic Mo which requires the presence of H_2S. A second difference in the behaviour of the two elements is that the transfer of authigenic Mo to the sediment can be accelerated through a particulate Mn–Fe oxyhydroxide shuttle, whilst authigenic U is unaffected by this process. Forming at the chemocline, particulate Mn-Fe-oxyhydroxides adsorb molybdate oxyanions during transit through the water column. These particles are reductively dissolved on reaching the sediment/water interface, releasing molybdate ions that either diffuse back into the water column or are scavenged by other minerals within the sediment. The latter process results in the rapid transfer of authigenic Mo to the sediment. The effectiveness of these shuttles largely depends on the size of the basin and the height and stability of its chemocline. The particulate shuttle is likely to be enhanced in smaller basins characterized by a deep or highly variable chemocline. Conversely, the particulate shuttle is likely to be less effective in larger basins characterized by a shallow or stable chemocline.

In a study of modern ocean anoxic environments, Algeo and Tribovillard (2009) found both U and Mo to show no enrichment in oxic facies. In suboxic facies modest enrichments occurred in both elements with Mo/U ratios of sediments distinctly lower than that of seawater. This is explained by the preferential uptake of U over Mo, as U^{6+} is reduced to U^{4+} at the Fe^{2+}:Fe^{3+} redox boundary prior to the onset of sulfate reduction which controls the accumulation of authigenic Mo. Strong enrichment of both elements were noted in anoxic facies. These display progressively higher Mo/U ratios as total authigenic concentrations increased.

In addition to both elements and the Mo/U ratio increasing as conditions become more reducing, the abundance of organic matter also exerts a significant control on Mo/U ratios. Concentrations of authigenic U are controlled by both redox and the abundance of organic matter. Organic matter is often preserved in anoxic environments, but sometimes it can accumulate under oxic-suboxic condition, where relatively low Mo/U ratios reflect the abundance of organic matter and the persistence of oxic or suboxic conditions. This effect was noted by Craigie (2015) in study of Jurassic organic-rich carbonates in central Saudi Arabia.

3.2.3.11 Remobilisation of Elements Associated with Anoxic Paleoenvironemnts

The aforementioned elements associated with the development of anoxic conditions are relatively 'mobile' compared with high field strength ones concentrated in detrital heavy minerals. Consequently, an understanding of depositional environments and the effects of post deposition-diagenesis should be sought before using inorganic geochemical data to interpret paleoredox conditions. The redox sensitive elements may be linked with several phases such as metal sulphides, pyrite, phosphate, sulfate, insoluble oxides and oxyhydroxides, organometallic complexes and organic matter. For this reason, the mobilisation and behaviour of these elements is highly variable and also dependent on pH, Eh and burial environment. These factors should not be ignored but, according to Tribovillard et al. (2006), sulphides are relatively stable in the absence of oxidizing agents and the elements concentrated in or coprecipitating with (iron-) sulphides typically do not move during diagenesis. This generally holds true for the elements Mo, V, Cd, Ni, Co, Cu and Pb which show varying tendency to be incorporated in pyrite. Based on the work of Tribovillard et al. (2006) and references therein, the following hierarchy is suggested for sensitivity to pyritization: Mo > Cu = Fe > Co > Ni > Mn > Zn > Cr = Pb > Cd.

A further complication factor, in the case of Zn, Pb and Cd, is that these elements may form soluble complexes with reduced S species, in addition to existing in insoluble sulphides (Huerta-Diaz and Morse 1992). For these reason, these elements may be remobilised and reprecipitated as solid sulphides or iron and manganese oxides.

Oxidation of previously anoxic sediments can result in the mobilisation of U but the effect is considered less pronounced for V, Cr, Cd, Mo, Cu and Ni (Morford et al. 2001; Thomson et al. 1998; Tribovillard et al. 2006).

In summary, mobilization of redox-sensitive elements during burial and diagenesis is relatively uncommon, but should not be completely ignored. For this reason, the use of geochemical data to identify flooding surfaces and anoxic events should be corroborated with available sedimentological, petrographic and/or palaeontological data. For example, if the examination of a core reveals the presence of authigenic pyrite occurring in an black organic rich mudrock, then elevations of Mo, Cu, Ni and Co associated with that bed are likely to be 'in situ' and unaffected by post depositional diagenesis.

3.2.4 Clay Minerals

The term "clay" refers to materials with particle sizes of less than 2 μm, and also to the family of minerals having similar chemical compositions and common crystal structural characteristics. They are hydrous aluminosilicates and have a number of distinct physical characteristics including plasticity, shrinkage under fire and under air-drying, cohesion and capacity of the surface to take decoration (Mockovciakova and Orolinova 2009; Odoma et al. 2013). Clay minerals are often major components of both clastic and carbonate sedimentary rocks and, though their structure, chemical formulae and major element geochemical compositions have been established for decades, very little information exists on the trace element composition of these minerals, and the principal controls on the absorption of these elements to clay mineral surfaces. An extensive literature search of 'standard' geological journals such as 'Chemical Geology', 'Geochimica et Cosmochimica Acta' and others (including text books on mineralogy) on the part of the author revealed very little, if any, information on the trace element compositions of clay minerals. This may be explained by the small size of clay mineral crystals (2 μm or less) making them more difficult to examine. It may also be explained by their limited application in chemostratigraphy studies, where most schemes are based on changes in source/provenance reflected by variations in heavy minerals. In other chemostratigraphy studies, changes in paleoredox are measured by abundances of trace elements linked with organic matter and redox sensitive minerals such as pyrite. As a general rule, it is very difficult to use the abundance of clay minerals to model changes in provenance as they are chemically/physically unstable compared with heavy minerals, and it is impossible to determine whether they are detrital or authigenic in origin based on geochemical data alone.

Some elements, such as Cr and V, are absorbed by clay minerals in anoxic paleoenvironments but, in normal circumstances, anoxic paleoenvironments can be recognized more clearly by increases in Mo and U. For these reasons, the trace element geochemistry of clay minerals has been largely overlooked. In fact, the most detailed information on this subject can be found in journals relating to chemical engineering and environmental science, where clay minerals have been utilized to remove heavy metal contaminants from aqueous samples. Uddin (2017) summarizes much of the research in this field from 2006 to 2016 and the reader is urged to consult this excellent work and references therein for more information. One unfortunate aspect of this research is that it tends to be concentrated on the use of clay minerals to remove 'toxic' elements such as As, meaning that the abilities of clay minerals to absorb other/less toxic elements are largely ignored. Another issue is that many studies utilize synthetic or modified clay minerals to remove trace metals, rather than naturally occurring clay minerals. In spite of this, Uddin (2017) documents research conducted on both naturally occurring and synthetically modified clay minerals, and is considered to represent one of the most studies on clay mineral trace element compositions in recent times.

Many types of clay minerals exist in nature, but the most common ones found in sedimentary rocks are kaolinite, illite, montmorillonite

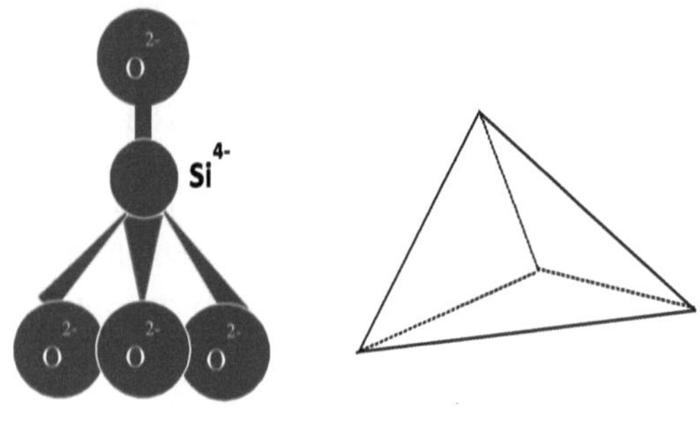

Fig. 3.1 Silica sheets consist of SiO_4^{2-} tetrahedra connected at three corners forming a hexagonal network in the same direction. This is called tetrahedral sheets (after Uddin 2017)

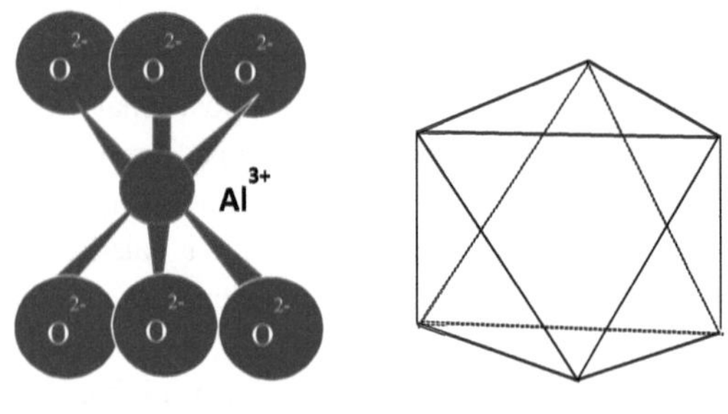

Fig. 3.2 Octahedron comprising central cation (Al^{3+}, Fe^{2+}, Mg^{2+}) surrounded by 6 oxygens (or hydroxyls) (after Uddin 2017)

(smectite), chlorite and vermiculite. The following paragraphs concentrate on these minerals.

Most clay minerals comprise layers containing silica and alumina sheets and are subdivided according to the type of layer structure. Tetrahedral and Octahedral sheets occur in clay minerals, with the former consisting of SiO_4^{2-} tetrahedra connected at three corners forming a hexagonal network in the same direction (Fig. 3.1). Octahedrons are defined by a central cation of Al^{3+}, Mg^{2+} or Fe^{2+} surrounded by 6 oxygens (or hydroxyls) (Weaver and Pollard 1973) (Fig. 3.2). The 1:1 clay mineral type is characterized by one tetrahedral sheet and one octrahedral sheet. By contrast, the 2:1 group comprises two silica tetrahedral sheets between which is an octrahedral sheet (Weaver and Pollard 1973).

With a 1:1 structure of one tetrahedral silica sheet linked to an octrahedral alumina sheet, kaolinite has the formula $Al_2Si_2O_5(OH)_4$. The octrahedral and tetrahedral sheets share a common plane of oxygen atoms (Fig. 3.3). As there is no interlayer swelling, the bonding between the layers is very strong. In an extensive review of the chemistry of clay minerals, Uddin (2017) documented 'typical' kaolinites as containing around 53% SiO_2, 2% Fe_2O_3, 43% Al_2O_3, 0.1% TiO_2 and 0.5%K_2O. Kaolinite has a very low cation exchange capacity, but Uddin (2017) records absorption of small quantities of Pb, Cd, Ni, Zn and Cu onto the structure of this mineral.

Montmorillonites (smectites) with the general formula $(Na,Ca)_{0.33}(Al,Mg)_2(Si_4O_{10})(OH)_2 \cdot nH_2O$, consist of 2:1 layered structures of an octahedral alumina sheet occurring between two opposing tetrahedral silica sheets (Fig. 3.3) (Brigatti and Galan 2013). Water and exchangeable ions are able to enter the structure of this mineral with relative ease as the bonding between the two silica sheets is very weak. This weak bonding explains the swelling capacity of this mineral (Fjar et al. 2008). The main geochemical components are SiO_2 (c. 65%), Fe_2O_3 (c. 2%), Al_2O_3 (c. 13%), CaO (c. 0.25%), NaO (c. 1%), K_2O (c. 1.5%) and MgO (c. 1%). With such an expanding lattice, montmorillonites have a considerably larger capacity to adsorb trace elements than other minerals. Uddin (2017) records the ability of montmorillonite to adsorb Cu, Pb, Hg, Cd, Zn, Ni, Co, Sr, Cs and Cr.

With the formula $(K,H_3O)(Al,Mg,Fe)_2(Si,Al)_4O_{10}[(OH)_2,(H_2O)]$, illite also has a 2:1 structure but the interlayers are bonded together with a K^+ ion to satisfy the charge and lock the structure (Fig. 3.3).The major element geochemistry of illite is dominated by SiO_2 (c. 62%), Fe_2O_3 (c. 5%), Al_2O_3 (c. 14%), CaO (c. 5%), NaO (c. 1%), K_2O (c. 3%), MgO (c. 2%) and TiO_2 (c. 0.7%) (Uddin 2017). The swelling capacity of illite is none to very slight, but the mineral is known to adsorb Pb, Cr and Cd (Uddin 2017). Strong links between Rb, K and illite have been noted in several studies (e.g. Craigie 2015) and this is not surprising given the closeness in ionic radii between the two elements.

Vermiculite has a 2:1 structure, medium shrink-swell capacity and the general formula $(Mg,Fe^{+2},Fe^{+3})_3[(Al,Si)_4O_{10}](OH)_2 \cdot 4H_2O$. This clay mineral has a high cation exchange capacity, in which the K+ ions of illite are replaced mainly by Mg^{2+} (Fig. 3.3). The major element geochemistry of vermiculite is documented by Uddin (2017) as follows: SiO_2 (c. 39%), Fe_2O_3 (c. 8%), Al_2O_3 (c. 12%), CaO (c. 3%), K_2O (c. 4%) and MgO (c. 20%). Uddin (2017) also recognizes the ability of this mineral to adsorb Cr, Cu, Ni, Co, Zn and Cd.

Chlorites are phyllosilicate minerals comprising regularly stacked and negatively charged 2:1 layers. They are characterised by Mg^{2+}, Al^{3+}, Fe^{3+}, and Fe^{2+} on octrahedral sites, with a single and positively charged interlayer octahedral sheet connected by H-bonds. The tetrahedral cations are Si^{4+} and Al^{3+} (Uddin 2017). Chlorite is a non-swelling clay and, as such, it is postulated to have a limited ability to adsorb trace elements. Despite an extensive literature search, the author could not find any information on the trace element compositions of this mineral, but Uddin (2017) documents the composition of a chlorite sample as follows: SiO_2 (27.4%), Fe_2O_3 (2.4%), Al_2O_3 (18.9%), MgO (34%).

Based on the results of tests on natural and synthetic clays, Uddin (2017) records the capacity of various clay minerals to adsorb heavy metals, but more precise estimates of the quantities of these elements and others in clay minerals are unavailable. Clearly, much more research is required on the trace element composition of clay minerals.

3.2.5 Elements Associated with Carbonates

A large number of carbonate minerals have been identified in nature, the most common of which are shown in the following list:

Calcite Group: Trigonal
Calcite $CaCO_3$
Gaspeite (Ni, Mg, Fe^{2+}) CO_3
Magnesite $MgCO_3$
Otavite $CdCO_3$
Rhodochrosite $MnCO_3$
Siderite $FeCO_3$

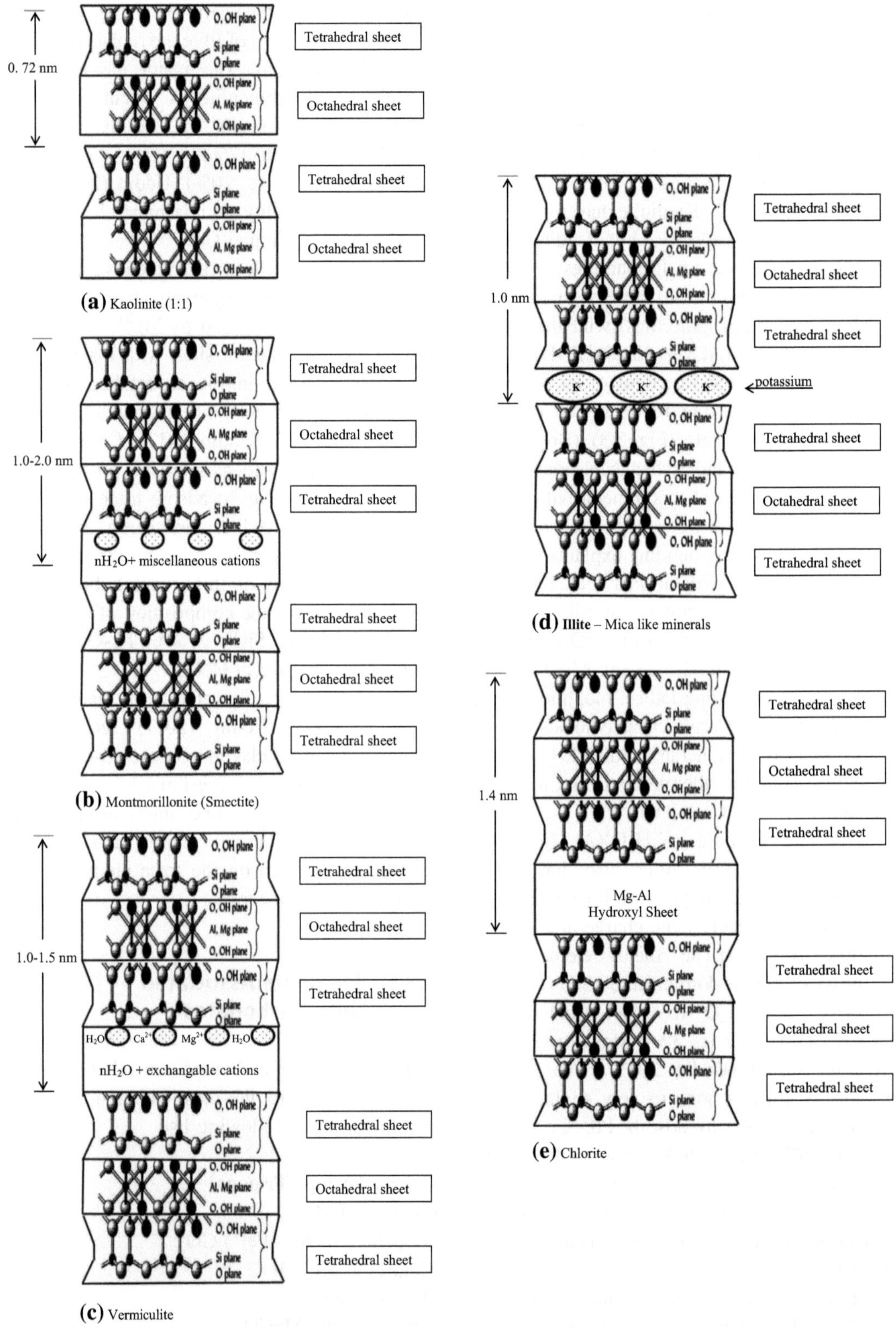

Fig. 3.3 Crystal structures in diagrammatic form, with respect to the following minerals: **a** Kaolinite, **b** Montmorillonite (Smectite), **d** Illite and **e** Chlorite (after Uddin 2017)

Smithsonite $ZnCO_3$
Spherocobaltite $CoCO_3$
Aragonite group: Orthorhombic
Aragonite $CaCO_3$
Cerussite $PbCO_3$
Strontianite $SrCO_3$
Witherite $BaCO_3$
Rutherfordine UO_2CO_3
Natrite Na_2CO_3
Dolomite group: Trigonal
Ankerite $CaFe(CO_3)_2$
Dolomite $CaMg(CO_3)_2$
Huntite $Mg_3Ca(CO_3)_4$
Minrecordite $CaZn(CO_3)_2$
Barytocite $BaCa(CO_3)_2$
Carbonate with hydroxide: Monoclinic
Azurite $Cu_3(CO_3)_2(OH)_2$
Hydrocerussite $Pb_3(CO_3)_2(OH)_2$
Malachite $Cu_2CO_3(OH)_2$
Rosasite $(Cu, Zn)_2CO_3(OH)_2$
Phosgenite $Pb_2(CO_3)Cl_2$
Hydrozincite $Zn_5(CO_3)_2(OH)_6$
Aurichalcite (Zn, Cu)5$(CO_3)_2(OH)_6$
Hydrated carbonates
Hydromagnesite $Mg_5(CO_3)_4(OH)_2.4H_2O$
Ikaite $CaCO_3{\cdot}6(H_2O)$
Lansfordite $MgCO_3{\cdot}5(H_2O)$
Monohydrocalcite $CaCO_3{\cdot}H_2O$
Natron $Na_2CO_3{\cdot}10(H_2O)$
Zellerite $Ca(UO_2)(CO_3)_2{\cdot}5(H_2O)$.

This list may seem long and the range of geochemical compositions almost endless, but most of these minerals are extremely rare. The most common carbonate minerals include calcite, dolomite and aragonite, the latter mineral being metastable at surface pressures and often being replaced by calcite. Calcite, dolomite and aragonite are the principal hosts for Ca and Mg. Other elements associated with carbonate minerals include Fe, Mn, Sr, Na, Ba, U, Cu, Co, Zn and Ni. Incorporation of these elements is thought to occur dominantly by substitution for Ca^{2+}, rather than interstitially between lattice planes, at site defects, as adsorbed cations, or within inclusions (Banner 1995; Dickson 1990; Moore and Wade 2013). Of the other carbonate minerals in this list, ankerite and siderite may be found in both carbonate and clastic sediments and are a major host of Fe. Given the abundance of calcite, dolomite and aragonite in carbonate sediments and sedimentary rocks, the following discussion centres on these minerals.

The magnitude of the distribution coefficient for the various carbonate minerals is largely determined by the size of the crystal lattice. Aragonite has an orthorhombic form capable of accommodating cations larger than Ca, such as Sr, Na, Ba and U. By contrast, the smaller unit cell of hexagonal calcite preferentially incorporates smaller cations, such as Mg, Fe and Mn (Moore and Wade 2013). This explains why the concentrations of Sr are in the range 7000–10,000 ppm in aragonite, but only 1000–3000 ppm in calcite. Similarly, Na is concentrated at levels of around 4000 ppm and 1000–2000 ppm in aragonite and calcite respectively. The element U occurs in concentrations lower than 3 ppm in most carbonates, but are typically an order of magnitude higher in aragonite (around 2.5 ppm compared with 0.05 ppm in calcite). Conversely, Fe and Mn exist in concentrations of less than 1 ppm in aragonite, but 1–2 ppm in calcite. Values of Mg are highly variable but are around 1000 ppm in aragonite, and 1000 ppm—4% in calcite.

Dolomites are known to contain lower proportions of Sr than calcite because Sr only partitions into the Ca sites, of which there are less than that of calcite. In general, therefore, the proportion of Sr increases with the content of Ca in dolomites (Moore and Wade 2013).

According to Moore and Wade (2013), the study of trace elements in carbonates initially sparked considerable interest. For instance, it was thought that the Sr and Mg contents of calcite cements might reveal aragonite and magnesium calcite precursors. However, this interest wained when it became apparent that there were many controls on the distribution of these elements such as Eh, pH, temperature, the chemistry of seawater and the influence of diagenetic fluids. The relative effects of these factors on the concentrations of trace elements in a given study are often very difficult to model. A further problem when studying whole rock geochemical data for carbonates, as is normally the case in chemostratigraphy studies, is that many of the trace elements may be associated with detrital minerals. For example, U can exist in small quantities in carbonate minerals, but occurs at much higher levels in heavy minerals, such as zircons, monazites and apatites.

In general, trace elements linked with carbonate minerals are often limited in their value to understand carbonate precipitation and diagenesis. In spite of this, they can be of some interest if used intelligently within a solid geologic and petrologic framework. One example of such a study was by Banner (1995) who constructed quantitative models of diagenesis using a combination of trace elements and isotopes. In a study of Upper Jurassic ooids in southern Arkansas. Moore et al. (1986) employed trace element and petrographic data to infer that these ooids were derived from both aragonite and magnesium calcite. The highest levels of Sr were yielded by inferred aragonite ooids, whilst inferred magnesium calcite ooids were characterised by low Sr but elevated Mg. In a study of Upper Jurassic carbonates in Spain, Coimbra et al. (2015) were able to employ a combination of elemental, isotopic, petrographic and sedimentological data to distinguish epicontinental from epioceanic sediments. The former were heavily influenced by post-depositional diagenesis, explaining their relatively high concentrations of Mg, Fe and Sr, compared with epioceanic carbonates which were largely unaffected by these processes. Modelling of changes in the chemistry of carbonate minerals can also be used for chemostratigraphic correlation and this is discussed in more detail in Chap. 5.

3.2.6 Drilling Mud Additives

Very little information has been published on the influence of drilling additives on inorganic geochemical data, possibly because data would not have been published in studies where this proved problematic. This issue must be given some consideration when analyzing cuttings and, to a lesser extent, core and sidewall core samples. As a very general rule, however, the distribution of elements are largely unaffected by the presence of drilling additives, provided the samples have been washed in liquid detergent and carefully 'picked' prior to analysis. The elements most commonly influenced by drilling additives are Ba, Ca and Sr. Of these, Ba is considered a proxy to the amount of drilling additive left in the sample after processing. Concentrations of greater than 20,000 ppm Ba should not be expected in cuttings samples, unless these have not been washed to remove additives. Where levels of this element exceed 20,000 ppm, reprocessing of the sample (including more rigorous washing and careful picking) is recommended. Relatively low levels of Ba (typically 200–1000 ppm) are expected in core samples but, as with cuttings, higher abundances than 20,000 ppm may indicate the presence of drilling additives. Even some core samples may have to be washed prior to analysis under such circumstances. Failure to reprocess samples with high Ba may result in the dilution of other elements, to the point at which they occur at levels below the limit of detection of the instrument. This may present a particular problem in XRF analysis, where the limits of detection are much higher than those of ICP.

The element Ba has been linked to barite cements and is often found in trace quantities in feldspars and clay minerals. Tribovillard et al. (2006) also mentioned that this element could be used as a paleoproductivity indicator in some studies, though not all. The reason that Ba is rarely used for chemostratigraphic purposes, however, is that it is most heavily concentrated drilling additives in the vast majority of core, sidewall core or cuttings samples.

Where a calcareous additive, such as marble, had been used during drilling, the data may show contamination of Ca and Sr. Careful consideration of data acquired from core/cuttings descriptions and wireline log signatures may be used to determine whether Ca and Sr occur 'in situ' (e.g. as calcite cement) or as additives. For example, if a set of samples were derived from quartz arenites with no recorded carbonate cement but produced 10% Ca, then it would be obvious that Ca would be associated with drilling additives. Reading mud logging and mud engineering reports should also give an indication of the amount and composition of the additives, though this information is often unavailable, particularly for historic wells. Another good way to determine whether Ca and Sr have been influenced by drilling additives is to consider the relationship between these elements and Ba. Given that Ba is considered a proxy for drilling additives in most studies, close links between Ba, Ca and Sr, revealed through statistical analysis or by plotting the data on binary graphs, may infer that Ca and Sr are also concentrated in additives.

In XRF studies, elevated proportions of drilling additive bearing Ca (i.e. where Ca > 9%), and less commonly, Ba and Sr, may result in dilution of low concentration trace elements to the point at which they occur at levels below the limit of detection. This 'dilution' effect can be resolved by using ratios for chemostratigraphic purposes, rather than individual elements. On a more serious note, a linear relationship may not necessary exist between the enrichment of Ca and the depletion of particular elements in XRF analysis. It is possible, for instance, that the element Th could be depleted by a factor of 50% where Ca values range from 10 to 18% but by more than 75% where Ca ranges from 18 to 25%. Given that most calibration methods developed for clastic sediments are based on Ca concentrations of 0–7%, the presence of high abundances of this element may result in non-linear dilution of others. In addition to this, some elements may be 'artificially' diluted at different proportions to others, adding further complexity to the data. These problems relating to enrichment of Ca and subsequent depletion of other elements are much more serious in XRF than ICP analysis, so the latter technique would be favoured under such circumstances.

Another problem noted by the author occurs when a bentonite drilling additive is employed. Bentonite is defined as an absorbant phyllosilicate clay mainly comprising montmorillonite. The main elements associated with bentonite include Na, Ca and/or K, though a number of trace elements may also be contained in this material. For example, contamination of Ti, Nb, Cr and other elements may occur, rendering them useless for chemostratigraphic purposes. Thankfully, additives of this composition are rarely used.

In summary, drilling additives rarely affect the quality of ICP and XRF data, except for the occasions when the samples have not been washed/prepared properly, or where Ca-bearing additives are used, or in the extremely rare occasions when a bentonite additive is employed during drilling. In completing chemostratigraphy projects since 1994, the author has only encountered problematic issues relating to drilling additives on a handful of occasions.

3.2.7 Summary of Element:Mineral Links

By taking the aforementioned information into account, the following summary details the most common mineralogical affinities for each element:

- Si = mainly quartz, but part of any silicate mineral
- Al and Ga = mainly clay minerals (particularly kaolinite, with less amounts associated with feldspars)
- K and Rb = K feldspars, micas and clay minerals (especially illite)
- Cs and Sc = clay minerals and feldspars
- V = mainly clay minerals. It is also noteworthy that V is adsorbed onto clay minerals in anoxic conditions.
- Ca = mainly calcite and dolomite, though high levels of Ca are also linked with gypsum and anhydrite. In addition to this less amounts of Ca may me associated with smectites and plagioclase feldspar
- Mg = mainly dolomite, calcite, and/or clay minerals (particularly chlorite)
- Fe and Mn = various clay and carbonate minerals, and pyrite

- Na = mainly plagioclase feldspar, though some Na is be associated with halite and/or clay minerals (e.g. smectite)
- Ti, Ta and Nb = titanomagnetite, magnetite, illmenite, rutile, anatase and/or sphene
- Th = heavy minerals, particularly monazite, zircon and apatite
- LREE and MREE = various mineralogical affinities. In general LREE are most abundant in clay minerals and feldspars, whilst MREE exist in heavy minerals, though this rules does not hold true in every study.
- Y and HREE = heavy minerals
- U = heavy minerals and organic matter, with redox influence
- Cr = heavy minerals such as chrome spinel
- Zr and Hf = zircon
- P = mainly biogenic phosphate and/or P-bearing heavy minerals such as apatite and monazite. However, some P may be concentrate in carbonate and clay minerals.
- Zn, Ni, Mo, Co and Cu = pyrite, Fe-oxyhydroxides, carbonate and/or clay minerals
- Ba = drilling additives (in the form of barite)
- Sr = mainly drilling additives and carbonate minerals, though some Sr may be associated with feldspars and clay minerals
- Be, W, Tl, Sn and Pb = uncertain mineralogical affinities.

Please note that this list should be treated with caution as, though these generalisations may hold true for vast majority of studies, they do not necessarily work everywhere.

3.3 Establishment of Element: Mineral Links

The previous section of this chapter describes the most common element:mineral links and controls on geochemistry and mineralogy in sedimentary basins. However, it is apparent that nearly all of the elements can have multiple mineralogical affinities. For example, the element U may be concentrated in organic matter or a variety of heavy minerals. Similarly K is often associated with K feldspars, but may also be found in clay minerals (particularly illite) and micas. Unless the interpreter has an adequate understanding of the mineralogical affinities of elements and the controls on geochemistry/mineralogy, then it is very difficult to propose a chemostratigraphic scheme with any degree of confidence. For example, a trend in K may be used to place chemostratigraphic boundaries if it is concentrated in K feldspar, reflecting a change in provenance. Conversely, if the same element is associated with authigenic illite, it may be wise to ignore trends in K for chemostratigraphic purposes. Information on probable element:mineral associations can be gleaned by comparing geochemical and well log data (i.e. in the form of sample descriptions and wireline log data). This information can also be obtained by considering geochemical and mineralogical datasets (e.g., heavy mineral, petrography, XRD data) acquired from the same samples, and by employing statistical and graphical techniques that utilize stoikiometric relationships to infer mineralogy.

Comparatively few references are included in the remainder of Chap. 3, mainly owing to the lack of published data on how element:mineral links were established in a given study. This is partly explained by the actions of most journals to discourage extensive outlines of methodology, preferring authors to concentrate on chemostratigraphic interpretations (i.e. the ‘final’ result). It is hoped that the following paragraphs will provide a workflow to define the mineralogical affinities of elements and principal controls on the distributions of elements and minerals in any given study. Most of this information has been gleaned by the author by working on a variety of chemostratigraphy projects since 1994.

Some programs have been developed by oil service companies to predict mineralogy, particularly bulk mineralogy, using inorganic geochemical data. An example of this is the ChemMin program developed by Chemostrat Ltd (www.chemostrat.com/mineral-modeling/). Unfortunately, most, if not all, of these software packages are not available for sale to the wider geological community but, even if they were, they should be viewed with considerable caution as they often yield spurious results unless they are calibrated with petrographic or XRD data. It is anticipated that these packages will be further developed in future and that they will become more widely available. For the purpose of this publication, however, they are given no further mention.

Although wireline log data and sample descriptions should never be ignored, it is necessary to compare the geochemical data with mineralogical data, and also employ statistical and graphical techniques to gain a more detailed understanding of the mineralogical affinities of elements, particularly the ones associated with accessory minerals.

3.3.1 Comparison with Well Log Data

Well log data in the form of core/cutting descriptions and wireline logs are available in nearly every chemostratigraphy project, except for those performed on field outcrop sections. Core and cuttings descriptions are normally highly subjective as terms such as 'abundant calcite cement' and 'common mica' are used to give a very general estimation of bulk mineralogy. In spite of the drawbacks of using these descriptions to establish element:mineral links, they should not be ignored for this purpose. For example, if cuttings samples are described as 'calcareous' and elevated values of Ca are associated with these, it may be concluded that Ca is concentrated in calcite. Similarly, gamma (including spectral gamma), density, neutron porosity, resistivity and sonic logs may be used to provide an indication of changes in bulk lithology and mineralogy with depth, particularly when used in conjunction with core and cuttings descriptions. Figure 3.4 shows that it was possible to determine 'bulk' lithology/mineralogy in a study of Triassic sediments in Saudi Arabia, using a combination of wireline log and geochemical data. The basal part of the interval (Jilh Formation) is dominated by argillaceous dolostones and dolomitic mudrocks. This is apparent from low GR readings, fast sonic log responses and relatively high concentrations of Mg. By contrast, the Minjur Formation is dominated by clastic sediments, yielding relatively slow sonic log and low Ca values. Within this section it is possible to differentiate sandstones from mudrocks, with the former producing the highest Si/Al but lowest GR values.

3.3.2 Comparison with Mineralogical Data

Mineralogical data take three main forms- petrographic, XRD and heavy mineral data. Of these, petrography and XRD data are normally more readily available. By comparing elemental data with those of petrography, it is possible to establish the mineralogical affinities of the most abundant elements and minerals. Figure 3.5 illustrates how, in a study of the Upper Ordovician Sarah Formation in northern Saudi Arabia, Craigie and Rees (2016) identified the mineralogical affinities of certain elements by comparing geochemical and mineralogical datasets acquired from the same samples. This diagram shows that the profile trends for quartz and Si are very similar, confirming that Si is predominantly linked with this mineral. Close matches are demonstrated between Na and plagioclase, and Ca and calcite, confirming these as probable element:mineral associations.

The mineralogical affinities of Fe and Mg are a bit more uncertain. Given the low abundances of other Fe- and Mg-bearing minerals in sandstones from this study, the most likely association is with chlorite. However, the profile plotted for this mineral shows little resemblance to that of either Fe or Mg. The reason for this poor 'match' may be genuinely related to the mineralogy/geochemistry of the sample. However, it is more likely to be explained by a lack of resolution of the petrographic data. Most of the values of chlorite fall in the range 0–5%, while the resolution of the petrographic data is normally in the region 5–15%, explaining the poor link between these profiles. The elements K and

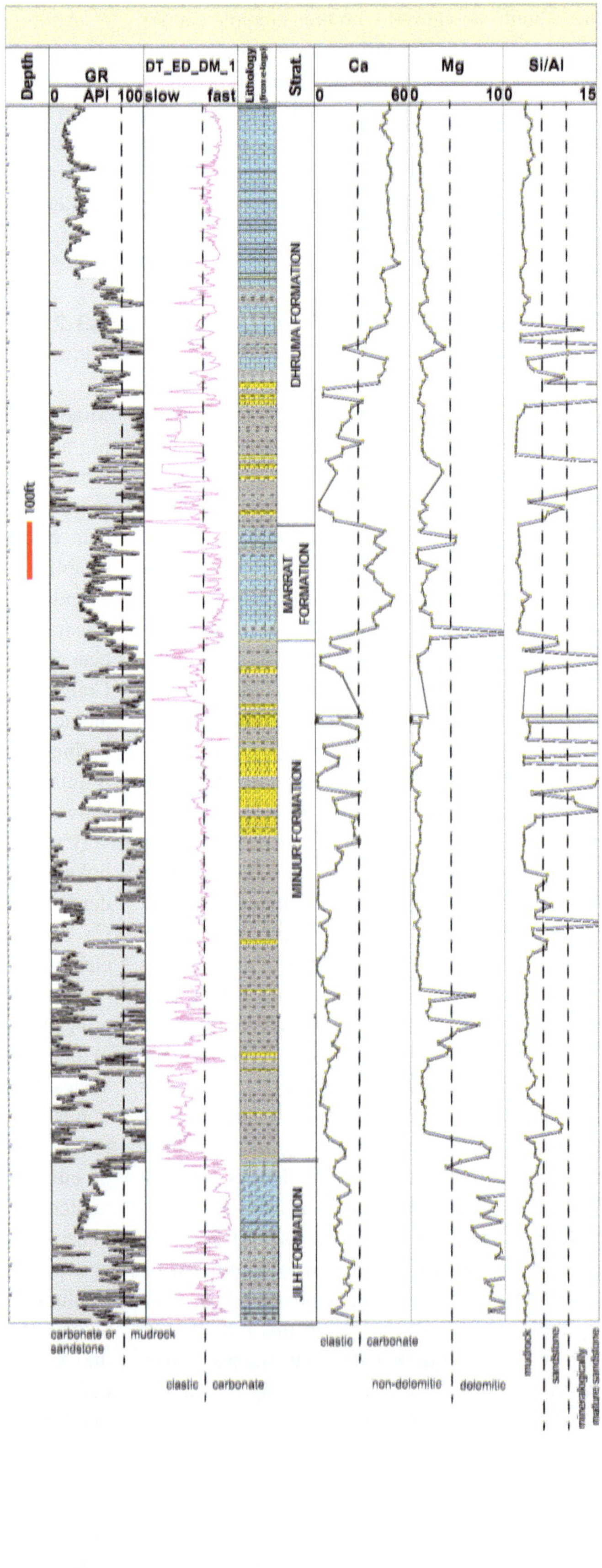

Fig. 3.4 Use of elog data and geochemical data to determine the lithology of samples analyzed from a well located in Saudi Arabia. All depths are log depths in feet and all samples were taken from cuttings material

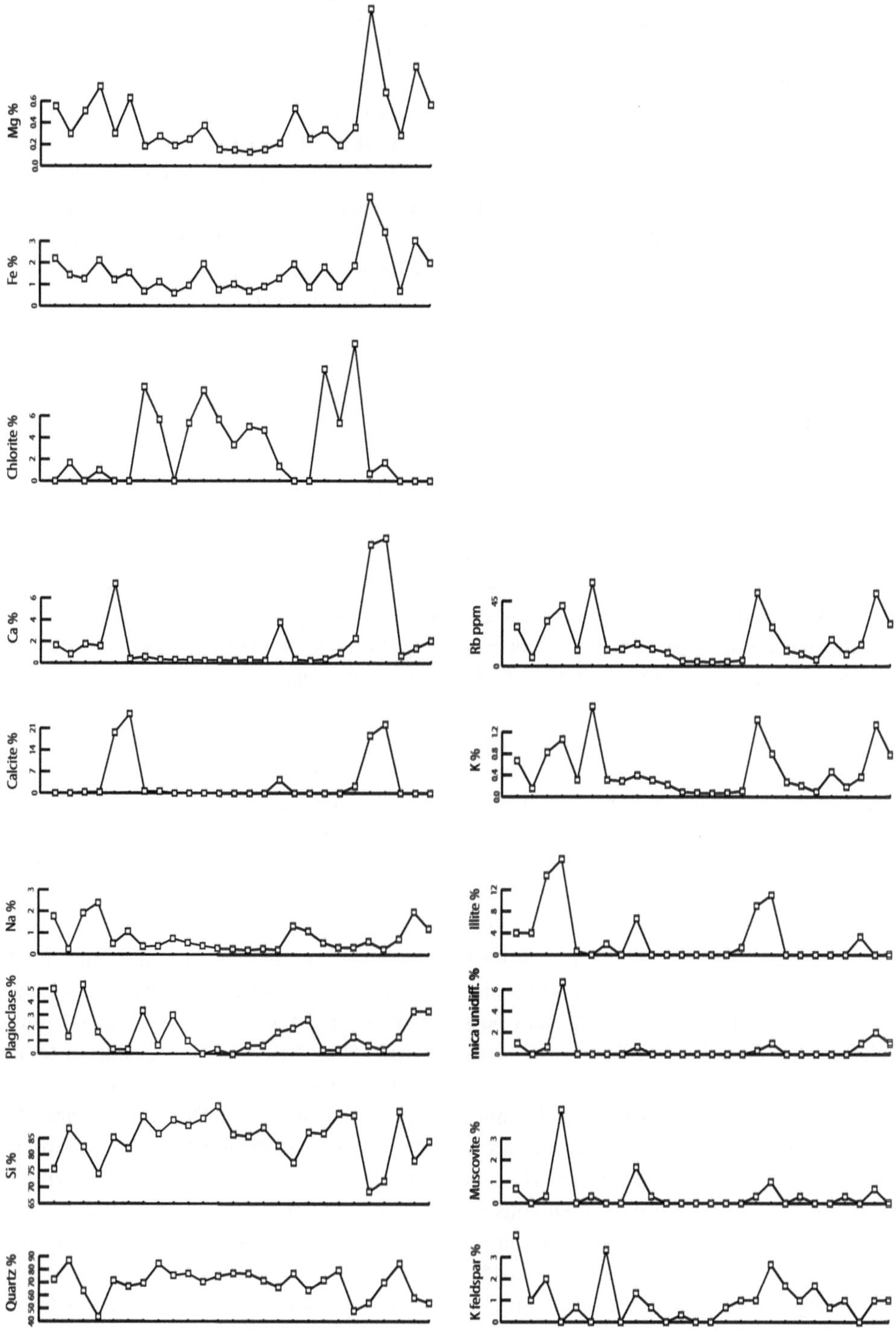

Fig. 3.5 Comparison of geochemical and mineralogical (petrographic) data acquired from equivalent samples. A careful examination of the trends enables the mineralogical affinities of some elements to be established. The data were generated in a study of the Sarah Formation of northwestern Saudi Arabia. Note that the samples were taken from multiple wells and no depth scale in implied from the profiles (after Craigie and Rees 2016)

Rb may have mixed mineralogical affinities, being associated with K feldspars, micas and clay minerals (mainly illite). In the present study, there is a roughly equal distribution of these minerals, hence the reason that the trends plotted for K and Rb do not match those displayed for K feldspar, mica or illite. The conclusion is that K and Rb probably occur in all three minerals in this particular study.

The principal drawback of using petrographic data is that it is only possible to establish, with any certainty, links between a relatively small array of elements and the most abundant minerals such as calcite, mica, K feldspar and quartz. A further difficulty is that the petrographic modal data are normally based on 300 point counts or less, meaning that there is always some uncertainty (typically 5–10% or more) regarding the resolution of the data and estimation of mineral concentrations. Moreover, considerable subjectivity may exist when recognizing minerals in thin section. For example, a degraded K feldspar grain may be described as a K feldspar by one interpreter, and as illite by another. It is recommended that the same petrographer analyses all of the samples in any given study to avoid such discrepancies but, even by adopting this approach, it is rarely possible to completely avoid such inconsistencies. Perhaps one of the greatest challenges of utilizing petrography is that it is often very difficult to use the technique on cuttings. In the case of reconstituted cuttings derived from drilling with a PDC bit, it is almost impossible to acquire mineralogical data via petrography. This may be a particular problem in studies where core or sidewall cores are unavailable.

Arguably a more accurate method of obtaining concentrations for the most abundant minerals is by employing the QEMSCAN technique. This is a fully automated process, with the output consisting of pixel-by-pixel compositional (both mineralogical and chemical) data (Jordens et al. 2016). The most significant advantages over petrographic analysis are that the subjectivity related to mineral identification is removed, and that the data are normally much more accurate and of a higher resolution. In addition to this, QEMSCAN can be applied, with an equal degree of success, to both core and cuttings samples (including reconstituted cuttings). The principal disadvantage of this technique is that it is normally very expensive and very few QEMSCAN instruments are available. Until recently, for instance, only two QEMSCAN analysers existed in the whole of the UK. It is hoped that QEMSCAN analysis becomes more widely available over time, as the wider geological community begins to appreciate the value of the technique. A further disadvantage of QEMSCAN is that it is difficult to distinguish some minerals with very similar chemical compositions, such as rutile and anatase. However, this is normally only an issue when acquiring data for accessory minerals. For analysing bulk mineralogy, QEMSCAN is generally favoured over standard traditional petrography.

In addition to petrography and QEMCSAN, it is possible to compare XRD and geochemical datasets to establish element:mineral links. As with petrography, this can be achieved by plotting profiles of elements and minerals, where data were derived from the same samples. Quantification of mineral abundances can be achieved using Reitveld Quantitative Analysis. A discussion on the details of this are beyond the scope of the present publication but, for more information, the reader is referred to Tamer (2013) and references therein. Next to petrography, XRD data is one of the most widely available techniques to provide data on the mineralogy of samples. The advantage of XRD is that it can be applied on a whole rock basis as well as to the clay fraction (i.e. <2 μm grain sizes). This technique is considered to be the most accurate method with respect to acquiring data for clay minerals. The disadvantage of XRD analysis is that, though the presence/absence of particular minerals is often recorded with a high degree of certainty, the quantitative data for mineral abundances may be of relatively low accuracy/resolution. Further complications may relate to instrument calibration, coelution and misidentification of 'peaks'.

As most chemostratigraphic schemes relate to changes in heavy mineral abundances comprising less than 1% of the total rock volume, petrographic and XRD data have limitations with respect to their usefulness. The acquisition of heavy mineral data for samples from which geochemical data has been obtained, is very useful in this regard. Figure 3.6 illustrates the comparison between geochemical and heavy mineral datasets acquired for Jurassic-Cretaceous sediments in the Bowser and Sustut basins, British Columbia, Canada (Ratcliffe et al. 2007). The results show an upward decrease in Cr roughly corresponding to a similar trend in the proportion of chrome spinel, suggesting that Cr is predominantly linked with chrome spinel in this study. The fact that Ni shows a similar trend may infer that this element has similar mineralogical affinities. This is not surprising as Ni is a common substitution in chrome spinel (Ratcliffe et al. 2007). Concentrations of Ce are relatively low and consistent throughout the Bowser Lake Group but are generally higher in the Sustut Group. Ratcliffe et al. (2007) explain this by the slightly higher levels of allanite in the latter stratigraphic interval. A significant inconsistency between the geochemical and heavy mineral datasets is that the trend in Zr does not match that of zircon. For example, higher Zr values, but lower zircon, are recorded in the Ritchie-Alger Assemblage than in the Todagin and Muskaboo Creek assemblages. This discrepancy probably relates to the size of the zircons. The heavy mineral analysis was performed on the 63–125 μm size fraction, but airfall zircons of a finer grade are known to occur in parts of the Bowser Group. Whole rock ICP analysis would result in acquisition of Zr data from both the 63–125 and <65 μm fractions, explaining the differences between the two datasets (Ratcliffe et al. 2007). Figure 3.6 also shows profiles plotted for Zr/Cr and Ce/Cr which were used by Ratcliffe et al. (2007) to differentiate the Bowser Lake and Sustut groups.

The aforementioned study highlights one significant drawback of comparing heavy mineral and geochemical data. As standard heavy mineral analysis is normally performed on the 63–125 μm size fraction, some heavy minerals may exist in grains of < 65 μm, explaining observed discrepancies between whole rock geochemical and heavy mineral datasets. A further drawback of heavy mineral analysis is that it is expensive and very few companies offer this service on a commercial basis. Consequently, this data is rarely available in the majority of chemostratigraphy studies. In recent years, Raman Spectroscopy (Bersani and Lottici 2010) has been developed to automate the process of heavy mineral analysis and QEMSCAN has also been used for this purpose (Jordens et al. 2016). It is hoped that these advanced technologies will be utilized on a more frequent basis in future heavy mineral studies.

One challenging aspect of many chemostratigraphy studies is the non-availability of mineralogical data, particularly heavy mineral data. For this reason, it is necessary to apply statistical and graphical techniques to establish the mineralogical affinities of elements. These are discussed in the following sections.

3.3.3 Statistical Techniques

The two most commonly used statistical techniques to establish element:mineral links are Correlation Coefficient (CC) and Principal Component Analysis (PCA). It is beyond the scope of the present publication to present the formulae used in CC and PCA, but more detailed information on CC is available through various websites (e.g. https://explorable.com/statistical-correlation, https://en.wikipedia.org/wiki/Correlation_and_dependence). The PCA technique is described in detail by Wold et al. (1987). As a general rule, CC scores of more than 0.7 infer a high level of correlation between two elements, 0.6–0.7 are moderate-high level, 0.5–0.6, a moderate, and below 0.5, a low level. The same rules can be applied to negative CC values. For example a CC score of less than −0.7 suggests that the two elements have a strong negative association.

Table 3.1 shows the results of CC applied to Triassic and Jurassic carbonate sediments in Triassic-Jurassic carbonates in the Rub' Al Khali Basin, SE Saudi Arabia. The technique was applied to almost every element in the range Na-U in the periodic table, but only a select group are shown to illustrate how it was possible to establish their mineralogical affinities in this dataset. CC values of greater than 0.7 are highlighted in this table. A high CC value of 0.77 is

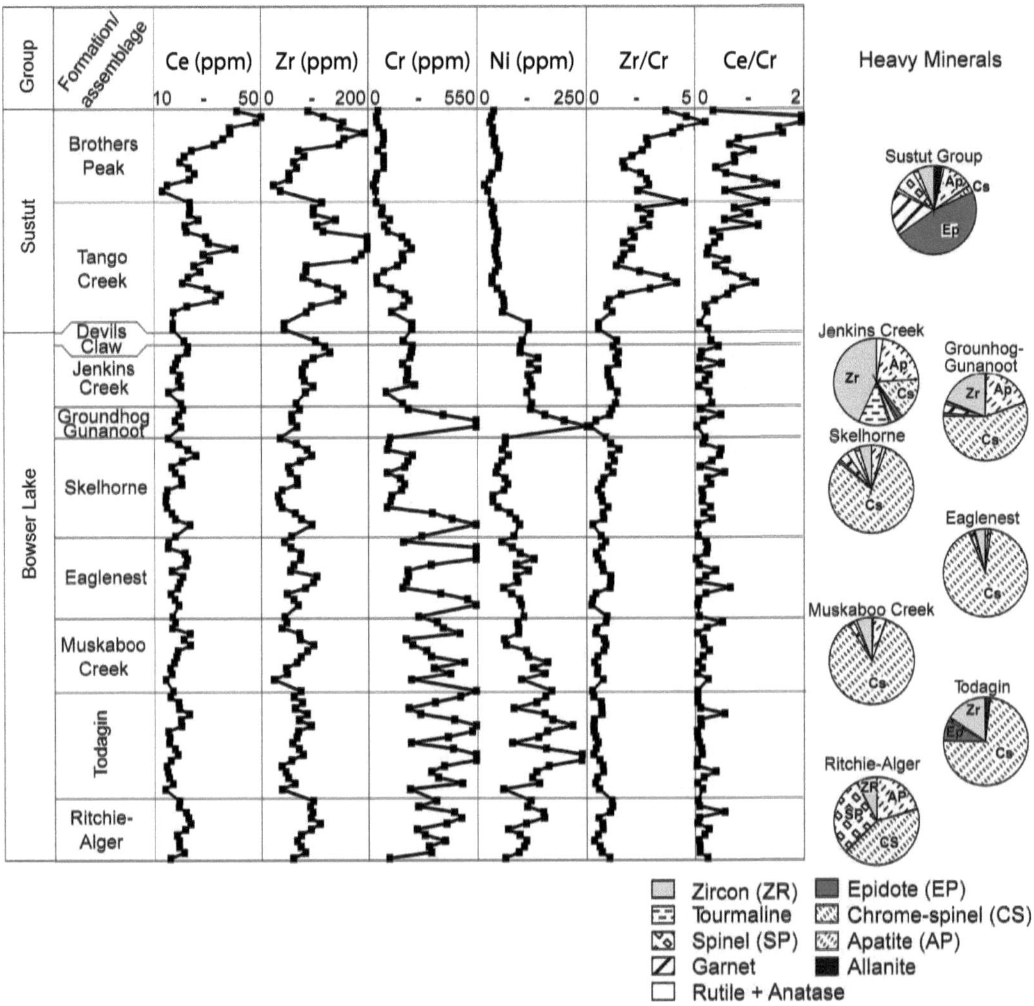

Fig. 3.6 Comparison between selected elements and the results of heavy mineral analysis in the Bowser and Sustut basins, British Columbia, Canada. Note that vertical position on profiles does not imply relative age of samples. The heavy mineral data is in the form of pie charts representing averages for each assemblage (after Ratcliffe et al. 2007)

produced between K and Al suggesting that K, along with Al, has the strongest mineralogical affinities with clay minerals. The high CC scores between Ti and Nb were expected, as both elements have similar mineralogical affinities, being associated with rutile, anatase, sphene and opaque heavy minerals (e.g. titanomagnetite, magnetite, ilmenite). The elements Zr and Hf also yield a high CC values as these are concentrated, almost exclusively, in zircon. The element Th may be linked with either zircon or the Ti- and Nb-bearing heavy minerals. However, this element produces a much closer association with Nb and Ti (CC of 0.79 and 0.86 respectively) than for Zr (CC of 0.66), inferring that most Th occurs in Ti and Nb-bearing heavy minerals in this study. Table 3.1 also illustrates a close association between Al and Ti, Nb and Th. This may seem strange as Al is concentrated, almost exclusively, in clay minerals, while Ti, Nb and Th are associated with heavy minerals. A probable reason for this apparent inconsistency is the that Ti-, Nb- and Th-bearing heavy minerals are likely to be concentrated, along with Al-bearing

Table 3.1 Results of correlation coefficient analysis for a select group of elements. See text for explanation of results

	Al	Si	Ti	Fe	Mn	Mg	Ca	Na	K	P	Hf	Nb	Th	Zr
Al	1.00	0.33	**0.92**	0.47	0.02	−0.16	−0.54	−0.05	**0.77**	0.35	0.56	**0.86**	**0.89**	0.54
Si	0.33	1.00	0.45	0.15	−0.25	−0.32	−0.89	−0.13	0.20	0.07	0.62	0.36	0.40	0.63
Ti	0.92	0.45	1.00	0.38	−0.06	−0.22	−0.61	−0.08	0.67	0.28	**0.77**	**0.92**	**0.86**	**0.75**
Fe	0.47	0.15	0.38	1.00	0.42	0.02	−0.36	−0.04	0.35	0.69	0.26	0.32	0.67	0.27
Mn	0.02	−0.25	−0.06	0.42	1.00	0.34	0.05	0.13	0.10	0.31	−0.12	−0.02	0.09	−0.12
Mg	−0.16	−0.32	−0.22	0.02	0.34	1.00	0.02	0.15	−0.03	0.03	−0.22	−0.18	−0.15	−0.22
Ca	−0.54	−0.89	−0.61	−0.36	0.05	0.02	1.00	−0.04	−0.41	−0.24	−0.66	−0.52	−0.61	−0.66
Na	−0.05	−0.13	−0.08	−0.04	0.13	0.15	−0.04	1.00	0.08	0.03	−0.09	−0.07	−0.07	−0.09
K	**0.77**	0.20	0.67	0.35	0.10	−0.03	−0.41	0.08	1.00	0.33	0.42	0.64	**0.70**	0.41
P	0.35	0.07	0.28	0.69	0.31	0.03	−0.24	0.03	0.33	1.00	0.21	0.23	0.55	0.22
Hf	0.56	0.62	**0.77**	0.26	−0.12	−0.22	−0.66	−0.09	0.42	0.21	1.00	0.67	0.68	**0.99**
Nb	**0.86**	0.36	**0.92**	0.32	−0.02	−0.18	−0.52	−0.07	0.64	0.23	0.67	1.00	**0.79**	0.65
Th	**0.89**	0.40	**0.86**	0.67	0.09	−0.15	−0.61	−0.07	**0.70**	0.55	0.68	**0.79**	1.00	0.66
Zr	0.54	0.63	**0.75**	0.27	−0.12	−0.22	−0.66	−0.09	0.41	0.22	**0.99**	0.65	0.66	1.00

clay minerals, in fine sand-silt and clay grade material. The fact that heavy minerals are largely associated with the fine sand-silt fraction was proven by Poldervaart (1955), Mange and Maurer (1992) and Morton and Hallsworth (1994).

Although CC is useful, one of the main problems associated with using it is that the presence of a high score of over 0.7 between two elements does not necessarily mean that they are concentrated in the same mineral(s). For example a high score between the elements K and Al could result from both elements being concentrated in the same grain size fraction rather than in the same minerals. K could be associated with illite, while Al may be linked with kaolinite, for example. Another potential problem, especially in relatively small datasets involving the analyses of less than 60 samples, is that of the 'one sample effect'. This refers to the fact that one or two samples containing anomalously high or low concentrations of particular elements/minerals may have an 'overriding' effect on the values of CC. For example the presence of one sample containing a high abundance of thoriferous zircons may result in a high CC score between Th and Zr. If data for that sample were to be removed and CC scores recalculated, the CC value between these two elements may be much lower. For this reason, CC should not be employed in isolation to establish the mineralogical affinities of elements.

One of the most commonly used statistical techniques for evaluating geochemical results is Principal Component Analysis (PCA), which is used to identify important element associations. The principal component scores for the original vectors (eigenvectors) indicate the association of certain elements and the relationship between some elements and mineralogy. The technique has been used in a chemostratigraphy study of Permian sedimentary rocks in New Mexico (Eisenberg and Harris 1995) and more recently in studies of carbonates and sandstones in Saudi Arabia (Craigie 2015, 2016, 2016a). In a study of Upper Jurassic strata in Spain, Coimbra (2015) also applied the technique to carbonates. The principal element associations predicted by the interpretation of PCA results from the aforementioned study of carbonate sediments on which CC was performed, are discussed in the following paragraphs.

EV1 (Eigenvector 1) and EV2 (Eigenvector 2) represent the first and second most important sources of statistical variance respectively. In this study, EV1 encompasses 38.15% of this variability, while EV2 related to 11.18%. When the EV1 and EV2 values are cross-plotted, the elements form several broad associations (Fig. 3.7a) as follows:

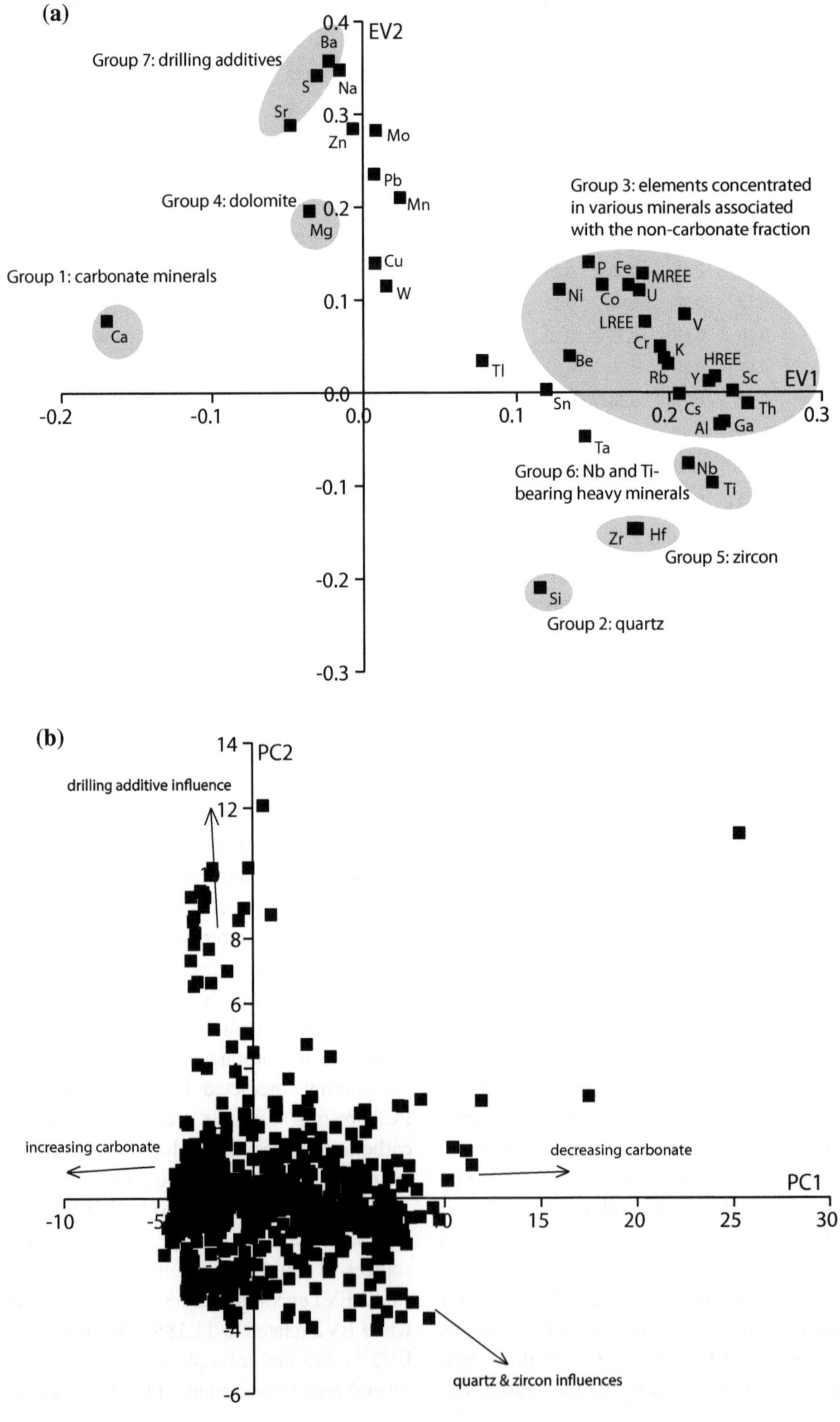

Fig. 3.7 Eigenvector (EV) and Principal Component (PC) crossplots used to aid the establishment of element:mineral linkages in a multiwell project on Triassic-Jurassic aged carbonates encountered in southwestern Saudi Arabia

- Group 1: includes Ca. This element is normally concentrated in carbonate minerals, particularly calcite.
- Group 2: includes Si. This element may be linked with a variety of minerals but is most commonly associated with quartz.
- Group 3: includes P, Ni, Co, Fe, U, LREE, MREE, HREE, Be, Cr, K, Rb, V, Cs, Y, Sc, Al, Ga and Th. These elements are concentrated in various minerals associated with the non-carbonate fraction.
- Group 4: includes Mg. This element is linked with dolomite.
- Group 5: includes Zr and Hf. These elements have strong mineralogical affinities with zircon.
- Group 6: includes Nb and Ti. These elements are concentrated in the heavy minerals rutile, anatase, sphene and/or opaque heavy minerals such as ilmenite, magnetite and titanomagnetite.
- Group 7: includes Ba, S and Sr. These elements are associated with barite drilling additives in both core and cuttings samples, but are more abundant in the latter.

Figure 3.7a shows that the EV1 axis is most strongly influenced by Ca-bearing carbonate minerals as Ca (group 1) plots with anomalously low values of EV1. By contrast, the EV2 parameter is affected by both grain size/quartz and drilling additives, with the group 2 (quartz) and 7 (drilling additives) elements plotting with anomalously low and high EV2 values respectively. The influences of carbonate, drilling additives and quartz/grain size/zircon are noted on the PC1:PC2 binary diagram (Fig. 3.7b).

EV2 and EV3 represent 11.18 and 8.59% of the variance of the geochemical dataset, respectively. When the EV2 and EV3 values are plotted, the elements form several broad associations (Fig. 3.8), as follows:

- Group 1: includes Ca. This element is normally concentrated in carbonate minerals, particularly calcite.
- Group 2: includes Si. This element is mainly associated with quartz.
- Group 3: includes Cs, Rb, K, Sn, Al and Ga. These elements are likely to have mineralogical affinities with clay minerals, micas and/or K feldspars.
- Group 4: includes Mg. This element is linked with dolomite.
- Group 5: includes Zr and Hf. These elements have strong mineralogical affinities with zircon.
- Group 6: includes Nb, Ta and Ti. These elements are concentrated in the heavy minerals rutile, anatase, sphene and/or opaque heavy minerals such as ilmenite, magnetite and titanomagnetite.
- Group 7: includes Y and HREE. These elements may be associated with a variety of heavy minerals.
- Group 8: includes P, Fe, LREE and MREE. The element P may be concentrated in biogenic phosphate and/or P-bearing heavy minerals such as apatites, monazites and xenotimes. The fact that LREE, MREE and Fe plot very close to P indicates that they have similar mineralogical affinities.
- Group 9: includes Ba. This element is associated with drilling additives.

Figure 3.8a shows that the EV2 axis is largely reflective of grain size and drilling additives, while EV3 is affected by the group 7 (heavy mineral association) and 5 elements (apatite and/or biogenic phosphate association), which plot with low values of this parameter. The influences of quartz/zircon, drilling additives and heavy minerals/biogenic phosphate are noted on the PC2:PC3 binary diagram (Fig. 3.8b).

Clearly, the EV1 axis is most strongly influenced by Ca and/or Mg, whilst EV2 is affected by quartz content (i.e. Si) and/or drilling additives (i.e. Ba). The PC values can, therefore, be used to distinguish samples that are enriched or depleted in carbonate minerals, detrital quartz (assuming Si is concentrated in this mineral rather than in biogenic Si) and drilling additives.

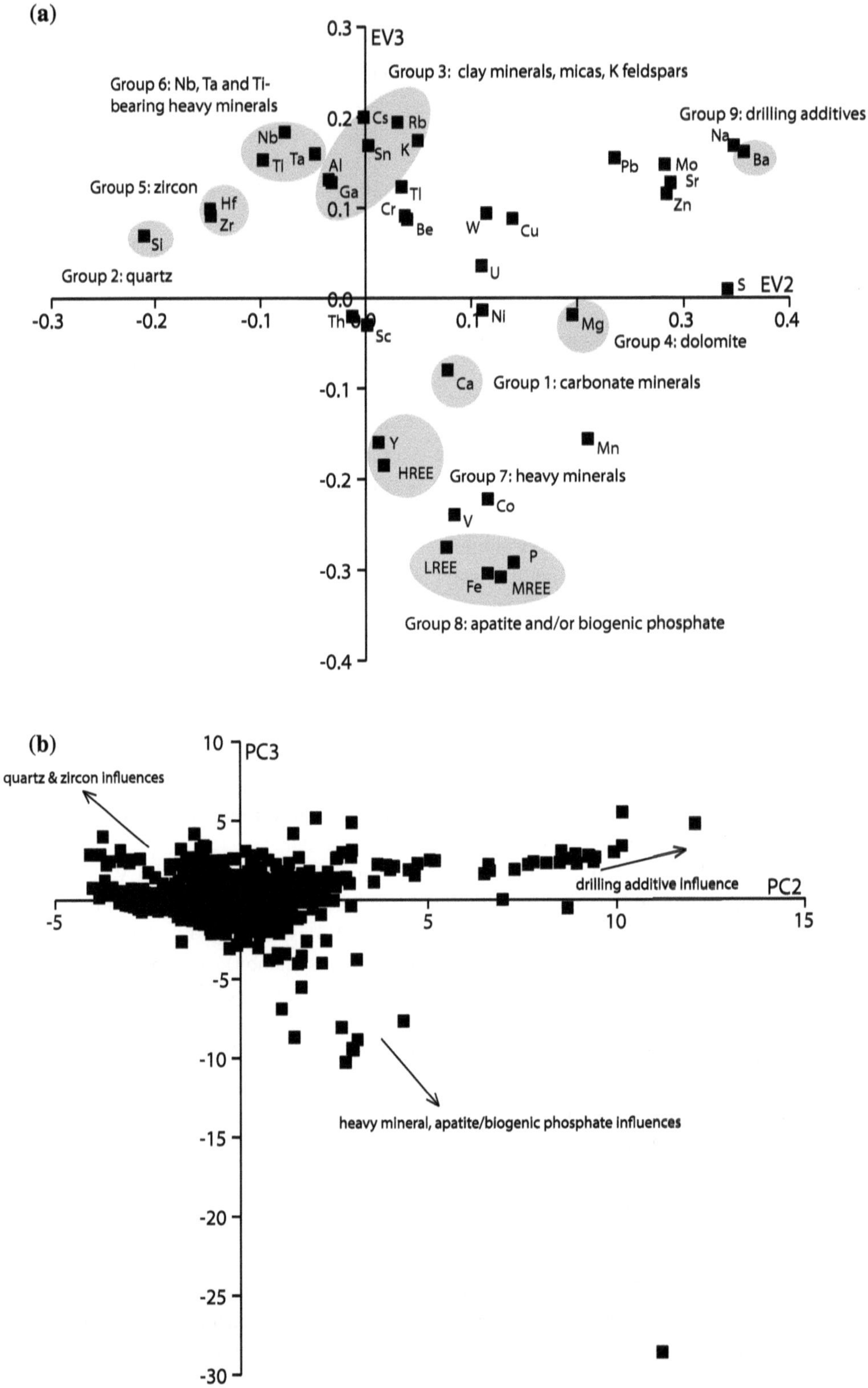

Fig. 3.8 Eigenvector (EV) and Principal Component (PC crossplots) used to aid the establishment of element:mineral linkages in a multiwell project on Triassic-Jurassic aged carbonates encountered in southwestern Saudi Arabia

In clastic sediments, values of Si, controlled by the proportion of detrital quartz, normally exert the principal control on EV1, whilst EV2 is often, though not always, influenced by levels of carbonate minerals (i.e. Ca, Mg, Mn) and/or drilling additives (i.e. Ba). In both clastic and carbonate sediments, the EV3 and EV4 parameters are often influenced by more subtle factors such as the proportion of accessory heavy or clay minerals. Consequently, EV1 versus EV2 crossplots often only provide information about the most abundant elements/minerals. A better understand of the mineralogical affinities of accessory minerals is often achieved by plotting EV2 versus EV3, PC2 versus PC3, EV3 versus EV4 and PC3 versus PC4 graphs. On some occasions, the EV5, EV6, PC5 and PC6 parameters are examined, but this is rarely necessary as, in nearly every study, EV5 and above encompass less than 4% of the variation of datasets.

As with other statistical techniques, the results and conclusion drawn from PCA should not be viewed in isolation. The fact that two elements plot in the same 'field' on EV plots does not necessarily mean that they are concentrated in the same minerals. It is possible that they are, for instance, linked to different minerals occurring in the same grain size fraction. In smaller datasets (i.e. less than 60 samples) it is possible that the observed elemental associations on EV crossplots relate to one or two samples yielding anomalously high or low concentrations of particular elements that are not reflective of the data as a whole. Another difficulty may relate to the recognition of 'fields' which can be highly subjective. For example, the element Ta plots almost as close to Al and Ga, as it does to Nb and Ti, in the EV2 versus EV3 crossplot (Fig. 3.8a). This may be explained by Ta occurring in similar grain size fractions as Al and Ga. However, it is highly unlikely that these elements would be concentrated in the same minerals as Ti, Ta and Nb have strong mineralogical affinities with heavy minerals, whilst Al and Ga are normally associated with clay minerals. In the absence of an existing knowledge of common element:mineral links in a given study, the interpretations of EV and PC crossplots may be misleading.

3.3.4 Graphical Evaluation

By plotting data on binary diagrams it is possible to recognize important positive and negative associations that may enable element:mineral links to be established, often with much more accuracy than statistical techniques. The binary diagrams presented in Figs. 3.9 and 3.10 are used to establish element:mineral links in samples from the same project on which PCA/EV was performed.

The Al versus Si binary diagram (Fig. 3.9a) shows both negative and positive associations between the two elements. The latter are explained by carbonate dilution, with an increase in the proportion of Ca resulting in a decrease in Si and Al, and vice versa. The samples that are depleted in Ca generally plot on the negative trend, reflecting the fact that Si and Al are mainly concentrated in quartz and clay minerals respectively. Figure 3.9b illustrates that the non-calcareous samples (< 8% Ca) do not plot on any specific trend on this Si versus Ca binary diagram. However, the other samples with greater than 8% Ca define negative trends, reflecting enrichment/depletion of carbonate minerals relative to non-carbonate ones. A positive trend exists between Mg and Ca (Fig. 3.9c), inferring that both elements are associated with carbonate minerals. However, the absence of a stronger trend may be taken as evidence that Ca and Mg are concentrated in different carbonate minerals. It is possible, for instance, that Ca is linked mainly with calcite, and Mg with dolomite. The Al versus K graph (Fig. 3.9d) illustrates that about 50% of the samples plot on a positive trend between the two elements. Because Al is mainly associated with clay minerals, it follows that the two elements have similar mineralogical affinities in these samples. The most commonly occurring clay mineral containing K is illite, so these samples are considered to be enriched in this mineral. The samples that do not plot on this trend generally yield low levels of K. These samples are characterised by relatively low proportions of K-bearing minerals and high concentrations of Al-bearing clay minerals (e.g., kaolinite). The strong positive trend defined by Al

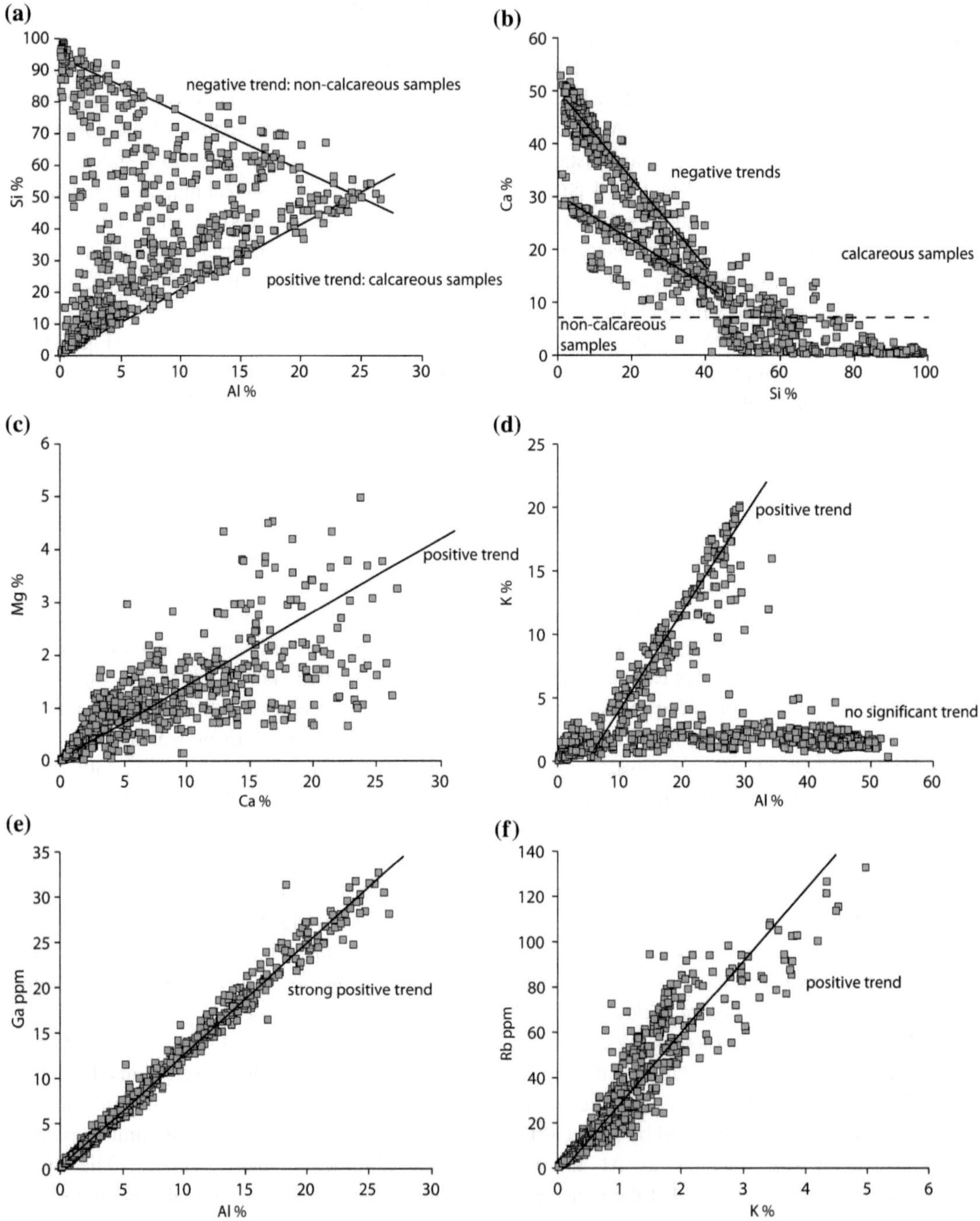

Fig. 3.9 Binary diagrams used to establish the mineralogical affinities of elements

and Ga (Fig. 3.9e) infers that both elements are associated with clay minerals. This is not surprising as Ga has a very similar ionic radius to Al and normally has similar mineralogical affinities. A positive trend is also displayed on the K versus Rb graph (Fig. 3.9f), which may be taken as evidence that these elements have similar mineralogical affinities, being mainly concentrated in in illite (from the petrographic data). Figure 3.10a displays a poorly developed positive relationship

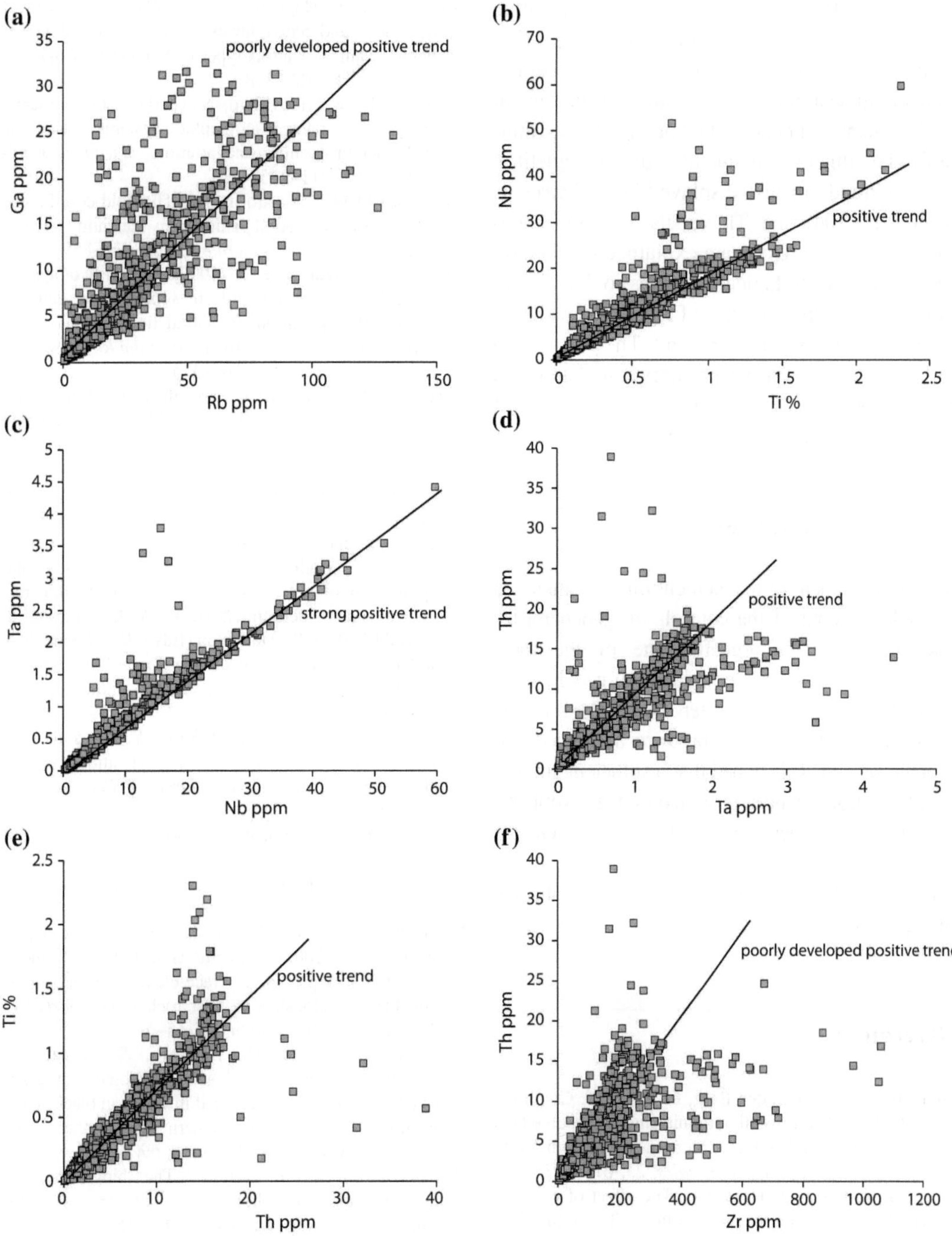

Fig. 3.10 Binary diagrams used to establish the mineralogical affinities of elements

between Rb and Ga. This is explained by the exclusivity of Ga in the clay fraction (regardless of clay type) and the more variable affinities of Rb (e.g. illite, mica K feldspars). However, the absence of a stronger relationship between the two elements suggests that at least some of the Rb is associated with other minerals such as K feldspars and micas.

Figure 3.10b and 3.10c show strong positive trends between Ti and Nb, and Nb and Ta, respectively. These elements have very similar mineralogical affinities as they are concentrated in rutile, anatase, sphene and/or opaque heavy minerals (i.e. ilmenite, magnetite, titanomagnetite). Positive trends are also displayed in the Ta versus Th (Fig. 3.10d) and Th versus Ti (Fig. 3.10e) binary diagrams. These are explained by Th having similar mineralogical affinities to Ta, Ti and Nb. However, the presence of a poorly developed positive trend between Zr and Th (Fig. 3.10f) indicates that, though Th is mainly associated with the Ti-, Nb- and Ta-bearing heavy minerals, it possesses a partial affinity with zircon.

3.4 Concluding Remarks

The establishment of the element:mineral links and an understanding of the controls on geochemistry and mineralogy is arguable one of the most important steps in the interpretation of geochemical data, though this is so often overlooked in the quest of the interpreter to produce a chemostratigraphic scheme. It is hoped that a sufficient amount of information has been provided on these subjects, and how the mineralogical affinities of elements can be established by comparing geochemical with mineralogical datasets, and by employing statistical and graphical techniques.

References

Achterberg, E. P., Van den Berg, C. M. G., & Colombo, C. (2003). High resolution monitoring of dissolved Cu and Co in coastal surface waters of the western North Sea. *Continental Shelf Reservoirs, 23,* 611–623.

Adam, J., & Green, T. H. (1984). The effect of pressure and temperature on the partitioning of Ti, Sr and REE between amphibole, clinopyroxene and basanitic melts. *Chemical Geology, 117,* 219–233.

Adelson, J. M., Helz, G. R., & Miller, C. V. (2001). Reconstructing the rise of recent coastal anoxia; molybdenum in Chesapeake Bay sediments. *Geochinica et Cosmochimica Acta, 65,* 237–252.

Algeo, T. J. (2004). Can marine anoxic events draw down the trace element inventory of seawater? *Geology, 32,* 1057–1060.

Algeo, T. J., & Maynard, J. B. (2004). Trace-element behaviour and redox facies in core shales of Upper Pennsylvanian Kansas-type cyclotherms. *Chemical Geology, 206,* 289–318.

Algeo, T. J., & Tribovillard, N. (2009). Environmental analysis of paleoceanographic systems based on molybdenum-uranium covariation. *Chemical Geology, V, 268,* 211–225.

Angel, R. J. (1997). Transformation of fivefold-coordinated silicon to octahedral silicon in calcium silicate, $CaSi_2O_5$. *American Mineralogist, 82,* 836–839.

Ayres, J. C., & Watson, E. B. (1993). Solubility of apatite, monazite, zircon and rutile in super-critical aqueous fluids with implications for subduction zone geochemistry. *Philosophical Transactions of the Royal Society, London, Series A, 335,* 365–375.

Balan, E., Neuville, D. R., Trocellier, P., Fritsch, E., Muller, J. P., & Calas, G. (2001). Metamictization and chemical durability of detrital zircon. *American Mineralogist, 86,* 1025–1033.

Banner, J. L. (1995). Application of the trace element and isotopic geochemistry of strontium to studies of carbonate diagenesis. *Sedimentology, 42,* 805–824.

Basu, B., & Molinaroli, E. (1991). Reliability and application of detrital opaque Fe-Ti oxide minerals in provenance determination. In A. C. Morton, S. P. Todd & P. D. W. Haughton (Eds.), *Developments in sedimentary provenance studies* (Vol. 57, pp 55–65). London: Geological Society of London Special Publication.

Bea, R. (1996). Residence of REE, Y, Th and U in granites and crustal protoliths: Implications for the chemistry of crustal melts. *Journal of Petrology, 37,* 521–552.

Belousova, E. A., Griffin, W. L., O'Reilly, S. Y., & Fisher, N. I. (2002a). Igneous zircon: Trace element composition as an indicator of source rock type. *Contributions to Mineralogy and Petrology, 143,* 602–622.

Belousova, E. A., Griffin, W. L., O'Reilly, S. Y., & Fisher, N. I. (2002b). Apatite as an indicator mineral for mineral exploration: Trace-element compositions and their relationship to host rock type. *Journal of Geochemical Exploration, 76,* 45–69.

Bellanca, A., Masetti, D., Neri, R., & Venezia, F. (1999). Geochemical and sedimentological evidence of productivity cycles recorded in Toarcian black shales from the Belluno Basin, Southern Alps, northern Italy. *Journal of Sedimentary Research, 69*(2), 466–476.

Benitez-Nelson, C. R. (2000). The biogeochemical cycling of phosphorus in marine systems. *Earth-Science Reviews, 51*(1), 109–135.

Bersani, D., & Lottici, P. P. (2010). Applications of Raman spectroscopy to gemmology. *Analytical and Bioanalytical Chemistry, 397,* 2631–2646.

Bragg, W. H. (1915). The structure of the spinel group of crystals. *Philosophical Magazine, 30,* 305–315.

Breit, G. N., & Wanty, R. B. (1991). Vanadium accumulation in carbonaceous rocks: A review of geochemical controls during deposition and diagenesis. *Chemical Geology, 91,* 83–97.

Brigatti, M. F., Galan, E., & Theng B. K. G. (2013). *Structure and mineralogy of clay minerals, Developments in Clay Science* (Vol. 5A, pp. 21–68). Amsterdam: Elsevier Press.

Buddington, A. F., & Lindsley, D. H. (1964). Iron-titanium oxide minerals and synthetic equivalents. *Journal of Petrology, 5,* 310–357.

Budzinski, H., & Tischendorf, G. (1989). Distribution of REE among minerals in the Hercynian posthinematic granites of Westerzgebirge-Vogtland, GDR. *Zritschrift fur Geologische Wissenschaften, 17,* 1019–1031.

Calvert, S. E., & Pedersen, T. F. (1993). Geochemistry of recent oxic and anoxic sediments: implications for the geological record. *Marine Geology, V, 113,* 67–88.

Calvert, S. E., & Pedersen, T. F. (1996). Sedimentary geochemistry of manganese: implications for the environment of formation of manganiferous black shales. *Economic Geology, 91,* 36–47.

Chang, L. L. Y., Howie, R. A., & Zussman, J. (1998). *Rock forming minerals, V. 5B: Non-silicates: Sulphates, Carbonates, Phosphates and Halides* (2nd ed). London: The Geological Society.

Coimbra, R., Immenhauser, A., Olóriz, F., Rodriguez-Galiano, V., & Chica-Olmo, M. (2015). New insights into geochemical behaviour in ancient marine carbonates (Upper Jurassic Ammonitica Rosso): Novel proxies for interpreting sea-level dynamics and palaeoceanography. *Sedimentology, 62,* 266–302.

Craigie, N. W. (1998). *Chemostratigraphy of Middle Devonian lacustrine sediments* (Unpublished PhD thesis). N.E. Scotland: University of Aberdeen.

Craigie, N. W. (2015). Applications of chemostratigraphy in Middle Jurassic unconventional reservoirs in eastern Saudi Arabia. *GeoArabia, 20*(2), 79–110.

Craigie, N. W., & Rees, A. J. (2016). Chemostratigraphy of glaciomarine sediments in the Sarah Formation, northwest Saudi arabia. *Journal of African Earth Sciences, 117,* 263–284.

Craigie, N. W., Rees, A., MacPherson, K., & Berman, S. (2016a). Chemostratigraphy of the Ordovician Sarah Formation, northwest Saudi Arabia: An integrated approach to reservoir correlation. *Marine and Petroleum Geology, 77,* 1056–1080.

Crusius, J., Calvert, S., Pedersen, T., & Sage, D. (1996). Rhenium and molybdenum enrichments in sediments as indicators of oxic, suboxic, and sulfidic conditions of deposition. *Earth and Planetary Science Letters, 145,* 65–78.

Crusius, J., & Thompson, J. (2000). Comparative behaviour of authigenic Re, Mo and U during reoxidation and subsequent long-term burial in marine sediments. *Geochimica et Cosmochimica Acta, 64,* 2233–2243.

Dasgupta, S., & Pal, S. (2005). Origin of grandite garnet in silicate granulites: mineral-fluid equilibria and petrogenetic grids. *Journal of Petrology, 46,* 1045–1076.

da Silva, A. L., Hotza, D., & Castro, R. H. (2017). Surface energy effects on the stability of anatase and rutile nanocrystals: A predictive diagram for Nb 2 O 5-doped-TiO 2. *Applied Surface Science, 393,* 103–109.

Deer, W. A., Howie, R. A., & Zussman, J. (1992). *The rock-forming minerals.* London: Longman.

Dick, H. J. B., & Bullen, T. (1984). Chromian spinel as a petrogenetic indicator in abyssal and alpine-type peridotites and spatially associated lavas. *Contributions to Mineralogy and Petrology, 86,* 54–76.

Dickson, J. A. D. (1990). Carbonate mineralogy and chemistry. In M. E. Tucker & V. P. Wright (Eds.), *Carbonate sedimentology* (pp. 284–313). Oxford: Blackwell Scientific.

Eisenberg, R.A. & Harris, P.M. (1995). Appplication of chemostratigraphy in multivariate statistical analysis to differentiating bounding stratigraphic surfaces. In Pause, P.H. and Candelaria, M.P. (eds.). *Carbonate Facies and Sequence Stratigraphy: Practical Applications of Carbonate Models*: Permian Basin section-SEPM Publication 95-36 / Permian Basin Graduate Center Publication 5–95, 83–102.

Erickson, B. E., & Helz, G. R. (2000). Molybdenum (VI) speciation in sulfidic waters: Stability and lability of thiomolybdates. *Geochimica et Cosmochimica Acta, 64,* 1149–1158.

Fernex, F., Féïvrier, G., Bebaïm, J., & Arnoux, A. (1992). Copper, lead and zinc trapping in Mediterranium deep-sea sediments: Probable coprecipitation with manganese and iron. *Chemical Geology, 98,* 293–308.

Fertl, W. H. (1983). *Gamma ray spectral logging: a new evaluation frontier. 5-applications in clastic reservoirs* (p. 79). March p: World Oil.

Fjar, E., Holt, R. M., Raaen, A. M., Risnes, R., & Horsrund, P. (2008). *Petroleum related rock mechanics. Developments in petroleum science* (2nd ed, Vol. 53). Amsterdam: Elsevier Press.

Fleischer, M., & Altschuler, Z. S. (1969). The relationship of the rare-earth composition of minerals to geological environment. *Geochimica et Cosmochimica Acta, 33* (6), 725–732.

Fleet, M. (1981). The structure of magnetite. *Acta Crystallographica. Section B, Structural Science, 37,* 917–920.

Förster, H. J. (2006). Composition and origin of intermediate solid solutions in the system thorite-xenotime-zircon-coffinite. *Lithos, 88*(1–4), 35–55.

Frost, B. R., Chamberlain, K. R., & Schumacher, J. C. (2000). Sphene (titanite): Phase relations and a role as a geochronometer. *Chemical Geology, 172,* 131–145.

Frost, B. R., & Lindsey, D. H. (1991). Occurrence of iron-titanium oxide minerals in igneous rock. *Review of Mineralogy, 25,* 433–486.

Goldberg, S., Su, C., & Forster, H. S. (1998). Sorption of moly on oxides, clay minerals and solid. In: Jenne, e. A. (Ed.), *Adsorption of metals by geomedia. Acadmemic Press, San Diego*, Chapter 19.

Gonçalves, G. O., Lana, C., Scholz, R., Buick, I. S., Gerdes, A., Kamo, S. L., et al. (2016). An assessment of monazite from the Itambé pegmatite district for use as U-Pb isotope reference material for microanalysis and implications for the origin of the "Moacyr" monazite. *Chemical Geology, 424,* 30–50.

Graeser, S., schwander, H., & Stalder, H. A. (1973). Asolid solution series between xenotime (YtPO4) and chernovite (YtAsO4). *Mineralogical Magazine, 39*, 145–151.

Grosjean, E., Adam, P., Connan, P., & Albrecht, P. (2004). Effects of weathering in nickel and vanadyl porphyrins of a Lower Toarcian shale of the Paris Basin. *Geochinical et Cosmochimica Acta, 60*, 3631–3642.

Hallsworth, C. R., Livingstone, A., & Morton, A. C. (1992). Detrital goldmanite from the Paleocene of the North Sea. *Mineralogical Magazine, 56*, 117–120.

Hallsworth, C. R., Morton, A. C., Claoue-Long, J. C., & Fanning, C. M. (2000). Carboniferous sand provenance in the Pennine Basin, UK: Constraints from heavy mineral and SHRIMP zircon age data. *Sedimentary Geology, 137*, 147–185.

Hartmann, L. A., Leite, J. A. D., Silva, I. C., Remus, M. V. D., McNaughton, N. J., Groves, & et al. (2000). Advances in SHRIMP geochronology and their impact on understanding the tectonic and metallogenic evolution of southern Brazil. *Australian Journal of Earth Sciences, 47*, 829–844.

Hastings, D. W., Emerson, S. R., & Mix, A. C. (1996). Vanadium in foraminiferal calcite as a tracer for changes in the areal extent of reducing sediments. *Paleoceanography, 11*(6), 665–678.

Heaman, L. M., Bowins, R., & Crocket, J. (1990). The chemical composition of igneous zircon suites: implications for geochemical tracer studies. *Geochimica et Cosmochimica Acta, 54*, 1597–1607.

Heatherington, C. J., & Harlov, D. E. (2008). Metasomatic thorite and uraninite inclusions inclusions in xenotime and monazite from granitic pegmatites, Hidra anorthosite massif, southwestern Norway: mechanics and fluid chemistry. *American Mineralogist, 93*, 806–820.

Heatherington, C. J., Jercinovic, M. J., Williams, M. L., & Mahan, K. (2008). Understanding geologic processes with xenotime: Composition, chronology, and a protocol for electron probe microanalysis. *Chemical Geology, 254*, 133–147.

Heinrich, W., Andrehs, G., & Franz, G. (1997). Monazite-xenotime miscibility gap thermometry. I. An empirical calibration. *Journal of Metamorphic Geology, 15*, 3–16.

Helz, G. R., Miller, C. V., Charnock, J. M., Mosselmans, J. L. W., Pattrick, R. A. D., Garner, C. D., et al. (1996). Mechanisms of molybdenum removal from the sea and its concentration in black shales: EXAFS evidences. *Geochimica et Cosmochimica Acta, 60*, 3631–3642.

Hisada, K., & Arai, S. (1993). Detrital chrome spinels in the Cretaceous Sanchu sandstone, central Jaman: Indicator of serpentinite protrusion into a fire-arc region. *Palaeogeography, Palaeoclimatology, Palaeoecology, 105*, 95–109.

Hisada, K., Sugiyama, M., Ueno, K., Charushiri, P., & Arai, S. (2004). Missing ophiolitic rocks along the Mae Yuam Fault as the Gondwana-Tethys divide in north-west Thailand. *The Isand Arc, 13*, 119–127.

Holland, H. D., & Gotfried, D. (1955). The effect of nuclear radiation on the structure of zircon. *Acta Crystallographica, 8*, 291–300.

Hoskin, P. W. O., & Ireland, T. R. (2000). Rare earth element chemistry of zircon and its use as a provenance indicator. *Geology, 28*, 627–630.

Huerta-Diaz, M. A., & Morse, J. W. (1992). Pyritization of trace metals in anoxic marine sediments. *Geochimica et Cosmochimica Acta, 56*(7), 2681–2702.

Hurst, A. (1990). Natural gamma-ray spectroscopy in hydrocarbon-bearing sandstones from the Norwegian Continental Shelf. In: A. Hurst, M. A. Lovell & A. C. Morton (Eds.), *Geological applications of wireline logs* (No. 48, pp. 211–222). London: Geological Society of London Special Publication.

Hurst, A., & Milodowski, T. (1994). Characterisation of clays in sandstones: Thorium content and spectral log data. In Paper S:-Transactions of the seventeenth European Formation Evaluation Symposium, October 11th–13th, 1994. SPWLA Aberdeen update.

Hurst A., & Morton, A. (2014). Provenance models: the role of sandstone mineral-chemical stratigraphy. In: R. Scott, H. Smyth, A. C. Morton & N. Richardson (Eds.), 2014. *Sediment provenance studies in hydrocarbon exploration and production* (386, pp. 7–26). London: Geological Society London, Special Publications.

Irvine, T. N. (1965). Chromian spinel as a petrogenetic indicator. Part 1. Theory. *Canadian Journal of Earth Sciences, 2*, 648–672.

Irvine, T. N. (1967). Chromian spinel as a petrogenetic indicator: Part 2. Petrologic applications. *Canadian Journal of Earth Sciences, 4*(1), 71–103.

Jarvis, I., Burnett, W. C., Nathan, Y., Almbaydin, F. S. M., Attia, A. K. M., Castro, L. N., & Serjani, A. (1994). Phosphorite geochemistry: state-of-the-art and environmental concerns. *Eclogae Geologicae Helvetiae, 87*(3), 643–700.

Johan, Z., & Johan, V. (2004). Accessory minerals of the Cinovec (Zinnwald) granite cupola, Czech Republic: Indicators of petrogenetic evolution. *Mineralogy and Petrology, 83*, 113–150.

Jonasson, R. G., Bancroft, G. M., & Boatner, L. A. (1988). Surface reactions of synthetic, end-member analogues of monazite, xenotime ansd rhabdophane, and evolution of natural waters. *Geochimica et Cosmochimica Acta, 52*, 767–770.

Jordens, A., Marion, C., Crammatikopoulos, T., & Waters, K. E. (2016). Understanding the effect of mineralogy on muscovite flotation using QEMSCAN. *International Journal of Mineral Processing, 155*, 6–12.

Kingsbury, J. A., Miller, C. F., Wooden, J. L., & Harrison, T. M. (1993). Monazite paragenesis and U-Pb systematics in rocks of the Mojave Desert, California, USA: Implications for thermochronometry. *Chemical Geology (including Isotope Geology), 110*, 147–167.

Klemme, S., Marshall, H. R., Jacob, D., Prowatke, S., & Ludwig, T. (2011). Trace-element partitioning and boron isotope practionation between white mica and tourmaline. *The Canadian Mineralogist, 49,* 165–176.

Klinkhammer, G. P., & Palmer, M. R. (1991). Uranium in the oceans: Where it goes and why. *Geochimica et Cosmochimica Acta, 55,* 1799–1806.

Knoche, R. L., Angel, R. J., Seifert, F., & Fliervoet, T. F. (1998). Complete substitution of Si for Ti in titanite $Ca(Ti_{1\text{-}x}Si_x)^{VI}Si^{IV}O^5$. *American Mineralogist, 83,* 1168–1175.

Kraal, P., Dijkstra, N., Behrends, T., & Slomp, C. P. (2017). Phosphorous burial in sediments of the sulfidic deep Black Sea: Key roles for adsorption by calcium carbonate and apatite authigenesis. *Geochimicla et Cosmochimica Acta, 204,* 140–158.

Mange, M. A., & Maurer, H. F. W. (1992). *Heavy minerals in colour.* London: Chapman and Hall.

Mange, M. A., & Morton, A. C. (2007). Geochemistry of heavy minerals. *Developemnts in Sedimentology, 58,* 345–391.

McConnel, D. (1973). *Apatite—its crystal chemistry, mineralogy, utilization and geologic and biologic occurrences.* New York: Springer.

McLennan, S. M. (2001). Relationships between the trace element composition of sedimentary rocks and upper continental crust. *Geochemistry. Geophysics, Geosystems* (G^3) *2*(2000GC000109).

Meyers, S. R., Sagemann, B. B., & Lyons, T. W. (2005). Organic carbon burial rate and the molybdenum proxy: Theoretical framework and application to Cenomanian-Turonian oceanic event 2. *Paleoceanography, 20.*

Mockoviakova, A., & Orolinova, Z. (2009). Adsorption properties of modified bentonite clay. *Chemical Technology, 1,* 113–130.

Morford, J. L., & Emerson, S. (1999). The geochemistry of redox sensitive trace metals in sediments. *Geochimica et Cosmochimica Acta, 63*(11), 1735–1750.

Moore, C. H., Chowdhury, A., & Heydari, E. (1986). Variation of ooid mineralogy in Jurassic Smackover limestone as control of ultimate diagenetic potential. *AAPG Bulletin, 70,* 622–628.

Moore, C. H., & Wade, W. J. (2013). *Developments in sedimentology, V. 76. Carbonate reservoirs, 2nd ed, Porosity and Diagenesis in a Sequence Stratigraphic framework.* Amsterdam: Elsevier.

Morford, J. L., Russell, A. D., & Emerson, S. (2001). Trace metal evidence for changes in the redox environment associated with the transition from terrigenous clay to diatomaceous sediments, Saanich Inlet, BC. *Marine Geology, 174,* 355–369.

Morse, J. W., & Luther, G. W., III. (1999). Chemical influences on trace metal-sulfide interactions in anoxic sediments. *Geochimica et Cosmochimica Acta, 63,* 3373–3378.

Morton, A. C., & Hallsworth, C. (1994). Processes controlling the composition of heavy mineral assemblages in sandstones. *Sedimentary Geology, 124,* 3–29.

Morton, A. C., & Yaxley, G. (2007). Detrital apatite geochemistry and its application in provenance studies. In: J. Arribas, S. Critelli & M. J. Johnsson (Eds.), *Sediment provenance and petrogenesis: perspectives from petrography and geochemistry* (Vol. 420, pp. 319–344). Geological Society of America, Special Paper.

Nadoll, P., Angerer, T., Mauk, J. L., French, D., & Walshe, J. (2014). The chemistry of hydrothermal magnetite: A review. *Ore Geology Reviews, 61,* 1–32.

Naimo, D., Adamo, P., Imperato, M., & Stanzione, D. (2005). Mineralogy and geochemistry of a marine sequence, Gulf of Salerno, Italy. *Quaternary International, 140,* 53–63.

Nameroff, T. J., Calvert, S. E., & Murray, J. W. (2004). Glacial-interglacial variability in the eastern tropical North Pacific oxygen minimum zone recorded by redox-sensitive trace metals. *Paleoceanography, 19.*

Nascimento, D. R., Jr., Sawakuchi, A. O., Guedes, C., Giannini, P. C. F., Grohmann, C. H., & Ferreira, M. P. (2015). Provenance of sands from the confluence of the Amazon and Madeira rivers based on detrital heavy mineral and luminescence of quartz and feldspar. *Sedimentary Geology, 316,* 1–12.

Nash, W. P. (1984). Phosphate minerals in terrestrial igneous and metamorphic rocks. In J. O. Nriagu & P. B. Moore (Eds.), *Phosphate minerals* (pp. 215–241). New York: Springer.

Neumann, H. (1985). Norges mineraler. *Norges geologiske undersokelse Bulletin, 68,* 278.

Nicholson, U., Poynter, S., Clift, P. D., & Macdonald, D. M. (2014). Tying catchment to basin in a giant sediment routing system: A source-to-sink study of the Neogene-Recent Amur River and its delta in the North Sakhalin Basin. In: R. Scott, H. Smyth, A. C. Morton & N. Richardson (Eds.), 2014. *Sediment provenance studies in hydrocarbon exploration and production* (386, pp. 7–26). Geological Society London, Special Publications.

Odomo, A. N., Obaje, N. G., Omada, J. I., Idakwo, S. O., & Erbacher, J. (2013). Paleoclimate reconstruction during Mamu formation (cretaceous) based on clay mineral distributions. *Journal of Applied Geology and Geophysics, 1,* 40–46.

Ondrej, M., Uher, O., Prešek, J., & Ozdin, D. (2007). Arsenian monazite-(Ce) and xenotime (Y), REE arsenates and carbonates from the Tisovec-Rejkovo rhyolite, Western Carpathians, Slovakia: composition and substitutions in the (REE, Y)XO_4 system (X = P, As, Si, Nb, S). *Lithos, 95,* 116–129.

Overstreet, W. (1967). *The geological occurrence of monazite.* US Geological Survey Professional Paper 530.

Owen, M. R. (1987). Hafnium content of detrital zircons, a new tool for provenance study. *Journal of Sedimentary Petrology, 57,* 824–830.

Piper, D. Z., & Perkins, R. B. (2004). A modern vs. Permian black shale—the hydrography, primary productivity, and water-column chemistry of deposition. *Chemical Geology, 206,* 177–197.

Plier, R., & Adams, J. A. S. (1962). The distribution of thorium and uranium in a Pennsylvanian weathering profile. *Geochimica et Cosmochimica Acta, 26,* 1137–1146.

Poldervaart, A. (1955). Zircon in rocks, 1: Sedimentary rocks. *American Journal of Science, C., 253,* 433–461.

Preston, J., Hartley, A., Hole, M., Buck, S., Bond, J., Mange, M., et al. (1998). Integrated whole-rock trace element geochemistry and heavy mineral studies: aids to correlation of continental red-bed reservoirs in the Beryl Field, UK North Sea. *Petroleum Geoscience, 4,* 7–16.

Pyle, J., & Spear, F. (1999). Yttrium zoning in garnet: Coupling of major and accessory phases during metamorphic reactions. *Geological Materials Research, 1*(6), 1–49.

Ratcliffe, K. T., Morton, A. C., Ritcey, D. H., & Evenchick, C. A. (2007). Whole-rock geochemistry and heavy mineral analysis as petroleum exploration tools in the Bowser Basin, British Columbia, Canada. *Bulletin of Canadian Petroleum Geology, 55*(4), 320–336.

Regenspurg, S., Margot-Roquier, C., Harfouche, M., Froidevaux, P., Steinmann, P., Junier, P., et al. (2010). Speciation of naturally-accumulated uranium in an organic-rich soil of an alpine region (Switzerland). *Geochimica et Cosmochimica Acta, 74,* 2082–2098.

Sabatina, N., Neri, R., Bellanca, A., Jenkyns, H. C., Masetti, D., & Scopellity, G. (2011). Petrography and high-resolution geochemical records of Lower Jurassic manganese-rich deposits from Monte Mangart, Julian Alps. *Palaeogeography, Palaeoclimatology, Palaeoecology, 299,* 97–109.

Saito, M. A., Moffet, J. W., Chisholm, S. W., & Waterbury, J. B. (2002). Cobalt limitation and uptake in Prochlococcus. *Limnology and Oceanography, 47,* 1629–1636.

Scheibe, C. (2002). *The application of whole-rock geochemistry and mineral chemistry to the correlation of Mesozoic reservoirs within the Alwyn Area, N. North Sea.* PhD thesis, University of Aberdeen.

Schneiderman, J. S. (1995). Detrital opaque oxides as provenance indicators in River Nile sediments. *Journal of Sedimentary Research, 65,* 668–674.

Schneideman, J.S. (1997). Potential paleoclimate indicators: composition of detrital pyroxene sediments. *journal of Sedimentary Research, 65,* 668–674.

Scott, M. R. (1968). Thorium and uranium concentrations and isotopic ratios in river sediments. *Earth and Planetary Science Letters, 4,* 245–252.

Scott, A. A., Smyth, H. R., Morton, A. C., & Richardson, N. (Eds.). 2014. *Sediment provenance studies in hydrocarbon exploration and production* (386). Geological Society London, Special Publication.

Serra, O., Baldwin, J., & Quirein, J. (1980). Theory, interpretation and practical application of natural gamma ray spectroscopy. In *SPWLA 21st Annual Logging Symposium Transactions,* paper Q.

Shannon, R.D., Prewitt, C.T. (1970). Effective ionic radii and crystal chemistry.*Journal of Inorganic and Nuclear Chemistry, 32*(5), 1427–1441.

She, Y. W., Song, X. Y., Yu, S. Y., & He, H. L. (2015). Variations of trace element concentration of magnetite and ilmenite from the Taihe layered intrusion, Emeishan large igneous province, SWChina: Implications for magmatic fractionation and origin of Fe-Ti-V oxide ore deposits. *Journal of Asian Earth Sciences, 113,* 1117–1131.

Smits, G. (1989). (U, Th)-bearing silicates in reefs of the Witwatersrand, South Africa. *Canadian Mineralogist, 27,* 643–655.

Spandler, C., Hammerli, J., Sha, P., Hilbert-Wolf, H., Hi, L., Roberts, E., et al. (2016). MKED1: A new titanite standard for in situ analysis of Sm-Nd isotopes and U = Pb geochronology. *Chemical Geology, 425,* 110–126.

Spear, F., & Pyle, J. (2002). Apatite, monazite and xenotime in metamorphic rocks. In M. L. Khon, J. Rakovan, & J. M. Hughes (Eds.), *Phosphates: Geochemical, Geobiological and Materials Importance. Reviews in Mineralogy and Geochemistry* (pp. 293–335). Washington: Mineralogical Society of America.

Tamer, M. (2013). Quantitative phase analysis based on Rietveld structure refinement for carbonate rocks. *Journal of Modern Physics., 4,* 1149–1157.

Thomson, J., Jarvis, I., Green, D. R. H., & Green, D. (1998). Oxidation fronts in Madeira Abyssal Plain turbidites: persistence of early diagenetic trace-element enrichments during burial, Site 950. In: P. P. E. Weaver, H. U. Schmincke, J. V. Firth, W. Duffield, (Eds.), In *Proceedings of ODP, Sci. Results* (Vol. 157, pp. 559–572). Ocean Drilling Program, College Station, TX.

Tieh, T. T., Ledger, E. R., & Row, M. W. (1980). Release of uranium from granitic rocks during in situ weathering and initial erosion (Central Texas). *Chemical Geology, 29,* 227–248.

Tiepolo, M., Oberti, R., & Vannucci, R. (2002). Trace-element incorporation in titanite: constraints from experimentally determined solid/liquid partition coefficients. *Chemical Geology, 191*(1), 105–119.

Trappe, J. (1998). Phanerozoic phosphorite depositional systems. *Lecture Notes in Earth Sciences, 76.*

Tribovillard, N., Algeo, T. J., Baudin, F., & Riboulleau, A. (2012). Analysis of marine environmental conditions based on molybdenum-uranium covariation-applications to Mesozoic paleoceanography. *Chemical Geology, 324,* 46–58.

Tribovillard, T., Algeo, T. J., Lyons, T., & Riboulleau, A. (2006). Trace metals as paleoredox and paleoproductivity proxies: An update. *Chemical Geology, 232,* 12–32.

Troitzsch, U., Ellis, D. J., Thompson, J., & Fitz-Gerald, J. (1999). Crystal structural changes in titanite along the join TiO-A1F. *European Journal of Mineralogy, 6,* 955–965.

Tyson, R. V., & Pearson, T. H. (1991). Modern and ancient continental shelf anoxia: An overview. In: R. V. Tyson & T. H.Pearson (Eds.), *Modern and ancient continental shelf anoxia* (Vol. 59, pp. 1–26). Geological Society Special Publication.

Uddin, M. K. (2017). A review on the adsorption of heavy metals by clay minerals, with special focus on the last decade. *Chemical Engineering Journal, 308,* 438–462.

Vallini, D., Rasmussen, B., Krapex, B., Fletcher, I. R., & McNaughton, N. J. (2002). Obtaining diagenetic ages from metamorphosed sedimentary rocks: U-Pb dating of unusually coarse xenotime cement in phosphatic sandstone. *Geology, 30*(12), 1083–1086.

Varva, G., Schmid, R., & Gebauer, D. (1999). Internal morphology, habit and U-Th-Pb micro-analysis of amphibolite-to-granulite facies zircons: geochronology of the Ivrea Zone (Southern Alps). *Contributions to Mineralogy and Petrology, 97,* 205–217.

Vorlicek, T. P., Kahn, M. D., Kasuza, Y., & Helz, G. R. (2004). Capture of molybdenum in pyrite-forming sediments: Role of lignand-induced reduction by polysulfides. *Geochimica et Cosmochimica Acta, 68,* 547–556.

Walderhaug, O., & Porten, K. W. (2007). Stability of detrital heavy minerals on the Norwegian continental shelf as a function of depth and temperature. *Journal of Sedimentary Research, 77,* 992–1002.

Wall, F., Niku-Paavola, V. N., Storey, C., Müller, A, & Jeffries, T. (2008). Xenotime-(Y) from carbonatite dykes at Lofdal, Namibia: Unusually low light to heavy REE ratio in carbonatite, and the first dating of xenotime overgrowths on zircon. *Canadian Mineralogst, 46.*

Wanty, R. B., & Goldhaber, R. (1992). Thermodynamics and kinetics of reactions involving vanadium in natural systems: accumulation of vanadium in sedimentary rock. *Geochimica et Cosmochimica Acta, 56,* 171–183.

Watson, E. B. (1976). Two = liquid partition coefficients: Experimental data and geochemical implications. *Contributions to Mineralogy and Petrology, 56,* 119–134.

Weaver, C. E., & Pollard, L. D. (1973). *Developments in sedimentology, the chemistry of clay minerals* (Vol. 15, pp. 1–213).

Wechsler, B.A., Lindsley, D.H. & Prewitt, C.T. (1984). Crystal structure and cation distribution in titanomagnetites (Fe3-XTiXO4). *American Mineralogist, 69,* 754–770.

Wedepohl, K. H. (1991). The composition of the upper Earth's crust and the natural cycles of selected metals. In E. Merian (Ed.), *Metals and their compounds in the environment* (pp. 3–17). Weinheim: VCH-Verlagsgesellschaft.

Wedepohl, K. H. (1971). Environmental influences on the chemical composition of shales and clays. In L. H. Ahrens, F. Press, S. K. Runcorn, & H. C. Urey (Eds.), *Physics and chemistry of the earth* (pp. 305–333). Oxford: Pergamon.

Wehrly, B., & Stumm, W. (1989). Vanadyl in natural waters: adsorption and hydrolysis promote oxygenation. *Geochimical et Cosmochimica Acta, 53,* 69–77.

Whitfield, M. (2002). Interactions between phytoplankton and trace metals in the ocean. *Advances in Marine Biology, 41,* 3–120.

Wignall, P. B., & Maynard, J. R. (1993). The sequence stratigraphy of transgressive black shales. In: B.J. Katz & L. M. Pratt (Eds.), *Source rocks in a sequence stratigraphic framework* (Vol. 37, pp. 35–47). AAPG Studies in Geology.

Wirth, S. B., Gilli, A., Niemann, H., Dahl, T. W., Ravasi, D., Sax, N., et al. (2013). Combining sedimentological, trace metal (Mn, Mo) and molecular evidence for reconstructing past water-column redox conditions: the example of meromictic Lake Cadagno (Swiss Alps). *Geochimica et Cosmochimica Acta, 120,* 220–238.

Wold, S., Esbensen, K., & Geladi, P. (1987). Principal componenet analysis. *Chemometrics and Intelligent Laborary Systems, 2,* 37–52.

Wright, A. M., Spain, D., & Ratcliffe, K. T. (2010). Application of inorganic whole rock geochemistry of shale resource plays. In *Paper presented at Canadian Unconventional Resources and International Petroleum Conference held in Calgary*, Alberta, 2010.

Zack, T., Kronz, A., Foley, S., & Rivers, T. (2002). Trace element abundances in rutiles from eclogites and associated garnet mica schists. *Chemical Geology, 184,* 97–122.

Zack, T., Von Eynatten, H., & Kronz, A. (2004). Rutile geochemistry and its potential use in quantitative provenance studies. *Sedimentary Geology, 171,* 37–58.

Zheng, Y., Anderson, R. F., van Geen, A., & Kuwabara, J. (2000). Authigenic molybdenum formation in marine sediments: A link to pore water sulphide in the Santa Barbara Basin. *Geochimica et Cosmochimica Acta, 64,* 4165–4178.

4 Production of Chemostratigraphic Correlation Schemes

Abstract

Producing a chemostratigraphic correlation scheme is probably the most challenging stage of any chemostratigraphy project. In order to avoid making interpretations based entirely on changes in grain size/lithology, it is recommended that separate schemes are proposed for sandstone and mudrock samples in studies of clastic sediments. In carbonates, no such differentiation is necessary and the data are treated as a whole. After making some final checks on data quality, profiles are plotted for each element, and then for Al-normalised data and geochemical ratios. Most chemostratigraphy projects involve the analysis of 40–55 elements (so profiles are plotted for more than 250 elements and ratios), but the majority of correlation schemes relate to variations in 4–12 key elements or ratios. Hierarchical schemes are developed, based on the recognition of correlative zones, subzones, divisions and subdivisions. In addition to being identified on element/ratio profiles, the geochemical characteristics of these chemozones can also be visualised on binary and ternary diagrams. Using histograms and DFA (Discriminant Function Analysis), it is possible to assign levels of statistical confidence to each chemozone. In recent years, chemostratigraphy has been used very often in conjunction with lithostratigraphy, biostratigraphy, sedimentology and seismic data. By employing such a multidisciplinary approach to reservoir correlation, it is possible to propose more robust correlations of higher resolution.

4.1 Introduction

Arguably the most exciting stage in any chemostratigraphy project involves the development of chemostratigraphic schemes, though this is also the most challenging. Choosing the correct elements and ratios to use for chemostratigraphic purposes and the recognition of 'trends in the dataset' are skills that often take years to develop. No textbook or academic paper can teach the individual all of the skills he/she requires to interpret geochemical data, or how to

N. Craigie, *Principles of Elemental Chemostratigraphy*, Advances in Oil and Gas Exploration & Production, https://doi.org/10.1007/978-3-319-71216-1_4

propose chemostratigraphic correlations, in any given study. It is hoped that the following paragraphs will, however, provide a basic grounding on interpretation techniques and information on common pitfalls that face chemostratigraphers on a regular basis.

It may surprise the reader that relatively few references are used in parts of this chapter. This is largely explained by the lack of information provided by authors on subjects relating to interpretive methodology. The reasons for this 'gap' in information are manifold but may be summarized as follows:

(a) Academic journals generally do not favour detailed explanations on analytical or interpretive methodologies, preferring the author to focus on 'the final result'.
(b) Comparatively few individuals are skilled in the interpretation of inorganic geochemical data and the development of chemostratigraphic correlation schemes. The perceived economic benefits of knowledge retention are apparent to individuals and companies alike—the theory being that too many chemostratigraphers may equate to dilution of stature and ultimate loss of income. Whilst these concerns may be justified in part, it is felt that chemostratigraphy can only grow as a subject and become more popular through the transfer of knowledge.
(c) Without mentioning specific publications or reports, some are written by individuals with little experience in the interpretation of inorganic geochemical data. Mistakes are made in many studies but are rarely rectified. Such errors are commonplace in both published and unpublished works, and often result in the unfair criticism of the technique (e.g. North et al. 2005). Publication of substandard work, and spurious correlations without sound scientific support, can only be to the detriment of chemostratigraphy as a whole. Note that no formal training courses in chemostratigraphy are available to the general public, or even the wider geological community, so an almost infinite number of possibilities exist for the inexperienced chemostratigrapher to misinterpret data.
(d) The numerous pitfalls in the interpretation of geochemical data are rarely documented in publications.

It is first necessary to determine the lithology of the analyzed samples before making final quality control checks, and then plotting profiles for individual elements and ratios. Key elements and ratios are chosen for chemostratigraphic purposes and hierarchical schemes developed. The final stage of many studies involves the integration of chemostratraphic, biostratigraphic and/or sedimentological data. Numerous studies have shown chemostratigraphy to be most effective when utilized as part of a multidisciplinary approach to reservoir correlation.

A discussion on the parameters that are commonly used to distinguish provenance is provided towards the end of this chapter. Given that most chemostratigraphic schemes are based on changes in this factor, the chapter would not be complete without such a discussion. The pitfalls associated with these models are, however, manifold and these are also mentioned.

4.2 Lithology Determination

After the mineralogical affinities of elements are established, the next step in chemostratigraphy projects is to plot the data in the form of profiles. In most studies conducted on clastic sediments, separate profiles are plotted for sandstone and mudrock lithologies. If this is not done, then the resulting trends in particular elements with depth may be entirely related to variations in grain size/lithology rather than to provenance (on which most chemostratigraphic correlations are based). The word mudrock, is used in the present study as an all-encompassing term to describe the following lithologies: claystone, shale, silty claystone, silty shale and siltstone. The lithology of core and field samples should be very obvious from sample descriptions, but this does not always hold true for cuttings, particularly where fragments take the form of 'reconstituted' materials. In such circumstances, it is important to examine cuttings descriptions in conjunction with wireline log responses and geochemical data acquired from major elements in order to establish the dominant lithology of each sample. For example, if the gamma API levels in a siliciclastic reservoir are below 50 through the interval from which a cuttings samples were taken, and the samples reveal Si/Al values of greater than 10, then it can be concluded that these take the form of sandstones, irrespective of whether this is obvious from sample descriptions or not. The following general rules may be employed to establish the bulk lithology of sedimentary rocks using geochemical data:

$$Si/Al < 5 = \text{claystone or shale}$$
$$Si/Al\ 5{-}6 = \text{siltstone}$$
$$Si/Al\ 6{-}8 = \text{argillaceous sandstone}$$
$$Si/Al > 8 = \text{non - argillaceous sandstone}$$
$$Ca\ 7{-}15\% = \text{calcareous}$$

Ca > 20% = limestone, dolomite, gypsum of anhydrite. Where samples are composed of either gypsum of anhydrite, high S contents would be observed in conjunction with elevated Ca

$$Ca \text{ in region of } 30{-}40\%, \text{ with Mg of } 20{-}28\% = \text{dolostone}$$

Although these rules can be useful, they should be applied with considerable caution as they may not hold true for every study. For example, in some studies argillaceous sandstones may yield Si/Al ratios of less than 5.

A multitude of elements and elemental ratios are used to classify igneous and metamorphic rocks, many of which are described by Rollinson (1993). As it is difficult to find a simple relationship between the mineralogy of sandstones and their chemical composition, however, it is difficult to use whole rock geochemical data to model the conventional mineralogical classification of sandstones based upon quartz-feldspar-lithic fragments. Instead, classification schemes proposed for this lithology normally differentiate between mature and immature sediments. In the 1970s and 1980s the most commonly used scheme to classify sandstones was proposed by Pettijohn et al. (1972). This is based on log (Si/Al) and log (Na/K) ratios with greywacke, litharenite, arkose, subarkose, sublitharenite and quartz arenite 'fields' identified in a crossplot of these two ratios (Fig. 4.1a). More recently, Herron (1988) modified the Pettijohn et al. (1972) diagram to use log (Fe/K) ratios along the y-axis instead of log (Na/K) (Fig. 4.1b). It was found that arkoses could be classified more successfully using the log (Fe/K) parameter. This is also a measure of mineral stability as ferromagnesium minerals tend to

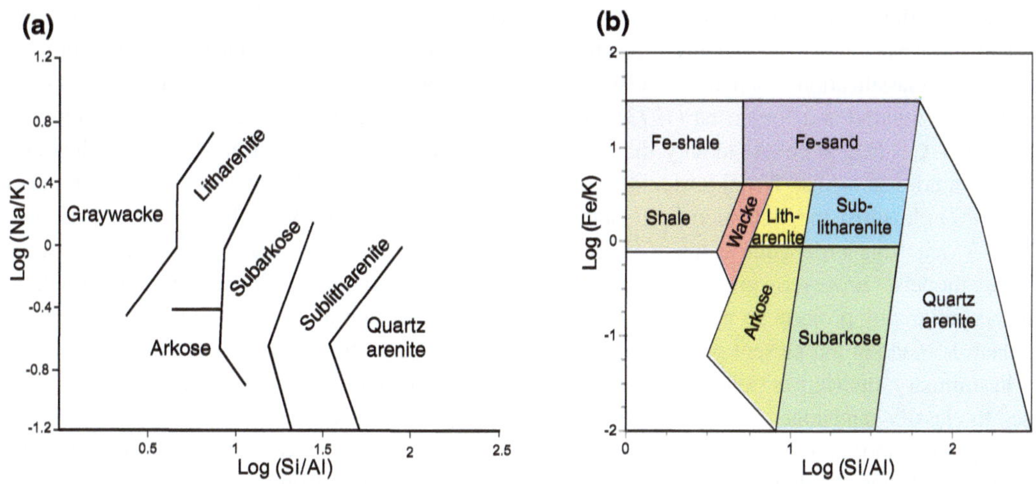

Fig. 4.1 Schemes used to classify sandstones based on the publications of **a** Pettijohn et al. (1972), and **b** Herron (1988)

be amongst the least stable minerals during weathering (Herron 1988; Rollinson 1993).

The work of Herron (1988) was published around 30 years ago but very few attempts have been made to utilize geochemical data to classify sandstones in more recent times and this is widely accepted as the most suitable scheme to use for this purpose. In spite of this, some caution should be applied when using the Herron (1988) scheme as the elements Fe and K may have multiple mineralogical affinities. For example a low log (Fe/K) value may be explained by the existence of K-bearing authigenic illite, making it difficult to differentiate the arkose from litharenite 'fields' on the log (Si/Al) versus log (Fe/K) crossplot.

Other scheme were proposed to classify sandstones, mostly in the 1970s and 1980s. These include the model of Blatt et al. (1972), who differentiated greywackes, lithic sandstones and arkoses, based on a ternary diagram of Na, Fe + Mg and K. This scheme was tested by Potter (1978), however, who found that the fields of this figure could not be duplicated in a study of recent big-river sediments, with the chemical composition of ancient sediments heavily influenced by diagenesis. Blatt et al. (1972) also suggested that Fe^{2+}/Fe^{3+} ratios could be used to subdivide greywackes and arkoses, but Potter (1978) could not prove this theory based on the study of recent big-river sediments, concluding that Fe^{2+}/Fe^{3+} values must be strongly controlled by the oxidation state of the sediment.

Unlike sandstones, no widely recognized chemical classification scheme exists for mudrocks. Englund and Jorgensen (1973) used a (K + Na + Ca)-(Mg + Fe)-Al ternary diagram to classify mudrock, whilst Wronkiewicz and Condie (1987) classified the Witwatersrand shales of South Africa using a triangular graph of Fe–K–Al. These models, however, are rarely used in chemostratigraphy projects so no further mention of them is made in the present publication.

In summary, the Herron (1988) scheme can be used to classify sandstones but should be ideally applied in collaboration with petrographic and/or XRD data. Where such mineralogical data are unavailable, this scheme should be utilized with caution, owing to uncertainties of element: mineral links and the effects of post depositional diagenesis. With the single exception of the Pettijohn et al. (1972), the other sandstone classification schemes are rarely used. In spite of these uncertainties associated with the classification of sediments, values of Si/Al, Ca and Mg should be used in conjunction with sample descriptions, XRD, petrography and wireline log data to differentiate sandstones, mudrocks and carbonates, whilst the Herron (1988) and Pettijohn et al. (1972) schemes may be employed tentatively to characterise sandstones.

4.3 Final Quality Control Checks

The quality of data received from the XRF or ICP laboratory should be checked before plotting it on a series of profiles, and the various checks are described in detail in Chap. 2. Additional quality control checks should, however, be implemented before attempting to identify chemozones and chemostratigraphic boundaries. In studies performed on clastic sediments, separate profiles should be plotted for sandstone and mudrock samples. These should include profiles for each major, trace and REE. In chemostratigraphy studies conducted on carbonates, it is generally impractical to plot data for separate lithologies, except under special circumstances. This is certainly true of well 2 in a study completed on the Jurassic Dhruma Formation of Saudi Arabia, where the lithologies comprise limestone, dolostone, calcareous mudrocks, dolomitic mudrocks and subordinate calcareous sandstones. Figure 4.2 shows profiles for major elements plotted for these lithologies in this well.

One of the final data quality steps should include an examination of major element data to ensure that the bulk lithology determined by sample descriptions and wireline log responses is consistent with the geochemical data. For example, if a group of cuttings samples are described as non-calcareous sandstones and that this is obvious from wireline log responses

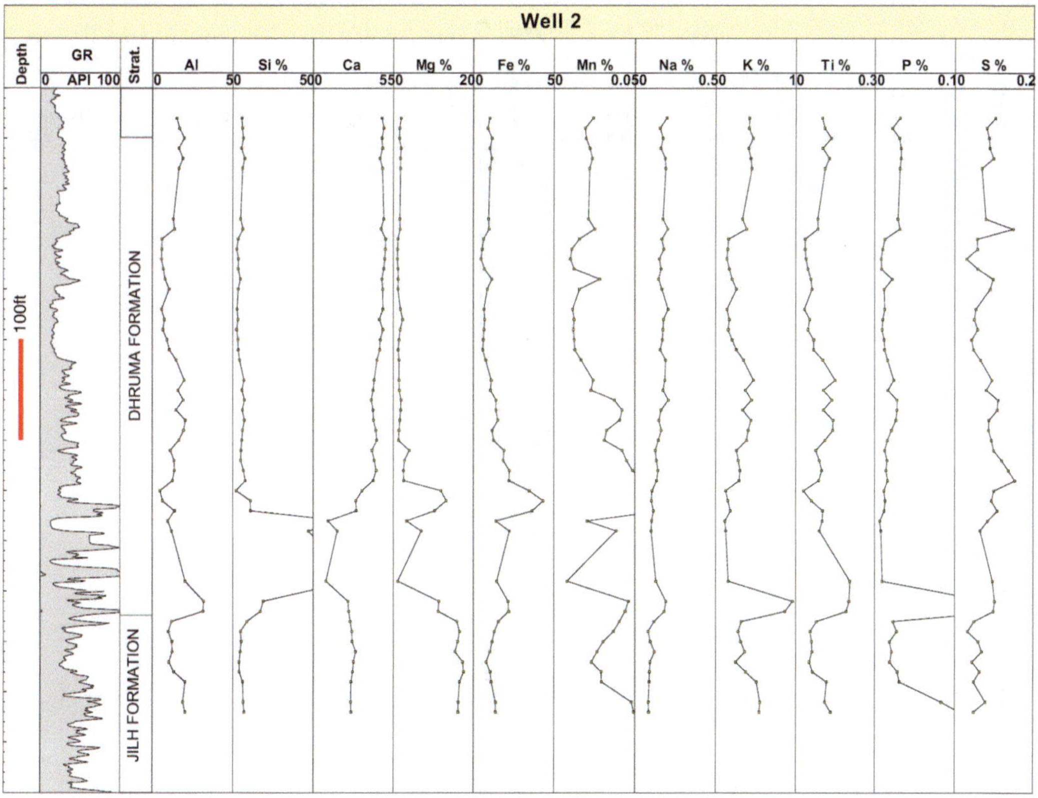

Fig. 4.2 Major element profiles plotted for well 2. All depths are log depths in feet and all samples were taken from cuttings material

(i.e. low gamma, low resistivity, low density and slow sonic), then it would come as a surprise if samples analysed from this interval were to yield Ca in the range 25–35%. If this were the case, then the high Ca values may either be explained by the presence of significant cavings within the cuttings, or by calcareous drilling additives, for example in the form of bicarbonate or marble. In order to determine which explanation is most likely, the interpreter should consult drilling reports to find out the materials added to the well during the drilling process. If drilling additives can be ruled out, then an examination of bulk lithologies immediately above the present study section may reveal that the cuttings contain an abundance of caved material from this overlying interval. In both scenarios, it may still be possible to use the resulting data for chemostratigraphic purposes if trends in other elements and ratios are consistent with the same trends observed in adjacent wells, where the presence of Ca bearing drilling additives or caved materials are not apparent. However, where samples have been heavily contaminated, they should not be used for chemostratigraphic purposes. Another, though less common, explanation exists for a poor match between the bulk lithology observed in samples and the geochemical data. It is possible that the analyzed core or cuttings samples have been mislabelled or have been collected from the wrong depths. Fortunately, inconsistencies of this nature are rare. If the presence of additives, cavings or flawed sample preparation procedures can be ruled out, the source of 'questionable' data may relate to malfunction of analytical equipment. If there are any doubts relating to this, then a thorough service of the instrument(s) is recommended.

An example of contamination of cuttings samples is illustrated in Fig. 4.3. This diagram

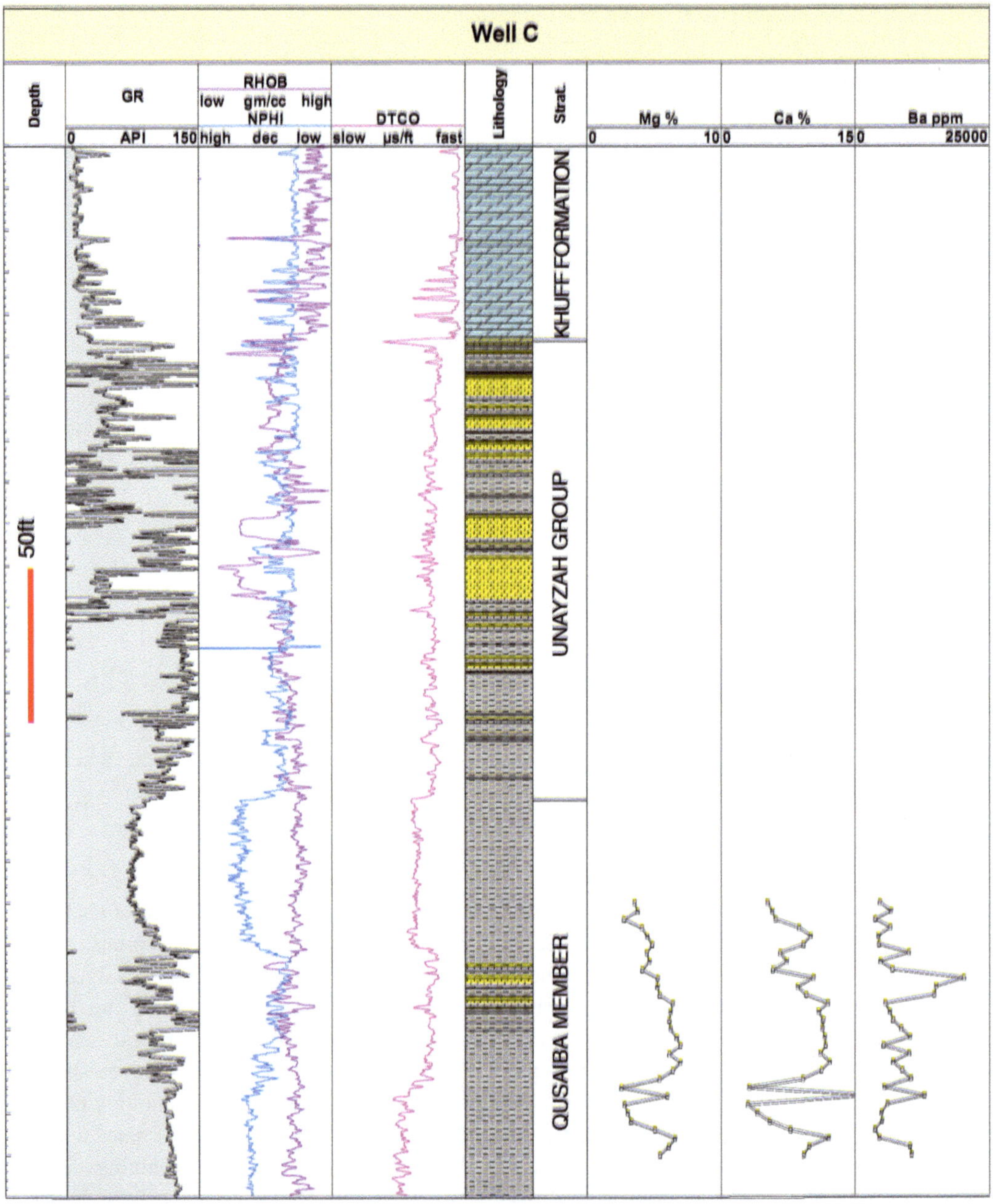

Fig. 4.3 The samples of well C are taken through the clastic lithologies of the Qusaiba Member. In adjacent wells these samples yield less than 2% Ca and Mg but in this well the samples are generally enriched in these elements. This may be explained by the presence of cavings from dolomites of the Khuff Formation. All depths are log depths in feet and all samples were taken from cuttings material

shows profiles plotted for a well penetrating siliciclastic sediments of the Silurian Qusaiba Formation in Saudi Arabia. The core and cuttings samples analysed in nearby wells produced less than 2% Ca and Mg, as was expected from this largely non-calcareous/non-dolomitic study section. However, the cuttings samples of well C generally produce Mg and Ca values of more than 5 and 10% respectively. No evidence of dolomite or calcite was mentioned in the sample descriptions and a careful examination of the wireline logs indicated the sediments to be non-calcareous/non-dolomitic. Furthermore, reading of the drilling report revealed that calcareous additives had not been used during the drilling of this well. In fact the contamination of Mg and Ca was explained by the presence of small quantities of 'caved' material from overlying dolostones of the Khuff Formation.

Another check on data quality involves the plotting of synthetic gamma ray profiles and comparing these with wireline derived gamma profiles from the same wells. According to Ellis and Singer (2007), the synthetic GR may be calculated as follows:

$$\text{Synthetic GR} = (16 \times K) + (8 \times U) + (4 \times Th)$$

In theory there should be a reasonably close match between the profiles plotted for synthetic and log derived GR, with some inconsistencies explained by poor depth control and the need for depth shifting prior to interpretation. Some caution should be exercised, however, when using these profiles as the values of synthetic gamma produced for cuttings samples represent a composite average over a c.10 ft interval, while wireline log data is normally acquired at c.1 ft spacings. Furthermore, disparities between these logs may be explained by fragments of one dominant lithology being selected for analysis in each cuttings sample, whilst the wireline logs may record the presence of two or more lithologies over a given 10 ft interval. In spite of these difficulties, synthetic Gamma Ray profiles should be plotted for every study interval with any inconsistencies investigated.

Figure 4.4 shows profiles for synthetic and wireline log gamma ray plotted for well 1. Clearly, there is a reasonably close match between the two, suggesting good depth control for both core and cuttings samples. The only exception is towards the base of the well (boxed area on Fig. 4.4) where there is a slight offset between the two logs. A depth shift of +20 ft for the cuttings data is suggesting in this zone.

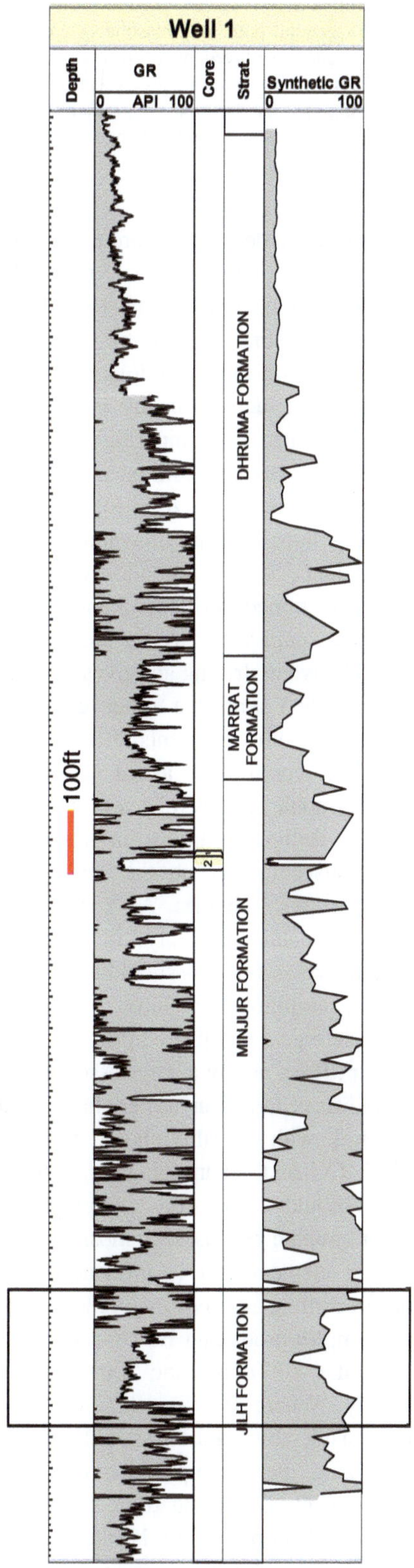

◀ **Fig. 4.4** Gamma and synthetic gamma profiles plotted for well 1. In general there is a reasonably close match between the trends displayed by these logs, suggesting good depth control. The only major exception to this is the zone highlighted by the box, where there is a slight offset. In this zone it is suggested adding a depth shift of +20 ft to the cuttings

In most studies, depth shifts are not applied to cuttings samples, owing to the aforementioned uncertainties relating to the comparison between synthetic and wireline log derived readings, but this is necessary on rare occasions. With regard to core materials, depth shifting is most accurately accomplished by comparing core gamma and wireline log derived gamma profiles, but a synthetic GR profile can be used for this purpose where core gamma data are unavailable.

The influence of drilling additives on data quality are mentioned in Chaps. 2 and 3, but it is important to compare the trends plotted for Ba (almost exclusive to drilling additives in core and cuttings samples) with the other elements in order to identify other elements affected by such additives. A commonly asked question in chemostratigraphic studies relates to whether the presence of drilling additives in samples can have a negative impact on the production of chemostratigraphic schemes. As a general rule, core samples contain relatively low quantities of drilling additives, while this should also hold true of cuttings, assuming they have been properly washed in detergent and picked prior to analysis. The element Ba is nearly always associated with drilling additives and, if any other element shows a strong link with Ba (this should be obvious using EV, CC and binary diagrams) then that element should not be employed for chemostratigraphic purposes. In most studies, Ba is the only element that is influenced by the presence of drilling additives, with most core and cuttings samples producing Ba concentrations in the range of < 1000 ppm and 500–25,000 ppm respectively. Where Ba concentrations exceed 25,000, considerable caution should be applied when utilizing these samples for chemostratigraphic purposes but, provided the samples have been processed properly, this is rarely the case. Other elements that may exist in the form of drilling additives include Sr and Ca where a carbonate drilling additive is employed. In studies of carbonate sediments, this may present a particular problem. In these circumstances, values of Ca should be used with caution as it may by uncertain whether the observed trends in this element relate to changes in lithology or the proportion of drilling additives. In order to determine this, a careful comparison of the geochemical data with wireline log data and sample descriptions is necessary, as is the reading of drilling reports, which should provide details on the composition of additives used at any given time during the drilling process.

Although the presence of small amounts of drilling additive in samples rarely causes chemostratigraphy projects to fail, on rare occasions companies use bentonite additives. This can have a more serious impact on the quality of data as the elements Ti, Ni, Nb, Cr and others may be contained within this material and these are used to produce chemostratigraphic schemes in many studies. Again, careful reading of drilling reports should reveal if any bentonite additives were used during the drilling of study wells.

4.4 Choice of Key Elements/Ratios

4.4.1 Introduction and Rationale Behind Choice of Key Elements and Ratios

Once these final checks on data quality have been made and the geochemical data plotted in the form of profiles for each well, the next step involves ratioing each element with respect to Al and then plotting Al-normalised profiles. The reasoning behind this is that the distribution of many elements is heavily influenced by variations in grain size. Figure 4.5 illustrates how changes in grain size can affect the distribution of certain elements in well D, penetrating the Silurian Qusaiba Formation in Saudi Arabia. The close association between the profiles plotted for Na, Si and Si/Al is explained by Si and Na being concentrated in quartz and plagioclase respectively, and that the distributions of both increase

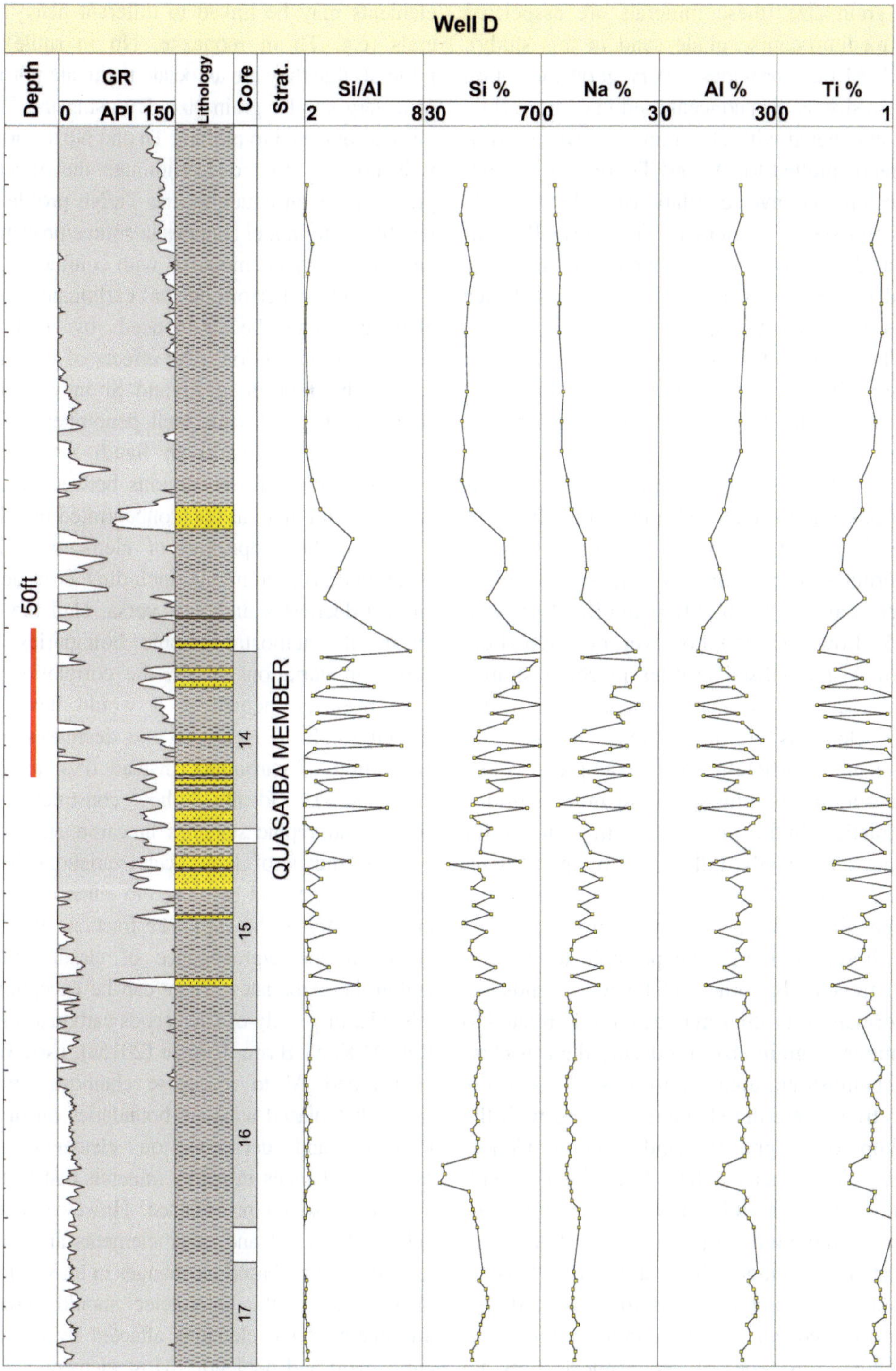

Fig. 4.5 This diagram illustrates the close association between the profiles plotted for Si, Si/Al and Na in well D, inferring that the distribution of the latter element generally increases with grain size. By contrast, the trends plotted for Al and Ti are very similar, suggesting an inverse relationship between Ti and grain size. All depths are log depths in feet

with grain size (these minerals are associated with medium-course grade sand in this study). The Si/Al ratio provides a very good indication of grain size as Si is concentrated in quartz, while Al is associated with clay minerals. By contrast, the trends plotted for Al and Ti are very similar, suggesting an inverse relationship between Ti and grain size. The element Ti is normally concentrated in heavy minerals which are generally found in the clay, silt and fine fractions, hence the association with Al.

Where individual elements show a close relationship with grain size they should not be used to construct chemostratigraphic schemes (unless they are utilized in the form of ratios such as Ti/Al). By ratioing elements to Al, it is often possible to reduce/eliminate the effects of grain size variations. Many authors have used Al-normalised data for this purpose, including Pearce et al. (2005a, b, 2010) and Ratcliffe et al. (2015). Lowey (2015), however, questioned the use of Al-normalised profiles in chemostratigraphy studies. Al-normalised data can be useful where elements are associated with silt and clay-sized particles—for example where Cr-bearing heavy minerals occur in the silt-clay size fraction. In this instance, profiles plotted for Cr may be heavily influenced by grain size, making it difficult to recognise trends reflecting changes in provenance. By using the Cr/Al ratio, however, it may be possible to identify such trends. In other instances, trends of Al-normalised profiles may be entirely related to variations in grain size, rendering them useless for chemostratigraphic purposes. This often holds true where the element is linked with the medium-coarse grained sand fraction. Consequently, Al normalised data should be used with caution and should only be applied for chemostratigraphic purposes where trends are unrelated to changes in grain size. This is obvious by comparing profiles plotted for Al-normalized, Si/Al and gamma ray log data.

Similarly, as series of geochemical ratios are plotted for the same reason—to eliminate the effects of grain size variations. For example, the elements Th and Nb are almost exclusively concentrated in heavy minerals. Though these elements may be linked to different heavy minerals (e.g. Th in monazite, Nb in rutile), the minerals that they are associated with are normally of a very similar grain size. By using the Th/Nb ratio, as opposed to plotting Th and Nb separately, it is possible to reduce/eliminate the effects of grain size. Consequently, the Th/Nb profile may be utilized to model changes in source/provenance and to identify chemozones with confidence.

In studies performed on carbonates, many elements may be influenced by carbonate concentration/dilution. The effects of this factor on the distribution of Zr and Sr in well 1 are shown in Fig. 4.6 (this well penetrates Triassic and Jurassic sediments in Saudi Arabia). An inverse relationship is obvious between Ca and Zr, indicating that, as Ca (concentrated in calcite) increases, the proportion of elements concentrated in detrital minerals (including Zr in detrital zircon) decreases and vice versa. Had the definition of chemostratigraphic boundaries been based on variations in Zr, the correlation may have been erroneous, as it would have been related solely to increases and decreases in the proportion of carbonate (instead of changes in provenance). Ideally, the construction of chemostratigraphic schemes in carbonates should be independent of large scale variations in carbonate content, and be related to either changes in the chemistry of the carbonate fraction, or to variations in source/provenance of clastic material within the carbonates. There can be exceptions to this rule. In a study of Cretaceous carbonates in the Rub Al'Khali Basin, Craigie (2015a), used values of Ca and Al to recognise chemostratigraphic zones, but placed subzone boundaries on upward increases and decreases on elemental ratios reflecting changes in heavy minerals distributions, and hence source/provenance. However, even if values of Ca, Al and other elements are used to recognize zones based on changes in bulk lithology in carbonates, the interpreter should have an awareness of the elements affected by carbonate enrichment and depletion. Most elements show a strong inverse relationship with Ca, but some are concentrated in carbonate minerals and may display a close association with this element. Figure 4.6 shows that very similar trends are displayed

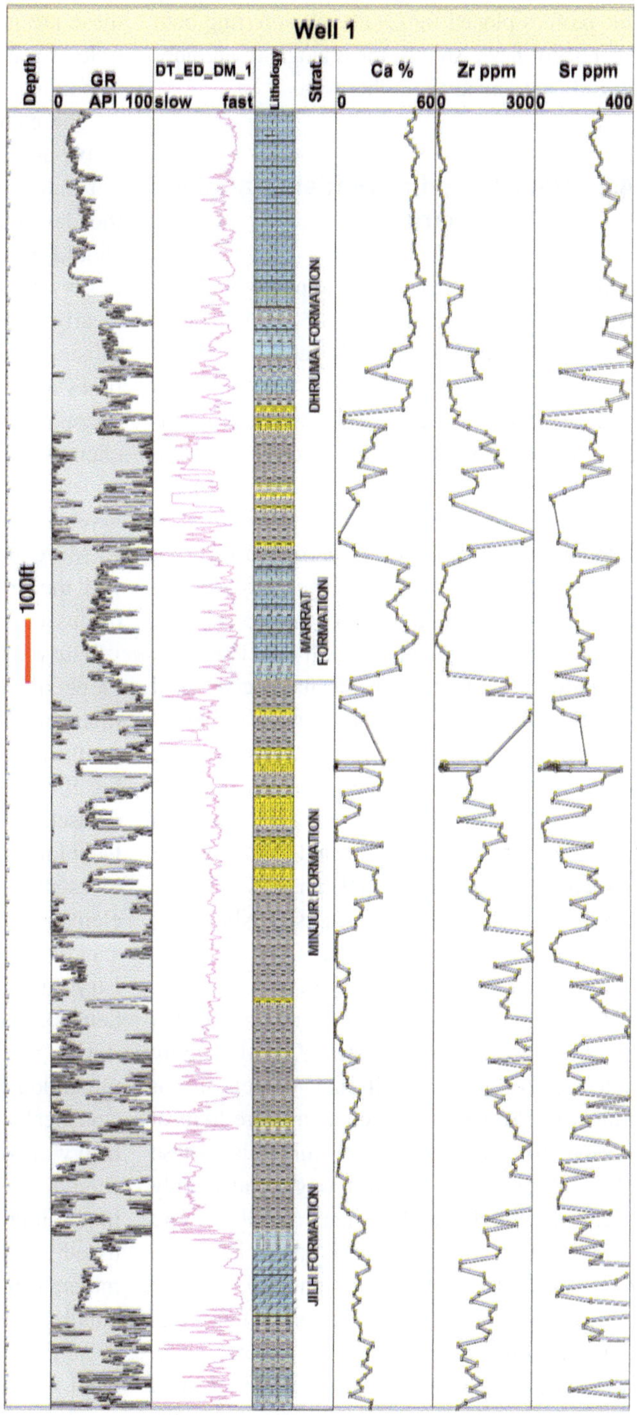

Fig. 4.6 Effects of carbonate dilution on the values of Zr in well 1. By contrast, the element Sr is mainly associated with carbonate minerals and generally increases with the proportion of Ca. All depths are log depths in feet and all of the samples take the form of cuttings

in the profiles plotted for Ca and Sr, inferring both elements to have strong mineralogical affinities with carbonate minerals in this well.

4.4.2 Ratios Used to Model Changes in Provenance

The choice of geochemical ratios to plot will largely depend on the nature and aim of the individual study. For example, if the study relates to fluvial sandstones of the northern North Sea, emphasis would be placed on the identification of ratios modeling changes in heavy mineral distributions (e.g. Zr/Nb could be plotted to model zircon/rutile). By contrast, more emphasis may be placed on the recognition of anoxic beds using ratios such as U/Th and Ni/Co, if the study were on organic-rich deep marine shales of the Kimmeridge Clay Formation in the central North Sea. However, as a first 'port of call' in nearly all studies conducted on clastic sediments, an attempt is made to model changes in source/provenance and then use that as a basis to recognize chemozones. For this purpose, the ratios normally involve elements associated with heavy minerals. These may include, but are not restricted to, the elements Zr, Hf, Th, U, Cr, Ti, Ta, Nb, P, Y and HREE. Ratios such as Zr/Nb, Zr/Th, Th/Nb, Ti/Ta and Cr/Nb are used in the identification of chemozones in most studies.

The benefits of using these elements are that they are generally very 'stable' and largely unaffected by post-depositional diagenesis or weathering. Some of the elements are almost exclusive to only one or two minerals. These include Zr and Hf, which are generally only associated with zircon. The elements Nb, Ti and Ta may have mixed mineralogical affinities, being associated with rutile anatase, sphene and/or opaque heavy minerals (e.g. titanomagnitite, magnetite, ilmenite). Cr is often concentrated in chrome spinel but may also be linked with clinopyroxenes and opaque heavy minerals. It is often more difficult to establish the precise mineralogical affinities of Th, Y and HREE but these are nearly always very stable and concentrated exclusively in heavy minerals, hence their importance in chemostratigraphic studies.

The element U is commonly associated with heavy minerals, particularly zircon and apatite, but may also be linked with organic matter and the development of anoxia. For this reason, U should only be used for chemostratigraphic purposes where its mineralogical affinities are clearly understood. Similarly, P often has strong mineralogical affinities with monazites, xenotimes and apatites but can also exist in biogenic phosphate. If this element is to be used to place chemostratigraphic boundaries, it is important to understand its mineralogical affinities in any given study.

More information on element:mineral links and the techniques used to establish these are discussed in more detail in Chap. 3.

Hurst and Morton (2014) criticised the technique of chemostratigraphy as element:mineral links are often not established, and that this can result in spurious correlations. This may be true but if equivalent geochemical and mineralogical datasets are compared and statistical/graphical techniques employed, then it should be possible to establish the mineralogical affinities of most elements or, at the very least, gain an understanding of the principal controls on geochemistry and mineralogy (see Chap. 3 for more details). In the absence of heavy mineral analysis, it may not be possible to establish the precise mineralogical affinities of every element, but statistical associations between elements and heavy minerals, or groups of heavy minerals, should be evident. For example, it is often impossible to determine whether Nb is concentrated in rutile, anatase, sphene or opaque heavy minerals but, the fact that this element is very 'stable' and is nearly always associated with these heavy minerals, means that it may to utilized in chemostratigraphic studies.

The following ratios, which generally reflect changes in source/provenance are normally plotted in the form of profiles:

Zr/Nb, Zr/Hf, Zr/Ti, Zr/Y, Zr/Ta, Zr/Yb, Zr/Th, Nb/Ti, Nb/Y, Nb/Ta, Nb/Yb, Nb/Th, Ti/Y, Ti/Ta, Ti/Yb, Ti/Th, Y/Ta, Y/Yb, Y/Th, Ta/Yb, Ta/Th, Yb/Th

Note that this list is by no means exhaustive and may include other elements and more complex ratios. For instance, Yb is selected because it is a high field strength HREE but other HREE such as Dy, Ho, Er, Tm and Lu could be used for this purpose. More complex ratios can also be included such as (Zr × Hf)/Ti, where it is necessary to emphasise the concentration of zircon over the Ti-bearing heavy minerals. Similarly, Nb, Ti and Ta often have similar mineralogical affinities, so the ratio (Nb × Ta × Ti)/Zr could be utilised where the intention is to emphasize the concentration/importance of Ti-, Nb- and Ta-bearing heavy minerals versus zircon.

Other ratios that may reflect either changes in provenance, depositional environment and/or weathering/diagenesis are as follows:

Zr/P, Ti/P, Ta/P, Y/P, Yb/P, Th/P, P/U, Zr/U, Ti/U, Ta/U, Y/U, Yb/U, Th/U, Gd/Zr, Gd/Ti, Gd/Y, Gd/Ta, Gd/Yb, Gd/Th, Gd/U, Gd/P, Na/K, Na/Sc, K/Sc

The usefulness of these ratios very much depends on the specific mineralogical affinities of P, U, Na and K. The elements P and U, for instance, may be concentrated in organic matter or heavy minerals. If they are associated with the latter, then variations in Zr/U, P/Ti and other ratios involving these elements may be very useful for chemostratigraphic purposes as they will be related to changes in provenance. Similarly, Na and K may be associated with feldspars, so the Na/K ratio may be employed to model the proportion of plagioclase versus K feldspar, again another potential indicator of changes in provenance. Conversely, the Na/K ratio may be less useful where K is related to authigenic illite. Ratios involving Sc should also be plotted but, as with Na and K, care should be taken to understand the controls on the distribution of this element before using such ratios for chemostratigraphic purposes. The element Gd is used in some of these ratios as it is a MREE. MREE may be linked with heavy minerals but, before Gd and ratios involving Gd are used to place chemostratigraphic boundaries, it should be established whether this element is concentrated in heavy minerals, feldspars or clay minerals.

Other ratios are less commonly used to define chemostratigraphic boundaries but are plotted as they may provide useful information on element: mineral links, weathering and depositional environment: These ratios are as follows:

K/Al, K/Rb, K/Cs, K/Ga, K/La, Rb/Cs, Rb/Ga, Rb/La, Cs/Ga, Cs/La, Ga/La.

The ratio K/Al normally occurs in the range 0–0.25, and greater than 0.28, where K is associated with illite and K feldspar respectively. High K/Al ratios may, therefore, be taken as evidence that K is predominantly linked with K feldpsar, the distributions of which may reflect changes in provenance. The elements K and Rb normally have very similar mineralogical affinities, being concentred in K feldspar, clay minerals (particularly illite) and micas. However, K/Rb ratios are higher in K feldspar than in illite and mica, so this parameter may be used to model the proportion of K feldspar versus that of illite. The elements Cs, La and Ga are normally found in clay minerals and feldspars, so ratios involving these elements may be used to model diagenetic trends or changes in depositional environment.

Note that data for some of the elements mentioned in this section, particularly Sc, Ta, MREE and HREE are routinely analysed by ICP-MS, but are often difficult to obtain by XRF.

4.4.3 Elements and Ratios Used to Model Changes in Paleoredox and the Abundance of Organic Matter

The elements affected by changes in paleoredox and/or the abundance of organic matter are normally Mn, Mo, Cu, Ni, Co, Zn, P and U. In many studies, data for these elements can be plotted in profile form, without the need to employ ratios, to identify changes in paleoredox and/or the abundance of organic matter (e.g. Craigie 2015b; Madhavaraju et al. 2015). In other studies, this may not be advisable as the distribution of the elements may be influenced by variations in the abundance of carbonate minerals, with high carbonate contents resulting in dilution. Under these circumstances, it may be advisable to use Al-normalized data (e.g. Mo/Al, Cu/Al etc.) or other ratios (e.g. U/Th) for this purpose. In addition to this, chondrite-normalised or shale average normalised data can be utilised (Tribovillard et al. 2012) Examples of ratios used to identify organic rock/anoxic paleoenvironments are illustrated in the studies of Hildred et al. (2011), Madhavaraju et al. (2015) and others. The mineralogical affinities of elements associated with changes in paleoredox and the concentration of organic matter are discussed in detail in Chap. 3.

4.4.4 Weathering Indices

Most chemostratigraphic schemes are based on elements and ratios associated with changes in provenance, but it is equally important to consider parameters that reflect the intensity of subaerial weather, not least because weathering surfaces often exist immediately below unconformities which may be correlative on local, subregional or regional scales.

4.4.4.1 Weathering Indices Involving Use of Major Element Data

Most of the commonly used weathering indices involve the use of major element data, particularly Al and the alkali metals (Na, Ca, K, Mg). Nesbitt and Young (1982) proposed the CIA (Chemical Index of Alteration) which is defined as follows:

$$\mathrm{CIA} = [\mathrm{Al}/(\mathrm{Al} + \mathrm{Ca}^{*} + \mathrm{Na} + \mathrm{K})]$$

where Ca* refers to the proportion of Ca concentrated in the siliciclastic fraction. Values of this ratio increase with the intensity of subaerial weathering, with unweathered igneous rocks and 'average' shales yielding CIA values of 50 or less and 70–75 respectively. By contrast residual clays with high kaolinite or gibbsite contents produce CIA values of up to 100 (Nesbitt and Young 1982). The basic theory behind this index is that Ca, Na and K are released from feldspar host minerals during weathering, causing a 'preferential' increase in Al and the CIA.

Another commonly used ratio used to measure the intensity of subaerial weathering, referred to in the present study as the WI (Weathering Index), is calculated as follows:

$$\mathrm{WI} = \mathrm{Al}/(\mathrm{Ca} + \mathrm{Mg} + \mathrm{Na} + \mathrm{K})$$

In this formula, Mg and Ca are assumed to exist in siliciclastic minerals. As with CIA, values of WI increase with the intensity of subaerial weathering. This index has been used by Retallack (1997) and Pearce et al. (2005a, b) to define zones of intense subaerial weathering.

One of the biggest challenges of using these ratios is that it is often very difficult, if not impossible, to correct for the proportion of Ca associated with carbonates and phosphates. Fedo et al. (1995) found that a value for silicate-bound Ca could be obtained by the following formula:

$$\begin{aligned}\mathrm{Mol\ Ca}_{(\mathrm{silicates})} &= \mathrm{mol\ Ca} - \mathrm{mol\ CO}_{2(\mathrm{calcite})} \\ &\quad - 0.5\ \mathrm{mol\ CO}_{2(\mathrm{dolomite})} \\ &\quad - 10/3\mathrm{mol\ P}_{(\mathrm{apatite})}\end{aligned}$$

This formula assumes that all P is concentrated in apatite and all inorganic carbon is linked with carbonates. Unfortunately, these assumptions normally result in overcorrection, with silicate bound Ca becoming negative (Bahlburg and Dobrzinski 2011; Garzanti and Resentini 2016). An alternative empirical correction method was proposed by McLennan (1993) to correct for apatite bound Ca in which Ca (silicates) = Na, if the number of Ca moles after correcting for Ca in phosphate, is greater than that of Na. More recently, Garzanti and Resentini (2016) used a different calculation based on the assumption that Ca and Na have similar levels of mobility in weathered sands (Garzanti et al. 2013). In this model $\alpha^{Al}\mathrm{Ca} = \alpha^{Al}\mathrm{Na}$ for those samples in which

$\alpha^{Al}Ca$ is lower than $\alpha^{Al}Na$, $\alpha^{Al}K$, $\alpha^{Al}Rb$, $\alpha^{Al}Sr$ and $\alpha^{Al}Ba$. In a study of volcaniclastic sands Garzanti and Resentini (2016) found this method to produce the best results as it did not lead to overcorrection.

Arguably the most accurate corrections can be achieved if quantitative petrographic heavy mineral data are obtained on the same samples that are analysed geochemically. In studies where such data exist, silicate bound Ca can be calculated by subtracting the amount of Ca associated with the optically determined concentration of calcite, dolomite and apatite. Calculations based on this data are rarely perfect, however, as petrographic analysis of thin sections cannot be undertaken on exactly the same fraction of the sample analysed geochemically. Furthermore, the precision of point counting methods are limited and heavy mineral analysis is normally performed on the < 32 μm fraction, meaning that at least some heavy mineral bound Ca could occur in grains of < 32 μm. After applying the aforementioned correction methods, Garzanti and Resentini (2016) still concluded that non-silicate bound Ca remained problematic, particularly in fluvial or marine sediments containing abundant extrabasinal and/or intrabasinal carbonate grains, and for sedimentary rocks that contain appreciable volumes of carbonate minerals precipitated as a cement or replacement of framework grains during early or late diagenesis.

In the case of WI, there are concerns about correcting for both Ca and Mg and, to date, no publication details correction factors for non-silicate bound Mg. Given the inherent difficulties associated with correcting for non-silicate bound Mg and Ca, it is suggested that no corrections are applied to either CIA or WI but alternative ratios should be employed where Ca and Mg exceed 7 and 4% respectively. Where values of Ca are greater than 7 % the following 'corrected' versions of CIA and WI could be applied:

$$\mathrm{CIA} = [\mathrm{Al}/(\mathrm{Al} + \mathrm{Na} + \mathrm{K})]$$
$$\mathrm{WI} = \mathrm{Al}/(\mathrm{Mg} + \mathrm{Na} + \mathrm{K})$$

Craigie et al. (2016a), for instance, applied the Al/(Mg + Na + K) ratio to identify weathered zones in Devonian, Carboniferous and Permian sediments in eastern Saudi Arabia. In addition to this, the WI formula could be recalculated as Al/(Na + K) where Mg concentrations are higher than 4%. Another potential problem may occur in samples containing halite-bound Na, though this mineral is generally of much lower abundance in clastic sediments.

According to Garzanti and Resentini (2016), other challenges to applying CIA include the existence of a provenance control on the values of this parameter which are often more significant than the effects of post depositional weathering. Furthermore, variations in grain size relating to hydraulic sorting may affect CIA values. Garzanti and Resentini (2016) point out that Al is mostly hosted in slow-settling fine grained and platy phyllosilicates, decreasing in abundance with increasing grain size and suspension depth. By contrast, Na, Ca and Sr occur mainly in faster-settling plagioclase, the values of which increase with grain size. The element K tends to remain constant, being linked with both phyllosilicates and K-feldspars.

A further complication with both CIA and WI relates to the assumption that Na, K and Ca originally occurred in detrital feldspars. This may hold true for many samples, but the existence of detrital clay minerals should not be ignored as these may contain significant quantities of all three elements.

Given the potential problems associated with non-silicate bound Ca and Mg, 'overriding' provenance signatures, hydraulic sorting/grain size and detrital heavy minerals, the WI and CIA indices often cannot be used in isolation to identify weathered zones, but they can be very useful when collaborated with sedimentological and wireline log data.

Other indices, though less commonly applied in chemostratigraphic studies, involving the application of major element data to define zones of weathering are as follows (Buggle et al. 2011; Harnois 1988; Fedo et al. 1995):

$$CPA\ (\text{Chemical proxy for alteration}) = [Al/Al + Na)] \times 100$$

$$CIW\ (\text{Chemical Index of Weathering}) = [Al/(Al + Na + Ca^*] \times 100$$

$$PIA\ (\text{PlagioclaseIndexofAlteration}) = [(Al-K)/(Al + Ca^* + Na-K)] \times 100$$

where Ca* refers to silicate bound Ca (Cox et al. 1995).

$$ICV\ (\text{Index of Compositional Variability} = (Fe + K + Na + Ca + Mg + Mn + Ti)/Al.$$

The drawbacks of using these ratios are similar to those applied to CIA and WI but, as these are rarely applied in chemostratigraphic studies, they are not discussed further in the present study. For more detailed information on the applications and relative merits of CPA, CIW and PIA, the reader is referred to the work of Buggle et al. (2011). For information on ICV, the most suitable references are Cox et al. (1995), and Tobia and Mustafa (2016).

4.4.4.2 Weathering Indices Involving Use of Trace Element Data

The application of "Sr-type" indices, including Rb/Sr or Ba/Sr, to identify weathered zones has become increasingly popular over the last 20 years (Chen et al. 1999; Ding et al. 2001; Tan et al. 2006; Bokhorst et al. 2009; Buggle et al. 2011). This is based on the theory that Sr can substitute for Ca in minerals and also shows an analogous behaviour to Ca in the weathering profile. Consequently, Sr is easily released into solution and mobilized in the course of weathering, whereas Rb or Ba can be regarded as relatively immobile during moderate weathering conditions, as a result of strong adsorption to clay minerals (Buggle et al. 2011). In a detailed evaluation of "Sr-type" indices, Buggle et al. (2011) concluded that these should only be used in carbonate free sediments as low and high Ba/Sr and Rb/Sr ratios are generally linked with elevated and depleted carbonate contents, respectively. This is not surprising given the strong mineralogical affinities of Sr with such minerals. A further drawback to using "Sr-type" indices is that Rb and Ba may undergo mobilization under extreme weathering conditions (Buggle et al. 2011 and references therein), making the Rb/Sr and Ba/Sr unreliable indicators of paleoweathering under such conditions. Possibly an even bigger problem of using these ratios in core and cuttings samples is that Ba and, to a lesser extent, Sr may be concentrated in drilling additives. For this reason, it is recommended that Rb/Sr and Ba/Sr ratios are not used to identify zones of weathering except, where it can be proven that the elements Ba and Sr do not occur in drilling additives.

In recent years, the Y/Ho ratio has been used to recognise weathered zones, with an inverse association apparent between values of this ratio and the intensity of weathering (Thompson et al. 2013; Babechuk et al. 2015). Changes in the Y/Ho ratio are known to track the progressive alteration of primary minerals to pedogenic clays and oxides. The mechanisms of fractionation are described by Thompson et al. (2013) and Babechuk et al. (2015), who suggest that Fe-(oxyhydr) oxides are the most significant minerals involved in Y/Ho fractionation, with maximum separation of Y from Ho occurring at intermediate pH ranges. Y and Ho have very similar ionic radii but Y has a lower affinity for Fe-oxyhydroxides and oxides in weathering profiles. However, this effect is thought to be suppressed in the presence of organic matter (Thompson et al. 2013). In addition to variations in pH and the pre-abundance of organic matter, the Y/Ho ratio is also likely to be influenced by variations in source/provenance. As with other weathering indices, the Y/Ho parameter should be used in conjunction with sedimentological and wireline log data to identify zones of weathering.

In summary, weathering indices can be used to identify weathering surfaces, some of which may be associated with unconformities. In spite of their usefulness, however, they may be influenced by factors other than weathering (e.g. provenance, hydraulic sorting), so they should always be applied in conjunction with sedimentological, wireline log, lithostratigraphic and/or biostratigraphic data, when attempting to identify zones of weathering.

4.4.5 Application of Ce and Eu Anomalies

REE anomaly data is rarely utilized in chemostratigraphy projects but is, nevertheless, noteworthy of mention in this chapter. Anomalies for all of the REE can be calculated but only the ones relating to Ce and Eu have been employed to any extent in chemostratigraphy studies. All REE form large trivalent (3+) ions, but Eu and Ce have additional valences. The element Eu forms 2+ ions, whilst Ce forms 4+ ions, leading to chemical reaction differences in how these ions can partition versus the 3+ REE.

4.4.5.1 Application of Ce Anomalies

Elderfield and Greaves (1982) first proposed the use of the Ce anomaly to identify changes in redox. Since that time a number of other papers have been published on that subject, including works by Liu et al. (1987), Wilde et al. (1996), Alibo and Nozaki (1998), Mongelli et al. (2015), Kocsis et al. (2016), and Tostevin et al. (2016). The Ce anomaly, generally quoted as Ce/Ce is based on the assumption of a "linear" decline in REE abundances with an increase of atomic number when the elements are normalised to some standard, for example, chondrites. The elements Ce and Eu are the only REE that deviate significantly from a line that declines towards the HREE. In order to calculate Ce/Ce, Wilde et al. (1996) used chondrite abundances of La, Ce and Sm of 0.332, 0.876 and 0.183 respectively (Haskin et al. 1968, 1971). An anomaly is identified where the normalized concentrations plots above or below a value calculated by the assumption of the straight line variation.

The order of LREE is La, Ce, Pr, Nd, Pm, Sm, so the calculation of Ce/Ce should be based on normalized Ce values that plot above or below the straight line extrapolation between La and Pr. This would result in the following formula for the this ratio:

$$Ce/Ce = [2Ce^{*}/(La^{*} + Pr^{*})]$$

where the superscript * implied the concentration normalized to chondrite. Given that data for Nd and Sm are normally available but this does not always hold true for Pr, Ce/Ce is more commonly calculated as follows (Wilde et al. 1996):

$$Ce/Ce = [2Ce^{*}/(La^{*} + Nd^{*})]$$

Based on the work of Haskin et al. (1968, 1971), the following chondrite values may be used in these Ce/Ce calculations:

$$Ce = 0.876$$
$$La = 0.332$$
$$Pr = 0.112$$
$$Nd = 0.60$$

Under oxic conditions Ce is less easily dissolved in seawater, so that oxic seawater and oxic sediments are depleted and enriched in Ce respectively. This results in organisms extracting phosphate from oxic seawater displaying a negative Ce/Ce anomaly, whilst Fe oxide-rich oxic sediment, such as red clay, produce a positive Ce/Ce anomaly. In suboxic seawater, Ce is mobilized from the sediments and released into the water column resulting in a less negative to a positive Ce/Ce value in the seawater. Ce is relatively depleted in anoxic sediments which show a negative Ce/Ce anomalies. As the remains of organisms extracting phosphate from dominantly anoxic waters would yield positive Ce/Ce anomalies and the host anoxic sediments would have negative anomalies, the resulting values of Ce/Ce may reflect a mix of such low and high Ce concentrations from two sources, and may not be a good indicator of paleoredox where whole-rock analysis is undertaken on core, cuttings or field samples (Wilde et al. 1996). Conversely, If analysis are undertaken separately on biogenic phosphate such as conodonts or fish bones (e.g. Kocsis et al. 2016) or on sediments not containing appreciable quantities of conodonts, inarticulate brachiopods and fossil fish (e.g. Berry et al. 1987), then it is possible to use Ce/Ce anomalies to identify anoxic paleoenvironments.

The mechanism of Ce uptake probably relates to oxidation of Ce(III) to Ce(IV) and subsequent incorporation into Mn oxyhydroxides. At neutral pH's the element is linked with Fe–Mn–Al–Ti-oxyhydroxide coatings on carbonate minerals (Wilde et al. 1996).

Apart form the aforementioned uncertainties relating to the presence/absence of Ce-bearing fossil material, the main reason the Ce/Ce index is not applied more frequently is that the element Ce is also concentrated in detrital siliciclastic minerals, so positive or negative Ce/Ce anomalies may simply reflect variations in the proportion of this material rather than any changes in paleoredox. This led Wilde et al. (1996) to suggest that the index is only valid during intervals of tectonic stability and in areas of low rates of sedimentation such as the relatively deep waters of the outer shelf and continental margins where the pycnocline and the lower mixed layers are anoxic.

4.4.5.2 Application of Eu Anomalies

The Eu anomaly is rarely applied in chemostratigraphy studies, but has the formula:

$$\mathrm{Eu/Eu} = [2\mathrm{Eu}^*/(\mathrm{Sm}^* + \mathrm{Gd}^*)]$$

where the superscript * implies the concentration normalized to chondrite. The following chondrite values, based on the publications of Haskin et al. (1968, 1971), are used in this equation:

$$\begin{aligned}\mathrm{Eu} &= 0.0685\ \mathrm{ppm}\\ \mathrm{Sm} &= 0.183\ \mathrm{ppm}\\ \mathrm{Gd} &= 0.252\ \mathrm{ppm}\end{aligned}$$

In general, enrichment or depletion of Eu relates to the tendency of this element to be incorporated into plagioclase. Eu has 2+ and 3+ states, with Eu^{2+} being adsorbed in plagioclase crystals preferentially over other minerals (Tang et al. 2017). Where plagioclase crystalizes in magma, most of the Eu will be incorporated into this mineral, causing a higher than expected concentration of Eu in this mineral versus other REE in the same mineral, resulting in a positive Eu anomaly. This will lead the rest of the magma to be relatively depleted in Eu. If this Eu-depleted magma then separates from its plagioclase crystals and solidifies, the resulting chemical composition will show a negative Eu anomaly (because the Eu is contained in the plagioclase left in the magma chamber). By contrast, a relatively positive Eu anomaly would occur if a magma accumulates plagioclase crystals before solidification (https://en.wikipedia.org/wiki/Europium_anomaly).

In siliciclastic sediments a positive Eu anomaly is usually caused by a relatively high proportion of plagioclase feldspar, but this is a huge generalisation as positive anomalies may result from post depositional diagenesis, with Eu enrichment in cements such a calcite. The susceptibility of Eu to post depositional diagenesis, together with the fact that this element may be concentrated in a range of minerals, not just plagioclase feldspar, means that the Eu/Eu parameter is rarely used for chemostratigraphic purposes and no further discussion of it is provided in the present publication. For further information on the applications of Eu/Eu, the reader is referred to the works of Tostevin et al. (2016), Tang et al. (2017), and references therein.

4.5 Application of Key Elements and Ratios to the Identification of Chemozones

The choice of which key elements and ratios to employ for chemostratigraphic purposes varies from study to study and the approach is slightly different for clastic and carbonate lithologies. Where clastic samples are analysed, care should be taken to avoid basing the chemostratigraphic scheme on changes in grain size/lithology. The distribution of some elements and minerals may be heavily biased towards certain grain sizes. For example, quartz and feldspars are generally more common in sandstones than mudrocks, so the elements Si, K and Na may be more abundant in the former lithologies. Conversely, Ga and Cs are more likely to be associated with clay minerals and, hence, have higher concentrations in mudrocks. Another reason for treating the two lithologies as separate 'entities' is that their mode of deposition can be quite different. For example, in studies of turbidites, sands may be deposited rapidly from a very specific source in the hinterland, whilst the geochemistry of mudrock samples may be different, reflecting a 'background' mix of material from more than one source. For these reasons, separate schemes should always be proposed for sandstone and mudrock lithologies. Difficulties may arise when interpreting data acquired from siltstones- these samples should be placed in either the sandstone or mudrock datasets but, based on the studies the author has completed over the last 20+ years, siltstones generally have closer similarities, both geochemically and mineralogically, with mudrocks.

In addition to plotting sandstone and mudrock data separately, the use of ratios, rather than individual elements, are preferred, as these are less likely to be influenced by changes in grain size. For example the elements Nb and Zr are almost exclusively linked with heavy minerals and often occur in similar grain size fractions, so the Nb/Zr ratio could be used to define chemozones under these circumstances. Plotting key element ratios can also avoid trends relating to carbonate dilution, where variations the abundance of carbonate minerals may result in increases and decreases in elements linked with the siliciclastic fraction. Carbonate may take the form of calcite cement or carbonate-bearing drilling additives. Assuming the elements concentrated in siliciclastic minerals are diluted and enriched equally, the use of ratios should allow the interpreter to 'see through' variations in carbonate concentrations, whether these relate to carbonate cements or drilling additives. The use of profiles for individual elements, however, should not be completely dismissed as trends in these elements may be unrelated to variations in grain size or carbonate contend, though the majority of chemostratigraphic schemes are based on ratios.

In studies undertaken on carbonates, grain size variations are unlikely to have the same impact on key elements and ratios, though the effects should not be ignored. Variations in the abundances of carbonate minerals, such as calcite and dolomite, however, are likely to present a more significant problem, hence the reason that ratios should be used to place chemostratigraphic boundaries in preference to individual elements.

It should never be assumed that the use of elements or ratios will eliminate the effects of grain size variations or carbonate content. For this reason, the trends of each element and/or ratio should be carefully compared with profiles plotted for Si, Al, Si/Al and Ca. A comparison with wireline logs such as the GR and sonic logs is also recommended. In general, any element or ratio displaying strong positive or negative associations with these parameters should not be used for chemostratigraphic purposes. In most studies, the use of geochemical ratios comprising elements associated with heavy minerals, eliminates the effects of changes in carbonate content or grain size, but this is not always the case.

It is advisable, particularly when studying clastic sediments, to initially attempt to build chemostratigraphic schemes based on changes in high field strength elements (Zr, Hf, Ti, Nb, Ta, Th, Y, Cr, REE) and ratios involving these elements (e.g. Zr/Nb, Nb/Th, Ti/Ta etc.). As these elements are almost exclusively concentrated in heavy minerals and are largely unaffected by post depositional weathering/diagenesis, they are considered to be most useful in chemostratigraphy studies. If these elements and ratios do not show any significant trends that can be used to define chemozones, the following elements (and ratios involving these elements) may be used: Na, P, U, Sc, K, MREE, LREE. These are used less commonly than the high field strength ones are they are more likely to be influenced by weathering/diagenesis and may have multiple mineralogical affinities. The element K, for instance, may be linked with K feldspars but it may also be concentrated in micas and clay minerals, whilst Na may be associated with plagioclase, smectite or halite. Variations in P can be used to infer changes in the abundance of P-bearing heavy minerals such as apatite and xenotime, but trends in this element should be interpreted with care as it may occur in biogenic phosphate. A number of heavy minerals are enriched in U (e.g. apatite, zircon), but so too is organic matter, meaning that trends in this element may reflect variations in the quantity of preserved organic matter rather than changes in source/provenance relating to heavy mineral distributions. The element Sc, as well as LREE and MREE may have multiple mineralogical affinities, being associated with clay minerals, heavy minerals and/or feldspars. For these reasons, Na, P, U, Sc, K, MREE, LREE should only be employed for chemostratigraphic purposes if their mineralogical affinities have been established through graphical/statistical analysis or by comparing mineralogical and geochemical data (see Chap. 3 for details).

Ideally, the use of unstable elements such as Fe, Cs and Ga should be avoided when proposing chemostratigraphic frameworks for clastic sediments, except in exceptional circumstances when data for other elements are unavailable or do not show trends that can be used to define chemozones. Even under these circumstances, it would be necessary to establish the mineralogical affinities of the elements before they could be used for chemostratigraphic purposes.

The use of redox-sensitive elements, such as U, V, Mo, Ni, Cu, Co and Zn, may be employed in study intervals, which include organic rich beds and/or sediments deposited in anoxic paleoenvironments and the mineralogical affinities of these elements and others are discussed in detail in Chap. 3.

The approach to the study of carbonates is very similar to that of clastics but, for practical reasons, the data for all lithologies are plotted together in the same set of profiles, rather than producing separate profiles for limestones, dolomites, calcareous mudrocks, argillaceous mudrocks, calcareous sandstones etc. The choice or ratios to plot will very much depend on the nature of the carbonates and whether they contain an appreciable amount of detrital material. In many studies of carbonates, it is possible to treat the data in the same way as for clastic sediments by identifying changes in provenance based on ratios relating to variations in heavy minerals, and to use that as a bases to identify chemozones. However, in studies where there is little or no detrital material within the samples, it may be necessary to produce a chemostratigraphic scheme based on changes in the chemistry of the carbonate fraction. For example, the Ca/Mg ratio may be used to model the proportion of calcite vs. dolomite, whereas variations in Ca/Sr, Sr/Mg, Ca/Zn, Ca/Ni and other ratios may reflect more subtle changes in mineralogy. However, when utilizing these ratios, care must be taken as they may reflect changes in post-depositional diagenesis that are unrelated to syn-depositional processes. It is necessary to compare the data with available sedimentological and petrographic data to avoid making erroneous correlations.

Profiles should be plotted for almost every element from Na–U and for a range of 50–200 or more ratios in each study section, though only a selection of 4–12 parameters are normally selected for chemostratigraphic purposes. Some elements and ratios may be influenced by changes in grain size or carbonate dilution. Others may be affected by depositional processes or post depositional diagenesis. In most studies a large number of profiles either show 'flat' trends or are very 'spikey' and cannot be used to identify chemozones. It is important to identify the elements/ratios that show significant trends that can be used to identify chemozones. The process of recognizing these trends is a skill that is difficult to teach, but is developed over years. Indeed, in the same way that some palaeontologists are able to identify fossils in the field more quickly than others, some chemostratigraphers have a natural ability to recognize 'trends' in profiles where others struggle. In spite of this, most chemostratigraphers are able to identify trends with some practice.

4.6 Development of Hierarchical Schemes

It is important to change the horizontal scales of profiles to recognize important trends. For example, in Fig. 4.7, profiles of P are plotted at different scales in well 2 to illustrate that, when a scale of 0–0.15 is used, it is clear that four samples occurring towards the base of this well yield high values of this element but the others plot on a 'flat' trend. By changing the scale to 0–0.04 however, it is possible to identify significant trends in the profile plotted for this element (Fig. 4.7). By altering the horizontal scale of profiles, trends can be visualized that would otherwise go unnoticed. The horizontal scales of profiles plotted for each element and ratio should be varied until the most suitable scale is found to enable the identification of trends and chemo-zone boundaries. In some studies, it is necessary to plot the same profile at two or three different scales to identify boundaries at different depths within the study section.

Where profiles are 'spikey', it may be possible to apply a '3 point moving average' through the data to eliminate variability which may reflect local/short term changes in grain size, provenance or depositional environment, that may not be of importance at a larger scale. Other moving averages (e.g. 2 or 5 point moving averages) can be applied but, in the experience of the author, a 3 point moving average generally works best. Figure 4.8 shows profiles plotted for Th/Y in well 6. Clearly, the data through the uncored sections are smooth, while data acquired for core samples are 'spikey' in character. In order to alleviate this problem, a 3 point moving average was applied to the cored interval and the data then replotted. In some instances, the application of such a moving average has little or no effect on the data but it is always worth applying this technique over intervals where there is extreme variability. A note of caution should be included, however, with respect to the application of moving averages, as the technique may produce spurious results where there are large sampling gaps in the data. Under these circumstances, it may be more appropriate to apply this technique over separate depth intervals.

One of the most common reasons for the production of erroneous correlations occurs when individuals interpret geochemical profiles in the same way as they would for wireline logs. When making lithostratigraphic correlations, trends, rather than specific values of wireline logs, are used to identify boundaries and zones that can be correlated between wells. For example, upward increases, decreases and 'ratty' signatures in the gamma ray may be diagnostic of a particular lithostratigraphic unit. Adopting a similar approach to chemostratigraphic studies is often problematic, as trends that look significant on initial inspection of the data may be entirely related to the presence of one or two samples yielding anomalous values of particular elements/ratios. For example, Fig. 4.9 illustrates Th/Y trends plotted for well 1. The left hand profile of Th/Y shows three trends represented by reference points 1, 2 and 3. However, the same data is plotted to the right of this, but with two samples removed at each reference point. The result is a 'flattening' of the profile shape at these points. The existence of trends may be entirely related to one or two samples representing a local change in a particular element or ratio that is not evident in adjacent wells. Furthermore, even if they are correlative between wells, their identification is based on a low number of samples, so these trends may be 'missed' in some wells depending on where the samples are taken and how many are collected. For this reason, it is important to recognize trends over a larger scale based on more than three samples when producing chemostratigraphic schemes.

Cut off values of key elements and ratios should be used to distinguish each chemozone (e.g. zone C1 = Zr/Nb ratios > 30, zone C2 = Zr/Nb < 30). Ideally the same cut off values should be applied to each study interval, particularly in field-scale studies where wells are rarely more than 10 miles apart. In studies performed on a subregional-regional scale (where wells are more than 15 miles apart) similar trends may be

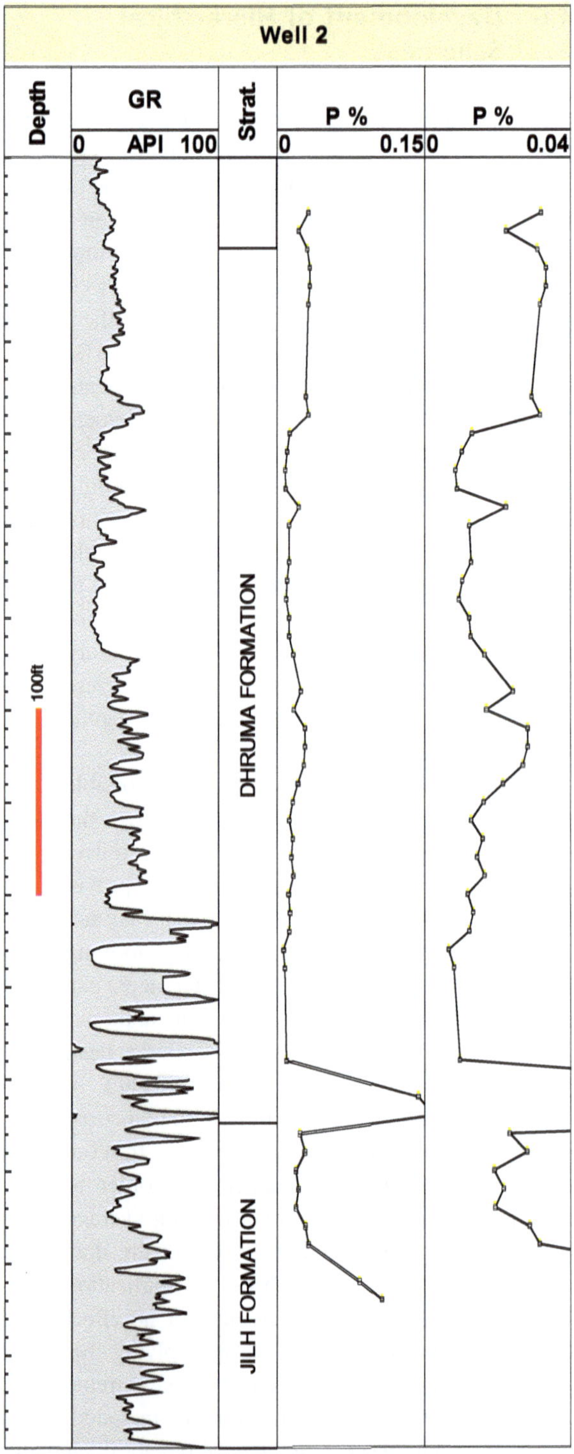

Fig. 4.7 The profiles for P are plotted at different scales in well 2 to illustrate the importance of changing the horizontal scales of profiles in order to recognize important 'trends'. All depths are log depths in feet and all samples take the form of cuttings

Fig. 4.8 The profile plotted for Th/Y in well 6 is relatively smooth over the non-cored intervals but 'spikey' over the sections of core. To reduce the 'variability' of this curve to enable trends to be more clearly visualized, a '3 point moving average' was applied though the cored interval. All depths are log depths in feet

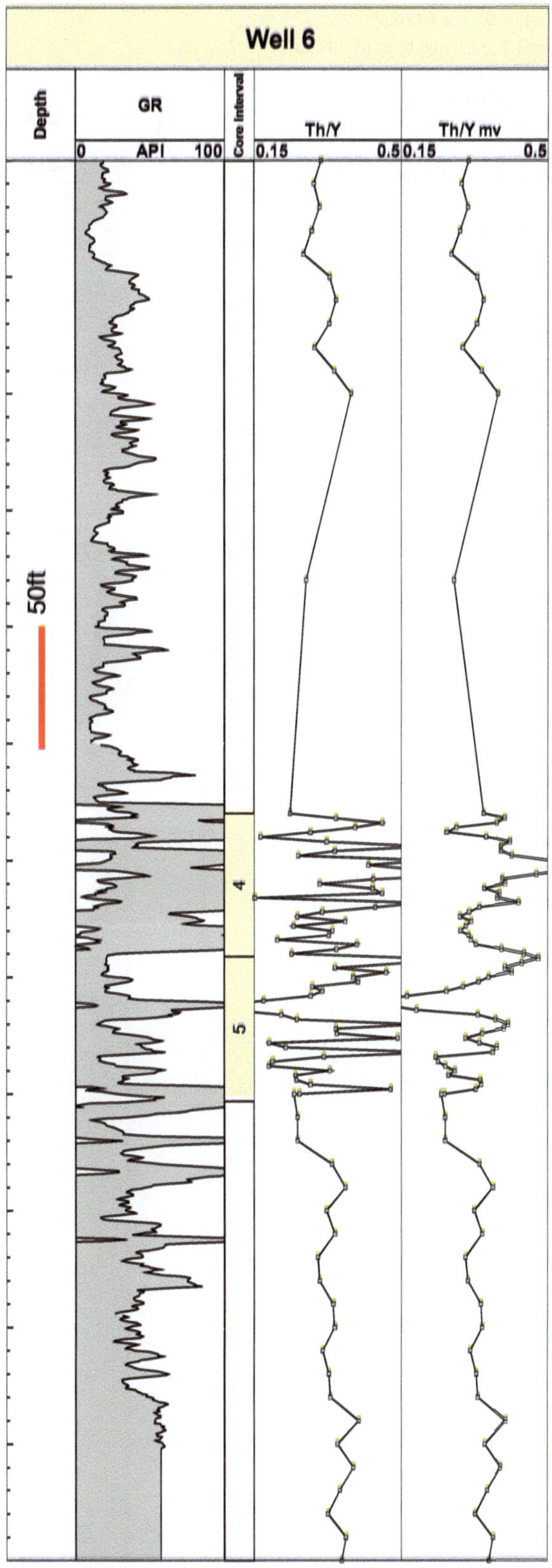

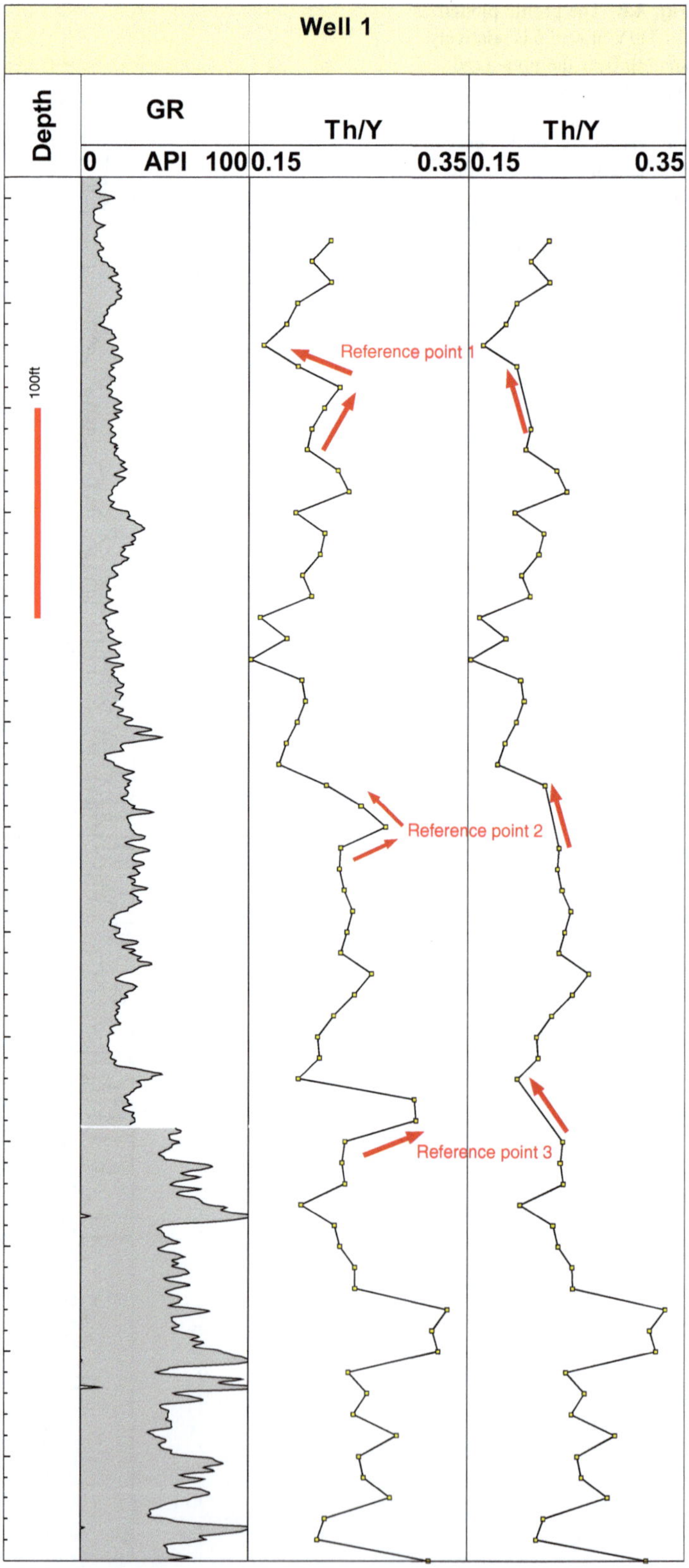

Fig. 4.9 An example from well 1 showing how trends in the key element ratio Th/Y (left column) may only be associated with one or two anomalous samples. When these anomalously high Th/Y samples are removed (i.e. two sample point at reference points 1, 2 and 3), the resultant trend plotted in the right column is very different. See text for further explanation. All depths are log depths in feet and all samples were taken from cuttings

recognized in wells that are distal to each other, but different cut offs may be ultilized to define the same chemozones. This is less common but may be explained by the effects of transportation and hydraulic sorting. For example, zircon is generally more resistant to chemical/physical attack than chrome spinel, so the Zr/Cr ratio (assuming Zr and Cr are concentrated in zircon and chrome spinel respectively) would be expected to be elevated in high-energy depositional environments. Variations in Zr/Cr may be also be influenced by changes in provenance, so it would be entirely plausible that similar trends would be observed in this ratio in two widely-spaced wells. Different cut-offs may be used to define the same chemozones, however, if different energy conditions persisted during the deposition of these chemozones in the two localities.

In nearly every chemostratigraphy study, the correct approach is to identify up to a maximum of 4 chemostratigraphic zones in the wells which have the greatest coverage of the stratigraphic section. Once these have been identified in two or three wells, an attempt should then be made to recognize the same zones in wells where stratigraphic sections are thinner or where the sample coverage is less. Once the correlation of zones has been achieved, the next step will be to identify subzones within each zone, divisions within each subzone and perhaps even subdivisions within each division. Nearly all chemostratigraphic schemes are based on the recognition of 3 or 4 levels of hierarchy, though the terms "unit" and "subunit" may be applied in the rare circumstances where more than four levels are required. To summarize, the following terms, in ascending hierarchical order, may be used in chemostratigraphic studies:

zones
subzones
divisions
subdivisions
units
subunits

Please note that there are no fixed number of zones, subzoned, divisions etc. in a given study as the number of chemozones very much depends on geochemical/mineralogical variability, the number of analysed samples and the vertical extent of the study interval(s).

An example of a chemostratigraphic scheme is presented for well 1 in a study of Triassic and Jurassic carbonates in southern Saudi Arabia (Fig. 4.10). Clearly, two principal zones are identified, with the basal zone, labelled C1 (identified by purple and blue coloured chemozones on Fig. 4.10), producing higher Th/Y ratios than in the overlying zone C2 (identified by orange colours). It is also noteworthy that Th/Cr ratios are generally higher in C1. Specific 'cut offs' are placed on the profiles plotted for both of these ratios in order to demonstrate that more than 70% of the C1 samples yield higher Th/Cr and Th/Y ratios than in C2, and vice versa. The chemostratigraphic subzones of C1 are labelled C1-1, C1-2, C1-3 and C1-4 in ascending stratigraphic order, with the former producing the highest concentrations of Mg. Subzones C1-2 and C1-4 are characterised by higher Th/Ta ratios than in the intervening subzone C1-3. A broad twofold subdivision is proposed for zone C2, with C2-1 occurring at the base and producing lower Y/Ta ratios than in the overlying subzone C2-2. With the exception of C1-1, divisions are identified in each subzone, with subdivisions present in division C1-2b. The details of the chemostratigraphic scheme are illustrated in Fig. 4.10 and summarised in Fig. 4.11.

Although chemozones are initially identified using geochemical profiles, it is often easier to visualise the differences between these when the data are plotted in the form of binary or ternary diagrams. Figure 4.12 shows an example of binary diagrams illustrating the differentiation of chemostratigraphic zones C1 and C2 in six closely spaced wells. These diagrams demonstrate that, though there is some 'overlap', most samples assigned to zone C1 yield Th/Y ratios greater than 0.26, while the C2 samples plot with Th/Y values that are lower than this. In wells 2, 4 and 5, it is also possible to use Th/Cr to differentiate the two zones, though these ratios cannot

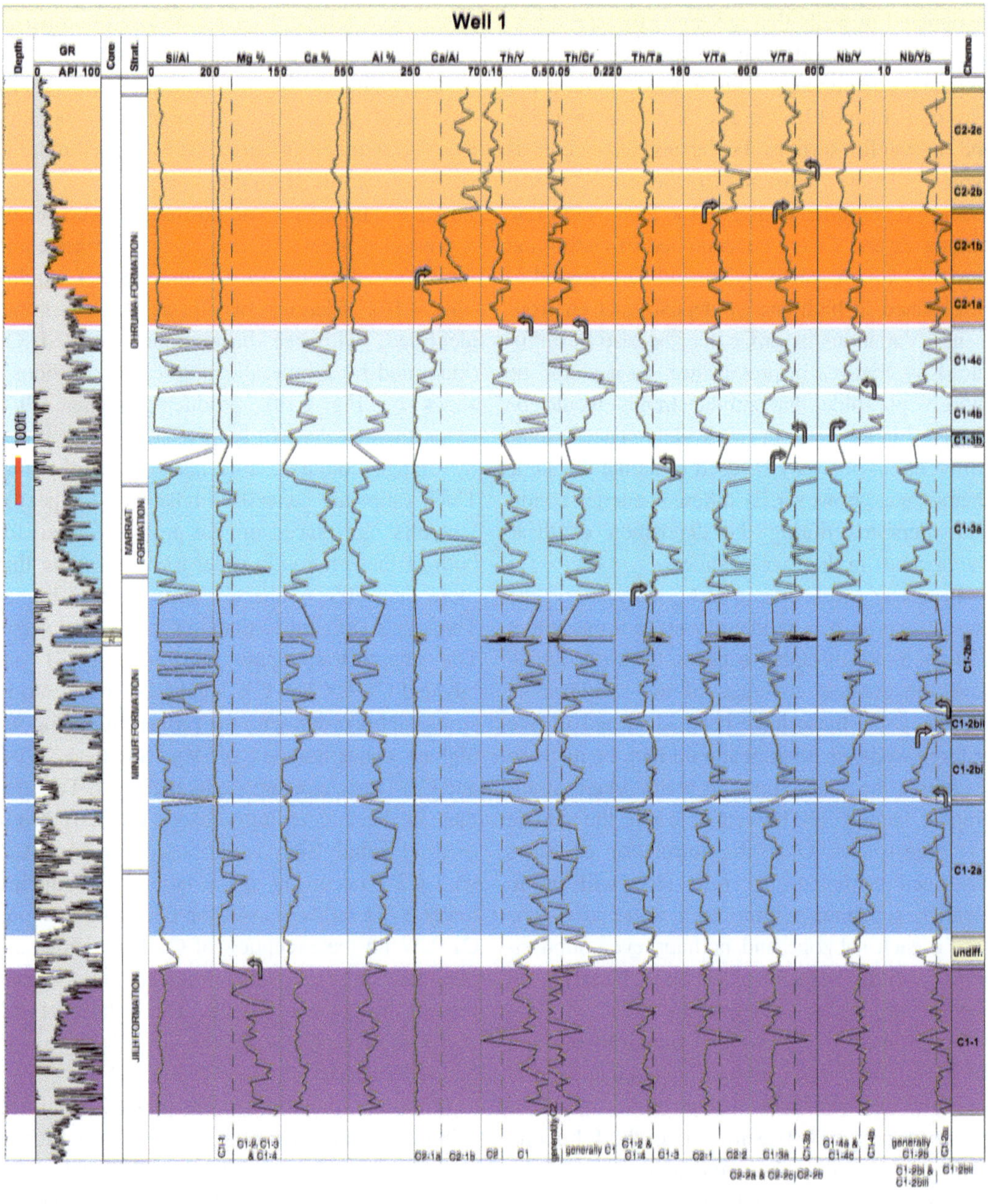

Fig. 4.10 Chemostratigraphic zonation proposed for well 1. All depths are log depths in feet

be employed for this purpose so easily in wells 1, 3 and 6. The reasons for this inconsistency are unknown but may relate to subtle differences in source/provenance between the locations of the wells. It can be concluded that Th/Y is a much more reliable parameter than Th/Cr for the purpose of differentiating the two zones in this study.

Figure 4.13 summarizes the chemostratigraphic correlation between wells 1, 2, 3, 4, 5 and 6. This correlation panel illustrates that subzone C1-1 is roughly equivalent to the Jilh Formation and is correlative between all six study wells. The overlying subzone C1-2 and associated divisions/subdivisions are only

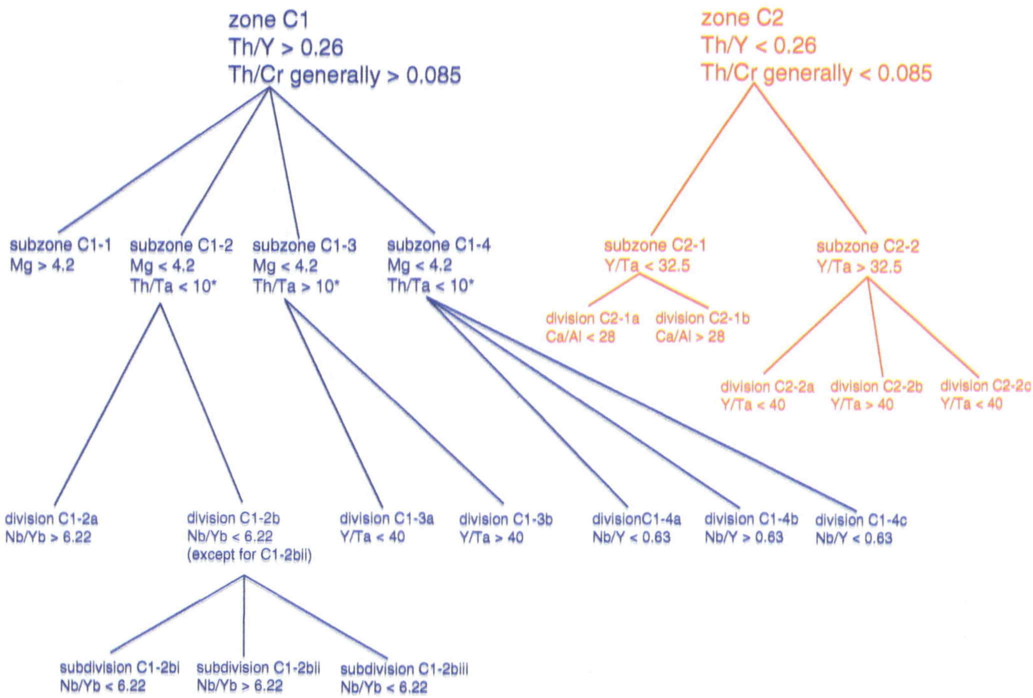

Fig. 4.11 Principal geochemical characteristics of zones, subzones, divisions and subdivisions proposed for a study of Triassic and Jurassic aged carbonates, Saudi Arabia

recognized in wells 1 and 5, though they may exist in the sample gap immediately above C1-1 in well 3. The absence of C1-2 in wells 2, 4 and 6, is most likely to be explained by erosion or non-deposition, with the areas around these wells probably representing topographic 'highs' during the deposition of C1-2. It is also noted that subzone C2-1 is only present in well 1. It is entirely possible that a higher level of accommodation space existed in the vicinity of this well than the others during the deposition of this subzone. Subzone C2-2a is identified in wells 2, 3 and 6, but is missing from 1, 3 and 4, probably as a result of erosion or non-deposition on a local scale. It is also noted that C1-3 cannot be further differentiated in wells 4 and 6, and the same holds true for C1-4 in well 2. The reasons for these inconsistencies are unclear but may relate to local variations in source/provenance that are not present on a larger scale.

4.7 Zonation and Correlation Confidence

Once the chemostratigraphic scheme has been produced, it is necessary to define levels of confidence for the placement of each chemostratigraphic boundary. This is best achieved by using a combination of statistical techniques and histograms. The statistical technique of Discriminant Function Analysis (DFA) can be used to define levels of statistical confidence for each chemozone and is best performed via a computer program such as UNISTAT or MATLABS. Table 4.1 presents the result of DFA applied at the zone level to the aforementioned study on Triassic and Jurassic carbonated in southern Saudi Arabia. This shows that the overall level of statistical confidence, with respect the identification of zones is 86.72%.

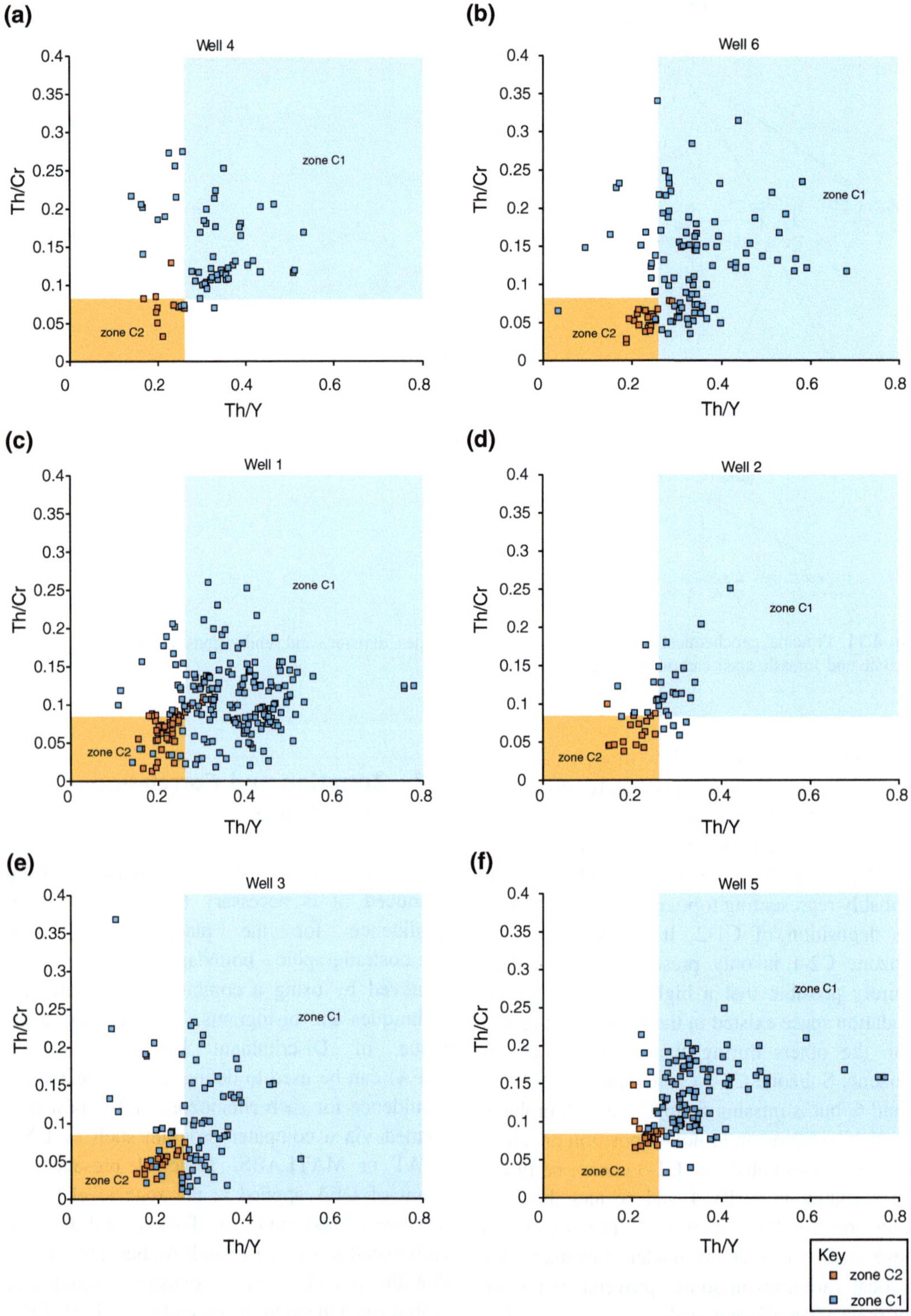

Fig. 4.12 Binary diagrams used to differentiate chemostratigraphic zones C1 and C2 in individual wells. Note that it is not possible to use the Th/Cr ratio to define the two zones in wells 1, 3 and 6

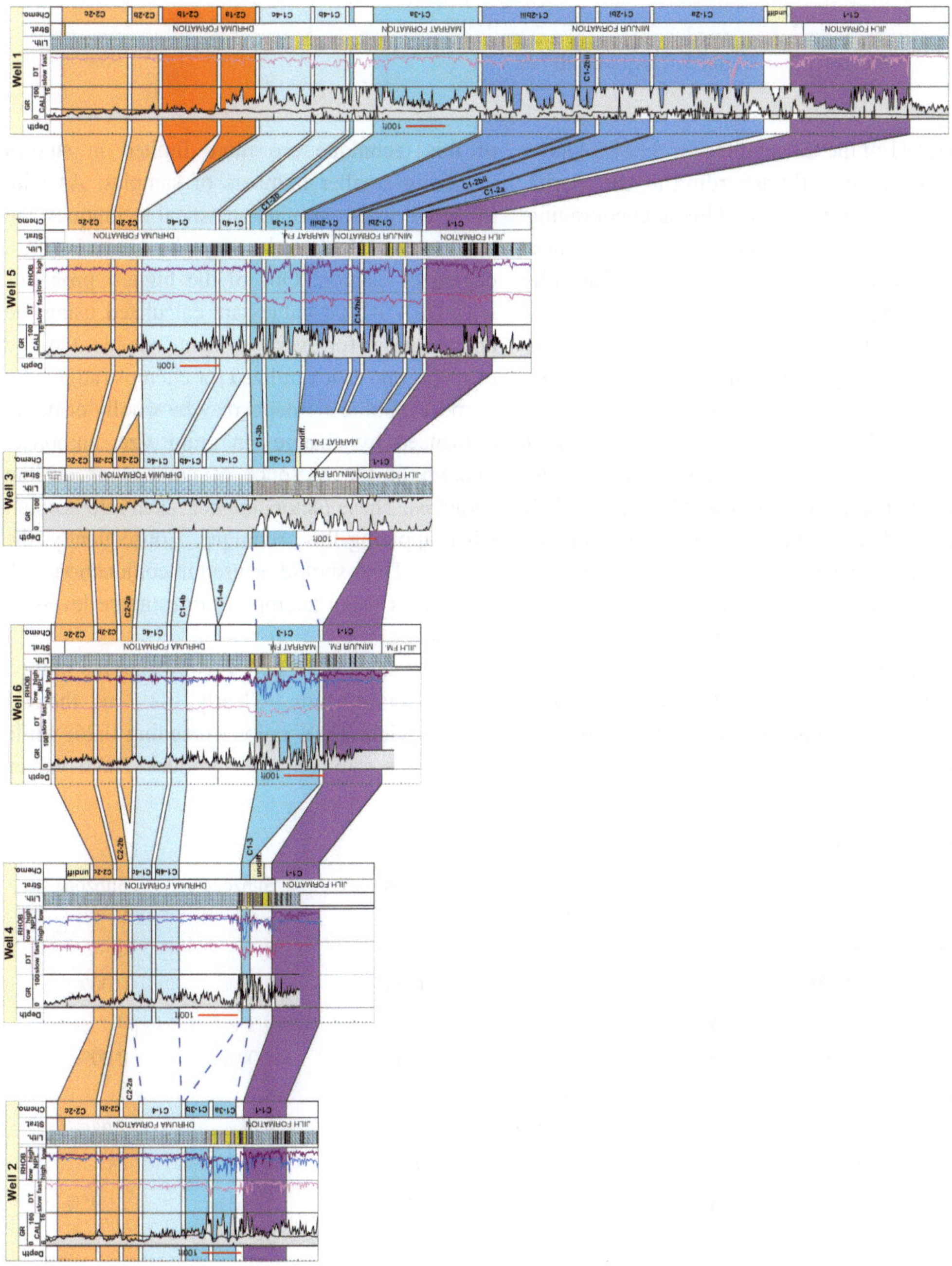

Fig. 4.13 Summary of proposed chemostratigraphic correlation with a lithology column and stratigraphic boundaries added for each well in a study of Triassic and Jurassic aged carbonate sediments, Saudi Arabia. All depths are log depths in feet

Table 4.1 Results of DFA applied at the zone level

	Zone C1	Zone C2
Zone C1	**541**	104
	83.88%	16.12%
Zone C2	0	**138**
	0.00%	**100.00%**
Correctly classified: 86.72%		

In detail, 541 of the C1 samples (or 83.88%) have been assigned correctly according to DFA, with 16.12% being misclassified. This is encouraging as any figure above 70% is taken to represent a high level of statistical confidence. All of the 138 zone C2 samples are deemed to be classified correctly. Table 4.2 shows the results of the same technique when applied at the subzone level. An overall confidence value of 74.76% is produced, with DFA values exceeding 80% in subzones C1-1, C1-2 and C2-1. Slightly lower DFA values, though still high, are linked to C1-3 and C2-2, while a 'moderate' level of statistical confidence (57.07%) is associated with C1-4. DFA was not applied at the division or subdivision levels, owing to the relatively low number of samples assigned to each division/subdivision. As a general rule, DFA works reasonably well at the zone and subzone levels when applied to projects comprising more than 500 samples. However, the applications of this technique are more limited in studies involving smaller numbers of samples. As with PCA, EV, CC and other statistical techniques, the results of DFA should always be viewed with a degree of caution. One of the biggest problems with DFA is that values are calculated using the entire dataset for every element, not just the key elements and ratios utilized for chemostratigraphic purposes. Some elements may be wholly or partly influenced by changes in grain size, carbonate dilution or the presence of cavings or drilling additives. These factors are not taken into account when applying this technique. Consequently, the results of DFA should be used in conjunction with histograms when attempting to estimate levels of confidence.

Figure 4.14 shows various histograms plotted for the same study. Clearly, 83% of the C1 samples yield Th/Y ratios exceeding 0.26, while 80% of the same samples produce Th/Cr values

Table 4.2 Results of DFA applied at the subzone level

	Subzone C1-1	Subzone C1-2	Subzone C1-3	Subzone C1-4	Subzone C2-1	Subzone C2-2
Subzone C1-1	**97**	7	0	0	0	0
	93.27%	6.73%	0.00%	0.00%	0.00%	0.00%
Subzone C1-2	0	**94**	11	4	0	0
	0.00%	**86.24%**	10.09%	3.67%	0.00%	0.00%
Subzone C1-3	8	11	**170**	22	9	7
	3.52%	4.85%	**74.89%**	9.69%	3.96%	3.08%
Subzone C1-4	1	3	9	**117**	30	45
	0.49%	1.46%	4.39%	**57.07%**	14.63%	21.95%
Subzone C2-1	0	0	0	0	**25**	0
	0.00%	0.00%	0.00%	0.00%	**100.00%**	0.00%
Subzone C2-2	0	0	0	0	33	**80**
	0.00%	0.00%	0.00%	0.00%	29.20%	**70.80%**
Correctly classified: 74.76%						

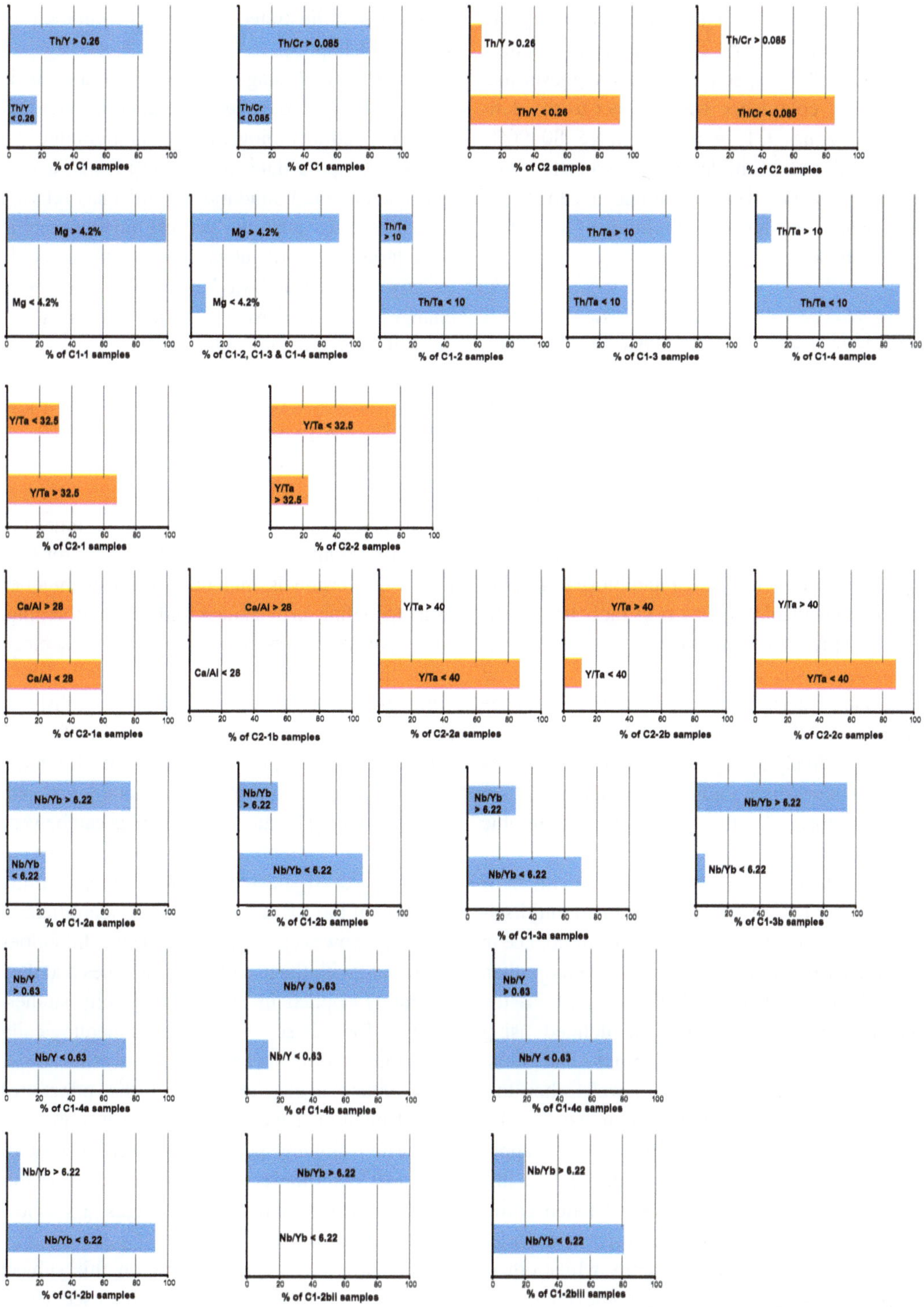

Fig. 4.14 Histograms used to define levels of confidence associated with each chemozone in a study of Triassic and Jurassic aged sediments, Saudi Arabia

of more than 0.085. Of the C2 samples, around 90% produce Th/Y and Th/Cr ratios of less than 0.26 and 0.085 respectively. This suggests that the level of confidence with respect to the definition of zones C1 is in the range 80–90%. Similar values of confidence are associated with each chemozone, with the single exception of C1-3, where the level of confidence is slightly lower at 62%.

4.8 Integration with Other Datasets

Failure to gain an understanding of paleoenvironments and to integrate available sedimentological and biostratigraphic data is one of the leading causes of failure with regard to chemostratigraphy projects. There are occasions when the technique has to be used in isolation (e.g. where sedimentological and biostratigraphic data are unavailable) but it works far better as part of a fully integrated approach to reservoir correlation. In some studies, there may be local diachroneity causing the same chemozones to be deposited at different times in different localities. This is less likely to happen in field-scale studies but may occur where study sections are more than 15 miles apart. Acquisition of biostratigraphic and sedimentological data may be used to determine whether chemozones are time-equivalent or not. Another reason for adopting this multidisciplinary approach is that the levels of confidence in any given scheme are improved where more than one technique is employed. For example, if the Hercynian unconformity can be defined using chemostratigraphy, lithostratigraphy and biostratigraphy in a particular set of wells, then this would be much more confident than using a single technique in isolation. A final reason to apply more than one correlation technique is that this approach usually results in more detailed correlation schemes. For instance, some boundaries may be defined biostratigraphically whilst others are placed using chemostratigraphic data. The techniques used to integrate chemostratigraphic, sedimentological and biostratigraphic data are most clearly illustrated by the case studies presented in Chap. 5.

In the initial stages of interpretation it is advisable to ignore existing correlations based on seismic, sedimentological, biostratigraphic or lithostratigraphic data as it is surprising easy to bias chemozone boundaries to 'fit' ones relating to a previous scheme, yet chemostratigraphic boundaries do not always have to coincide exactly with ones based on other schemes as changes in source/provenance may occur at any point in a given study interval.

4.9 Provenance and Tectonic Setting Determination

Given that most chemostratigraphic schemes are based on changes in provenance, it is appropriate that a discussion is included on the techniques used to define provenance and tectonic setting. Many papers have been published on this subject, particularly in the late 1980s and early 1990s when the use of geochemical data to establish the provenance of clastic sediments was one of the 'in vogue' research topics. When used intelligently, and with collaboration with mineralogical and sedimentological data, these schemes can be very useful. For example, it may be possible to establish 'sand fairways' and future drilling locations if the source of the sand is identified. In practice, however, there are many difficulties associated with the use of these schemes and they are rarely applied in chemostratigraphy studies, except for those involving mineralogically immature sediments derived from first order sedimentary cycle, and where the effects of post-depositional weathering/diagenesis are limited. The following paragraphs describe some of the most commonly used provenance discrimination schemes and the challenges associated with using them.

One of the first publications on this subject was written by Bhatia (1983) who used (Fe + Mg) versus Ti, and (Fe + Mg) versus Al/Si crossplots, to identify sandstones derived from

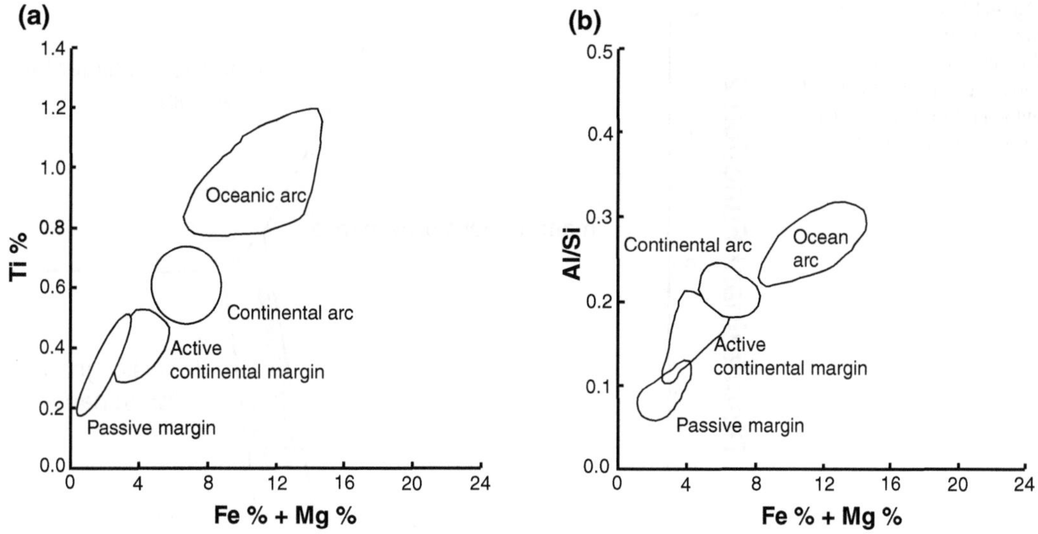

Fig. 4.15 Binary diagrams used to di_erentiate provenance/tectonic setting (after Bhatia 1983)

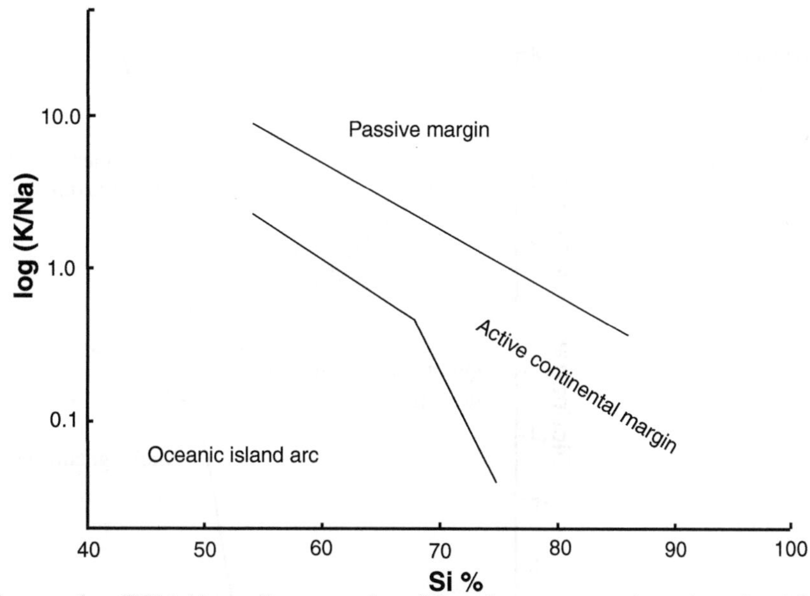

Fig. 4.16 Si versus log (K/Na) binary diagram used to differentiate provenance/tectonic setting (after Roser and Korsch 1986)

passive margin, active continental margin, continental arc and oceanic arc settings (Fig. 4.15). This was followed by the study of Roser and Korsch (1986) who employed a Si versus log (K/Na) binary diagram to distinguish passive margin, active continental margin and island arc fields in sandstone and mudrock samples (Fig. 4.16). A more commonly used scheme was proposed, for the same lithologies, by Roser and Korsch (1988) based on the following discriminant functions:

$$\mathrm{F1} = 30.638\ \mathrm{Ti/Al} - 12.541\ \mathrm{Fe/Al} + 7.329\ \mathrm{Mg/Al} + 12.031\ \mathrm{Na/Al} + 35.402\ \mathrm{K/Al} - 6.382$$

$$\mathrm{F2} = 56.500\ \mathrm{Ti/Al} - 10.879\ \mathrm{Fe/Al} + 30.875\ \mathrm{Mg/Al} - 5.40\ - 5.40\ 4\mathrm{Na/Al} + 11.112\mathrm{K/Al} - 3.89$$

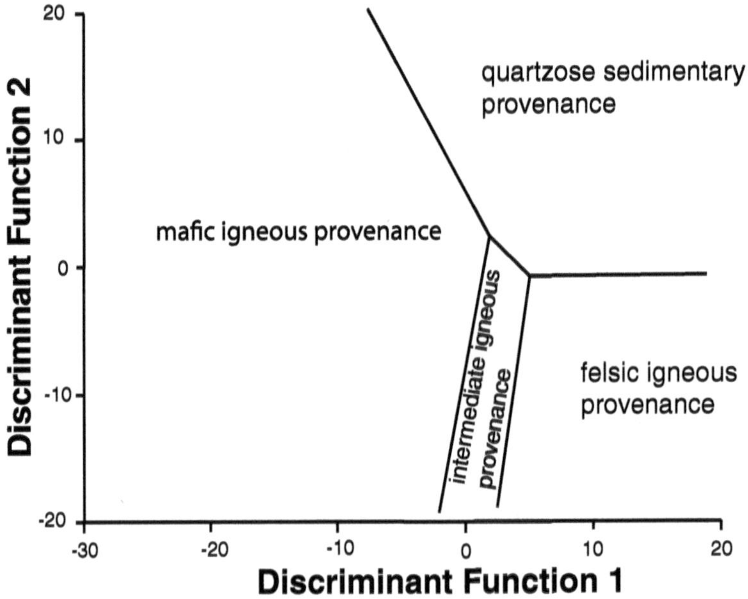

Fig. 4.17 Discriminant function diagram used to differentiate provenance signatures in sandstone and mudrock lithologies (after Roser and Korsch 1988)

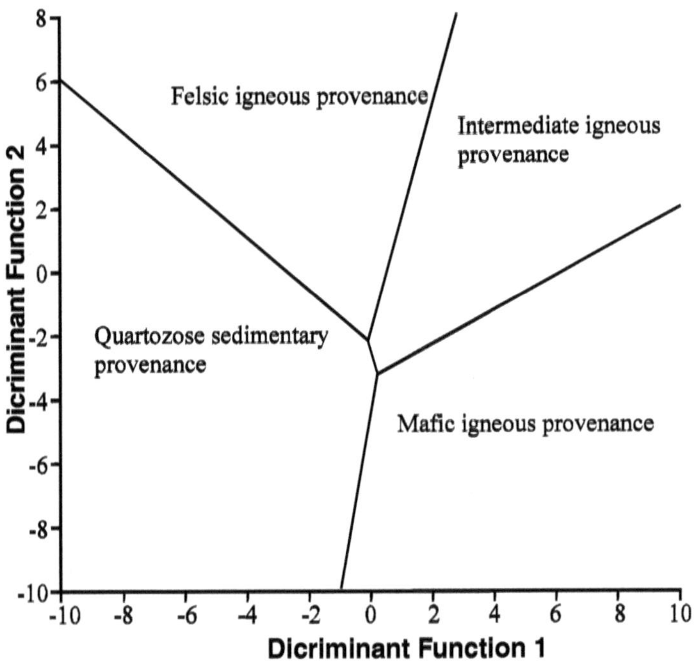

Fig. 4.18 Discriminant function binary diagram used to differentiate provenance/tectonic setting (after Roser and Korsch 1988)

Details of this scheme are presented in the binary diagram of Fig. 4.17. In addition to this, Roser and Korsch (1988) also proposed a scheme based on variations in Ti, Al, Fe, Ca, Na and K. The following discriminant functions are plotted on a binary diagram (Fig. 4.18):

$$\begin{aligned} F1 &= -1.773\ Ti + 0.607\ Al + 0.76\ Fe - 1.5\ Mg \\ &\quad + 0.616\ Ca + 0.509\ Na - 1.224\ K - 9.09 \\ F2 &= 0.445\ Ti + 0.007\ Al - 0.25\ Fe - 1.142\ Mg \\ &\quad + 0.438\ Ca + 1.475\ Na + 1.426\ K - 6.861 \end{aligned}$$

Bhatia and Cook (1986) used a La/Sc versus Ti/Zr crossplot (Fig. 4.19a), a Sc–Th–La ternary graph (Fig. 4.19b), and a Zr/10-Sc–Th ternary diagram (Fig. 4.19c), to discriminate greywackes derived from passive margin, active continental margin, continental island arc and oceanic island-arc sources.

The main problem associated with the use of these diagrams is that they do not take the following factors into account:

(a) Hydraulic sorting and reworking
The actions of transportation, hydraulic sorting and reworking often result in an increase in quartz, and hence Si. This is normally accompanied by a decrease in the proportion of physically and chemically unstable grains (e.g. Mg-bearing olivines and pyroxenes). Failure to take account of these actors may result in erroneous interpretations. For example, the provenance determination crossplots of Roser and Korsch (1988) are based on variations in a number of elements including Al, K, Mg and Ti. Given that Ti-bearing heavy minerals are much more resistant to chemical/physical attack than most Mg, K and Na-bearing minerals, the discriminant function parameters shown

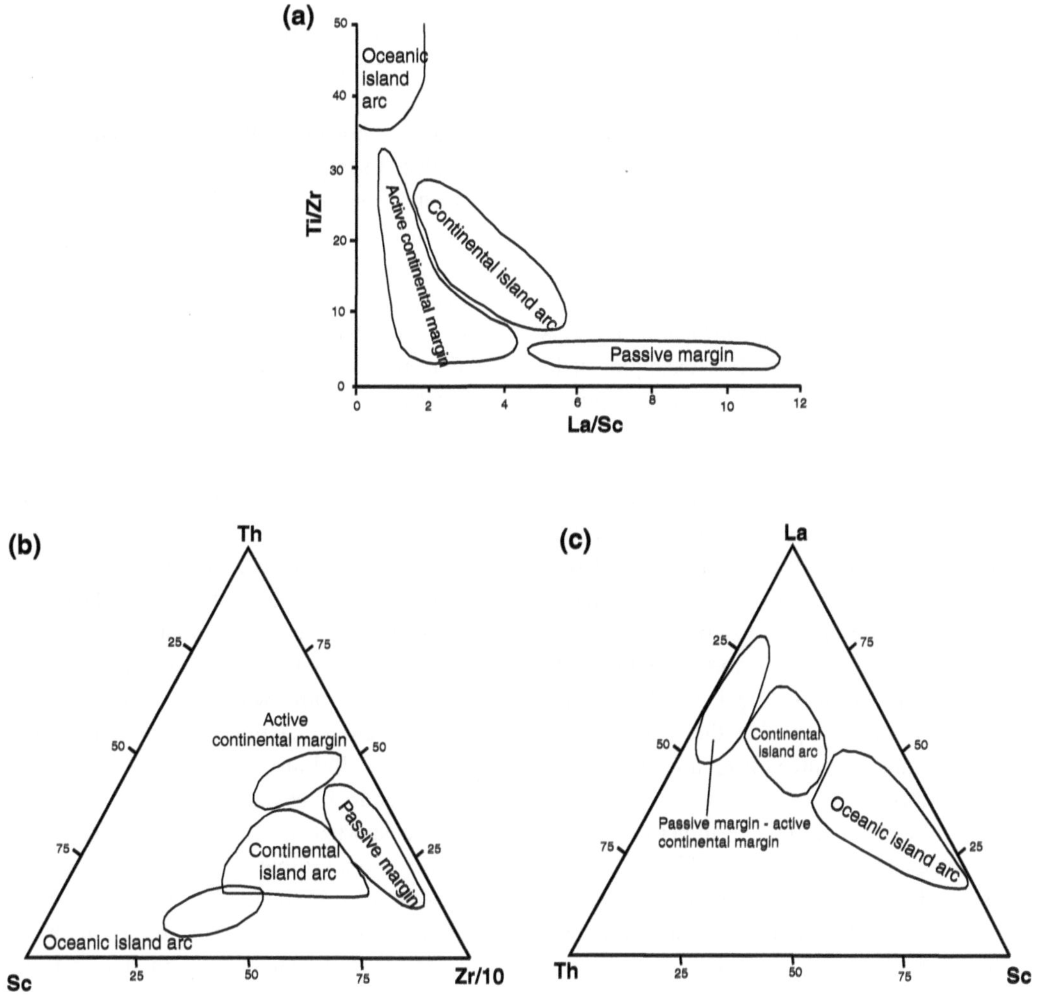

Fig. 4.19 Binary diagrams used to differentiate tectonic setting/provenance in greywackes (after Bhatia and Cook 1986)

in Figs. 4.16 and 4.17 may simply reflect variations in hydraulic sorting, transportation and reworking, rather than provenance.

(b) Grain size variations
Variations in grain size, often related to hydraulic sorting, transportation, reworking etc., may exert a significantly greater control over the distribution of elements and minerals than changes in source/provenance. Using the Sc–Th–La ternary diagram of Bhatia and Cook (1986), for example, the element Th is almost certainly linked with heavy minerals, being associated with fine sand-silt grade fractions. By contrast, Sc and La may be relate to a variety of grain sizes, depending on their mineralogical affinities. They may be linked with heavy minerals occurring in the fine sand-silt fraction, clay minerals of clay grade, or feldspars in medium-coarse grade sand material. The unintended consequences of using provenance determination diagrams are that different interpretations may result from different grain sizes, irrespective of changes in provenance.

(c) Post depositional weathering/diagenesis
The effects of post depositional diagenesis should not be ignored. For example, sediments derived from a felsic igneous provenance would be expected to contain a high proportion of K-bearing feldspars. If the sediments were then subjected to intense subaerial weathering/diagenesis, however, this could result in dissolution of K feldspars and consequential loss of K. In this scenario, it is entirely possible that samples from these sediments would plot in the quartzose sedimentary provenance 'field' of Roser and Korsch (1988) in spite of the fact that they were originally sourced from a felsic igneous provenance.

(d) The influence of carbonate minerals
Some schemes (e.g. Roser and Korsch 1988) use variations in Ca to predict changes in provenance but make the assumption that Ca is associated with siliciclastic minerals. Unfortunately, this element is often linked with carbonate minerals, meaning that erroneous interpretations may be made unless the presence of Ca-bearing carbonate is corrected for. Given the inherent difficulties associated with such corrections, it is recommended that the interpreter avoids using provenance determination diagrams involving the use of Ca unless carbonate minerals are absent from the study samples.

(e) Mixing of sediment sources
The mixing of material from more than one source/provenance can result in confusion and erroneous interpretations. For example, using the log (K/Na) versus Si discrimination diagram of Roser and Korsch (1986) it is possible to differentiate sandstones and mudrocks derived from island arc, active continental margin and passive margin settings. If sediments were derived from the mixing of island arc and passive margin sources, however, this may result in the analysed samples plotting in the active continental margin field. The clastic sediments of many basins have been derived from mixed sources and reworked several times, making the application of provenance determination diagrams useless.

(f) Transportation from one tectonic environment to another
Geochemical discrimination diagrams for sedimentary rocks are often based on the underlying assumption that a close relationship exists between plate tectonic setting and sediment provenance. This often holds true and the main success of the technique involves studies of immature sediments containing significant volumes of lithic fragments from which provenance, and hence tectonic setting, may be identified. However, Rollinson (1993) recognised a major area of uncertainty where sediments are transported from their tectonic setting of origin into a sedimentary basin in a different tectonic environment. For this reason, users of discrimination diagrams for sedimentary rocks must be cautious in their claims for original tectonic setting of a sedimentary basin (Rollinson 1993).

(g) Use of provenance discrimination diagrams in old rocks
Rollinson (1993) points out some potential pitfalls of the application of provenance discrimination diagrams in very old rocks. The term 'very old" is not defined by Rollinson (1993) but is presumed to refer to rocks of Cambrian and Pre Cambrian age and sediments derived from these rocks. It is probable that the trace element composition of mantle source regions has changed with time and the melt may have been less fractionated early in the history of the Earth. Given that higher mantle temperatures occurred in the Archaean, higher degrees of mantle melting and a greater probability of crustal melting should be considered (Pearce et al. 1984; Rollinson 1993).

Many workers have now realised the weaknesses of using the aforementioned schemes to discriminate the provenance and tectonic setting of clastic sediments (e.g. Rollinson 1993; Armstrong-Altrin 2009; Blanco et al. 2011; Caracciolo et al. 2012; Ghosh et al. 2012; Zaid 2012; Verma and Armstrong-Altrin 2013; Tobia and Mustafa 2016). Almost all of these schemes were devised prior to 1995, but are still used for provenance determination in more recent studies (e.g. Akarish and El-Gahary 2011; Tobia and Mustafa 2016; Yassin and Abdullatif 2017). The scheme proposed by Verma and Armstrong-Altrin (2013) is one of the few exceptions to this rule and is considered a vast improvement on previous ones, as it employs separate discriminant functions for different concentrations of Si. For samples with Si in the range 35–65%, the following discriminant functions are applied:

$$
\begin{aligned}
\text{Discriminant Function 1} = {} & (-0.263 \times (\text{Ti/Si})) \\
& + (0.604 \times (\text{Al/Si})) \\
& + (-1.725 \times (\text{Fe/Si})) \\
& + (0.660 \times (\text{Mn/Si})) \\
& + (2.191 \times (\text{Mg/Si})) \\
& + (0.144 \times (\text{Ca/Si})) \\
& + (-1.304 \times (\text{Na/Si})) \\
& + (0.054 \times (\text{K/Si})) \\
& + (-0.330 \times (\text{P/Si})) \\
& + 1.588
\end{aligned}
$$

$$
\begin{aligned}
\text{Discriminant Function 2} = {} & (-1.196 \times (\text{Ti/Si})) \\
& + (1.064 \times (\text{Al/Si})) \\
& + (0.030 \times (\text{Fe/Si})) \\
& + (0.436 \times (\text{Mn/Si})) \\
& + (0.838 \times (\text{Mg/Si})) \\
& + (-0.407 \times (\text{Ca/Si})) \\
& + (1.021 \times (\text{Na/Si})) \\
& + (-1.706 \times (\text{K/Si})) \\
& + \left(-0.126 \times (\text{P/Si})_{\text{adj}}\right) \\
& - 1.068
\end{aligned}
$$

For samples with Si in the range 35–65%, the following discriminant functions are applied:

$$
\begin{aligned}
\text{Discriminant Function 1} = {} & (0.608 \times (\text{Ti/Si})) \\
& + (-1.854 \times (\text{Al/Si})) \\
& + (0.299 \times (\text{Fe/Si})) \\
& + (-0.550 \times (\text{Mn/Si})) \\
& + (0.120 \times (\text{Mg/Si})) \\
& + (0.194 \times (\text{Ca/Si})) \\
& + (-1.510 \times (\text{Na/Si})) \\
& + (1.941 \times (\text{K/Si})) \\
& + (0.003 \times (\text{P/Si})) \\
& + 1.588
\end{aligned}
$$

$$
\begin{aligned}
\text{Discriminant Function 2} = {} & (-0.554 \times (\text{Ti/Si})) \\
& + (-0.995 \times (\text{Al/Si})) \\
& + (1.765 \times (\text{Fe/Si})) \\
& + (-1.391 \times (\text{Mn/Si})) \\
& + (-1.034 \times \text{In}(\text{Mg/Si})) \\
& + (0.225 \times (\text{Ca/Si})) \\
& + (0.713 \times (\text{Na/Si})) \\
& + (0.330 \times (\text{K/Si})) \\
& + (0.637 \times (\text{P/Si})) \\
& - 3.631
\end{aligned}
$$

Figure 4.20 shows the binary diagram used by Verma and Armstrong-Altrin (2013) to differentiate high and low silica clastic samples using these discriminant function parameters. Given that separate distriminant functions have been employed for these two types of samples, this is considered an improvement on earlier schemes which are more likely to be influenced by variations in grain size/lithology, sorting,

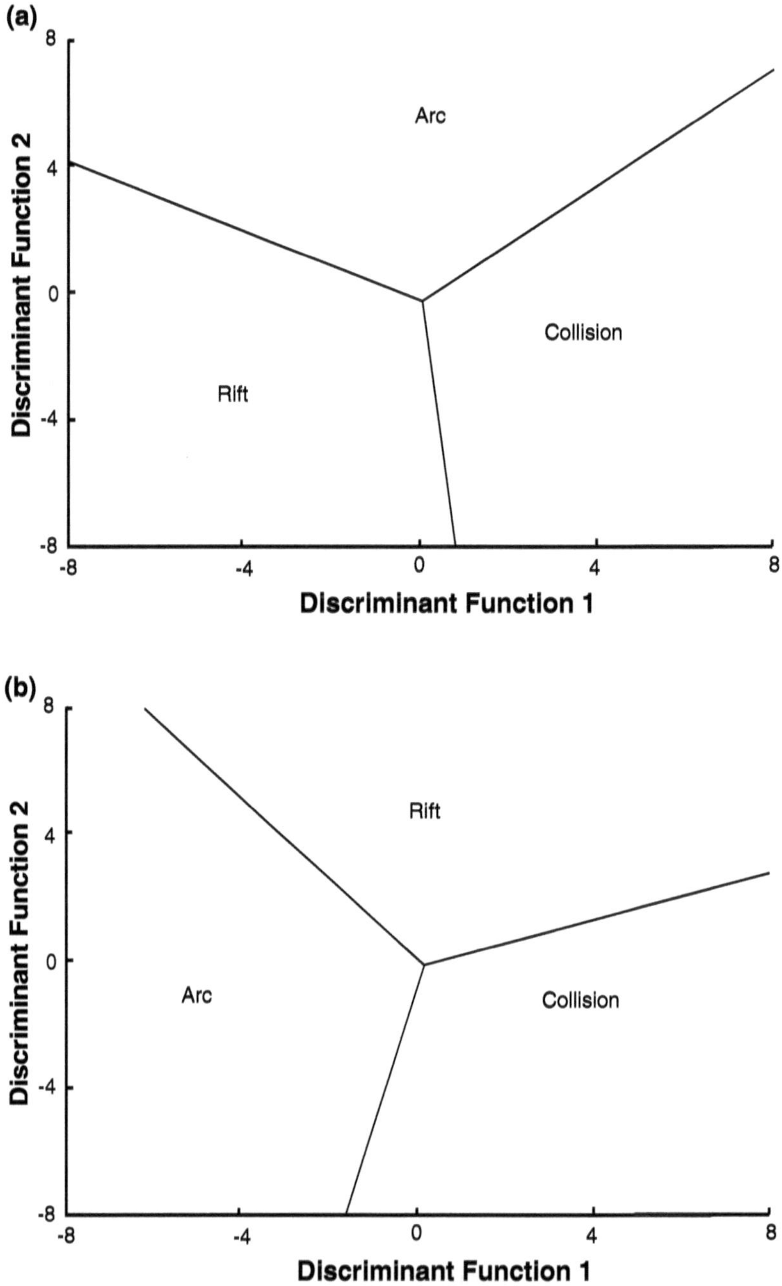

Fig. 4.20 Discriminant Function binary diagrams used to distinguish sediments sourced from arc rift and collision settings in **a** high silica samples, and **b** low silica samples (after Verma and Armstrong-Altrin 2013)

transportation etc. (e.g. Bhatia 1983; Roser and Korsch 1986). The Verma and Armstrong-Altrin (2013) scheme was used by Armstrong-Altrin et al. (2014), and Tobia and Mustafa (2016), in recent studies of sedimentary provenance. In spite of this improvement, this scheme is by no means perfect and factors such as carbonate content, grain size, sorting, transportation, weathering/diagenesis, still need to be considered before using it.

In summary, provenance discrimination diagrams may be used effectively where the material is textually/mineralogically immature, it has been derived from first order sedimentary cycle, is largely unaffected by post-depositional weathering/diagenesis and where it has been transported from a single source, rather than multiple sources. Unfortunately, these requirements are rarely satisfied in the vast majority of chemostratigraphy studies, making the aforementioned provenance discrimination schemes of little use. Furthermore, most chemostratigraphic schemes are based on very subtle changes in source/provenance beyond the resolution of these schemes. For example, Roser and Korsch (1988) used variations in various major elements to identify mafic igneous, quatrzose sedimentary, intermediate igneous and felsic igneous provenance fields. In most chemostratigraphy studies, all of the samples are likely to plot in only one of these 'fields', with the chemostratigraphic zonations/correlations based on more subtle variations in key elements/ratios, probably reflecting minor changes in the distribution of heavy mineral assemblages.

Another issue with most chemostratigraphic studies is that they tend to be restricted to a limited number of wells or field localities. It may be a difficult task to even 'type' the sediment source (e.g. where samples are mineralogically mature, affected by post depositional weathering etc.) and a near impossibility to identify the exact location of the source, unless the study is conducted on a regional/subregional scale where the sediment sources of the hinterland have been distinguished with clarity. In the case of the North Sea, for example, much of the clastic sediments were originally sourced from Cambrian-Precambrian metamorphic basement rocks from the Scottish mainland to the West, but the mineralogical composition of the basement is highly variable in this region, making it very difficult to determine the precise location of the original source of the sediments. This is further complicated by the fact that many of the sediments have been reworked several times and have been subjected to numerous phases of transportation and post-depositional weathering/diagenesis.

For more information on the application of provenance determination schemes, the reader is referred to publications by Rollinson (1993), Akarish and El-Gahary (2011), Verma and Armstrong-Altrin (2013), Armstrong-Altrin et al. (2014), Tobia and Mustafa (2016), and Yassin and Abdullatif (2017).

4.10 Criticisms of Chemostratigraphy

Chemostratigraphy is now a popular and well-established correlation technique but some individuals and organisations still criticise it, normally without much foundation. Sweeping statements such as "chemostratigraphy does not work" are still heard in the corridors of oil and service companies, even with the thousands of publications on the subject proving the contrary (Ramkumar 2015). The basis of the criticism often relates to an individual with little/no prior experience of using the technique trying, and usually failing, to produce a viable correlation scheme. Alternatively, individuals or companies may commission third parties to complete a chemostratigraphic study on a particular field, only to discover that the resulting schemes are spurious at best (probably owing to inexperience on the part of the interpreter). The failure of a minority of chemostratigraphic studies can be explained by the following:

1. Poor sampling strategy.
   This may include samples being taken from the wrong lithologies, at the wrong depths, the wrong type of material being sampled, a lack of available samples etc.
2. Problems with sample preparation.
   Arguably, this is the biggest reason why some chemostratigraphy projects fail, particularly with regard to studies performed on cutting samples.
3. Problems associated with analysis and inadequate data QC.
   Without proper QC of the data, it is impossible to ascertain whether the quality of data is adequate but, in the absence of good

quality data, it is not possible to produce an acceptable chemostratigraphic correlation.

4. Lack of attempt to establish element:mineral links.
   Chemostratigraphic interpretation should not commence until these are established along with the controls on geochemistry and mineralogy.
5. Errors made whilst interpreting the data.
   These are manifold but include failure to take into account changes in grain size/lithology, carbonate content, drilling additives, weathering/diagenesis. In addition to this, plotting data at the wrong horizontal and vertical scales (e.g. North et al. 2005) may inhibit the interpreter to recognize trends that could be utilized for chemostratigraphic purposes. Other studies may have been successful had a '3 point moving average' been applied to the data, or where key elements and ratios had been chosen with more care. Failure to identify chemozones with distinct "cut offs" in the values of key elements/ratios can also be a source of failure.
6. Correlation confidence.
   This can be achieved by employing binary/ternary diagrams in conjunction with DFA and histograms but, in some studies, none of these techniques are employed.
7. Lack of integration with other datasets.
   Some chemostratigraphy projects are performed on a 'stand-alone' basis, but the technique works far better as part of a multidisciplinary approach to reservoir correlation. The skill of the chemostratigrapher is not just to understand the geochemical data and to produce a correlation scheme, but to have enough 'all round' geological knowledge to integrate this with available biostratigraphic and sedimentological data.

Most critics of the technique only convey their thoughts and concerns through conversation, but there are two notable exceptions. The publication of North et al. (2005) is highly critical of chemostratigraphy as a correlation tool. This involved a study of Upper Cretaceous deltaic sediments in the Book Cliffs, Utah, United States. Failure to use the chemostratigraphic data to recognize previously defined parasequences, led North et al. (2005) to conclude "the use of whole-rock geochemistry proved notably inferior to detailed sedimentological analysis in subdividing the succession." Though they recognized internal variability within each field outcrop section, they did not use these variations to identify chemozones within each paprasequence that could have been used for correlation/correlation. Perhaps a more pragmatic approach would have been to apply a fully integrated methodology to zonation/correlation, utilizing sedimentology to identify larger scale parasequences, and chemostratigraphy to recognize chemozones within each parasequence. The fact that North et al. (2005) could not identify chemozones at parasequence boundaries should not have been viewed as a failure of chemostratigraphy as a zonation/correlation technique, because changes in source/provenance can occur at any depth within individual sequences as well as at sequence boundaries. More careful analysis of the data, for instance, may have resulted in the identification of two or more chemozones within each parasequence.

Failure to plot data at realistic scales may have also contributed to the difficulties North et al. (2005) faced in this study. For example, profiles are plotted for Cr at a scale from 0 to 150 ppm, when data for this element exists mainly in the range 2–20 ppm. This makes it almost impossible for the reader to visualise trends in the profiles plotted for each element. The short vertical scales used by the authors also make it difficult to identify these trends. It is noted that some of the somewhat limited number of profiles displayed by North et al. (2005) show considerable variability, but trends may have been more apparent had a three-point moving average been applied to the dataset.

In a more recent paper, Hurst and Morton (2014) claim that "for the most part, successful applications of this technique are based on mudstone-rich successions….", yet this claim is without foundation. In fact, most chemostratigraphy projects are completed on sandstone reservoirs. Many of these involve the analysis of both mudrock and sandstone lithologies, but the focus is usually on the latter, as mudrocks are normally thin, volumetrically unimportant or even non-existent in conventional reservoir intervals (e.g. Ratcliffe et al. 2004, 2007; Sahraeyan et al. 2015; Craigie and Rees 2016; Craigie et al. 2016a, b; Yassin and Abdullatiff 2017).

Hurst and Morton (2014) go on to mention "in cases where this technique has been applied to sandstones, it is generally calibrated using heavy mineral analysis." This holds true for some studies (e.g. Ratcliffe et al. 2004, 2007) but not for most (e.g. Holmes et al. 2015; Craigie and Rees 2016; Craigie et al. 2016a, b; Yassin and Abdullatif 2017). In fact, heavy mineral analysis is extremely expensive in comparison to chemostratigraphy and the low number of experts in this field, particularly those offering this service on a commercial basis, has meant that almost all chemostratigraphy projects are now performed in the absence of this data. Instead, element:mineral links are normally established by employing statistical and graphical techniques, in addition to comparing geochemical, petrographic and/or XRD datasets (Chap. 3).

Another questionable claim by Hurst and Morton (2014) is that "whole-rock geochemical studies (sometimes referred to as chemostratigraphy), by contrast, quantify the abundance of major and trace elements in sandstone, but provide no information on the distribution and location of the elements in minerals." Again this sweeping statement is largely unfounded. Some elements are almost exclusive to particular minerals, such as Zr and Hf being linked with zircons. Other high field strength elements (e.g. Th, Y, HREE, Ta, Cr, Ti) are nearly always concentrated in heavy minerals and, by employing the statistical and graphical techniques detailed in Chap. 3, it is often possible to establish the mineralogical affinities of elements with a high degree of confidence. Even in cases where this is not possible (e.g. Th may be associated with apatite, zircon and monazite), these elements can still be used to define chemostratigraphic boundaries, as they are normally resistant to chemical/physical attack and reflect the distribution of heavy minerals (and hence, changes in provenance).

4.11 Concluding Remarks

The steps required to produce a chemostratigraphic interpretation are manifold but are summarized in Fig. 4.21. Note that some of these steps (e.g. sandstone classification, provenance determination) are not required in every study. Assuming the correct sample preparation, analytical procedures (Chap. 2) and modelling of element:mineral links (Chap. 3) have been completed, the steps shown in this figure should be followed to ensure the production of a robust chemostratigraphic correlation. Chemostratigraphy rarely fails as a correlation tool but, where it does, this is often explained by complacency on the part of the interpreter and/or failure to follow the correct procedures with respect to sample preparation, analysis, and interpretation (Fig. 4.21).

It is hoped that this chapter will serve as a general guide to the production of chemostratigraphic schemes for the beginner and experienced chemostratigrapher alike. It is also hoped that chemostratigraphy may become more widely taught within both academia and industry. Though the technique is now very well established, many of the analytical and interpretation techniques outlined in Chaps. 2–4 are not well documented. This often results in inexperienced interpreters making spurious correlations, which can result in chemostratigraphy being given 'bad press'. Consequently, this publication may serve as a teaching aid, so that the many 'pitfalls' are avoided in future studies.

Many studies in recent times involve the integration of sedimentological, biostratigraphic, lithostratigraphic and/or seismic data. Section 4.8 discusses this very briefly but this is best explained by the case studies detailed in Chap. 5.

1. Lithology determination- split samples into sandstone and mudrock lithologies (clastics only).
2. Sandstone classification (optional).
3. Final quality control checks.
4. Plot profiles for major elements, trace elements and REE.
5. Plot profiles for Al-normalized major elements, trace elements and REE.
6. Plot profiles for ratios.
7. Choose suitable horizontal and vertical scales for profiles.
8. If necessary, apply a '3 point moving average' to the data and re-plot it (core and field outcrop samples only).
9. Identify 4-14 'key' elements and/or ratios to be use for chemostratigraphic puposes.
10. Ensure key elements show significant trends that can be used to identify chemozones..
11. Ensure that key elements are not influenced by grain size, weathering/diagenesis, carbonate content or drilling additives
12. Identify 2 or 3 chemostratigraphic zones and check if these correlate between study sections.
13. Identify subzones (usually 2-4) in each zone.
14. Identify divisions (usually 2-4) in each subzone.
15. Identify subdivisions in each division, and units in each subdivision (not necessary in most studies)
16. Use weathering indices to model changes in the intensity of subaerial weathering (not necessary in some studies)
17. Integrate available lithostratigraphic, sedimentological, biostratigraphic and seismic data. At this stage it may be necessary to revise the placement of some chemostratigraphic boundaries.
18. Plot data on binary and ternary diagrams to illustrate the differentiation of chemozones in each study section.
19. Plot data on histograms to show the level of confidence associated with each chemozone
20. Employ DFA to assign a level of statistical confidence to each chemozone.
21. Determination of provenance and tectonic setting (not necessary for most chemostratigraphy studies).

Fig. 4.21 Summary of steps to employ in chemostratigraphy projects (after data acquisition, data QC, establishment of element:mineral links)

References

Akarish, I. M., & El-Gahary, A. M. (2011). Provenance and source area weathering derived from the geochemistry of Pre-Cenomanian sandstones, East Sinai, Egypt. *Journal of Applied Sciences, 11*(17), 3070–3088.

Alibo, D. S., & Nozaki, Y. (1998). Rare earth elements in seawater: Particle association, shale-normalisation, and Ce oxidation. *Geochimica et Cosmochimica Acta, 62,* 363–372.

Armstrong-Altrin, J. S. (2009). Provenance of sands from Cazones, Acapulco and Bahia Kino beaches, Mexico. *Revista Mexicana de Ciencias Geologicas, 26,* 764–782.

Armstrong-Altrin, J. S., Nagarajan, R., Lee, Y. I., Kasper-Zubillaga, J. J., & Córdoba-Saldaňa, L. P. (2014). Geochemistry of sands along the San Nicoláa and San Carlos beaches, Gulf of California, Mexico: Implications for provenance. *Turkish Journal of Earth Sciences, 23,* 533–558.

Babechuk, M. G., Widdowson, M., Murphy, M., & Kamber, B. S. (2015). A combined Y/Ho, high field strength element (HFSE) and Nd isotope perspective on basalt weathering, Deccan Traps, India. *Chemical Geology, 386,* 25–41.

Bahlburg, H., & Dobrzinski, N. (2011). A review of the chemical index of alteration (CIA) and its application to the study of Neoproterozoic glacial deposits and climate transitions. In: E. Arnaud, G. P. Halverson, & G. Shields-Zhou (Eds.), *The geological record of neoproterozoic glaciations* (Vol. 36, pp. 81–92). Geological Society London, Memoirs.

Berry, W. B. N., Quinby-Hunt, M. S., Wilde, P., & Orth, C. J. (1987). *Use of the cerium anomaly in black shales—Climatic interpretation in the Ordovician-Silurian boundary interval, Dob's Linn,*

Scotland (Vol. 19, 587p). Geological Society of America Annual Meeting

Bhatia, M. R. (1983). Plate tectonics and geochemical composition of sandstones. *Journal of Geology, 91,* 611–627.

Bhatia, M. R., & Cook, K. A. W. (1986). Trace element characteristics of graywackes and tectonic discrimination of sedimentary basins. *Contributions to Mineralogy and Petrology, 92,* 181–193.

Blanco, G., Germs, G. J. B., Rajesh, H. M., Chemale, F., Jr., Dussin, I. A., & Justino, D. (2011). Provenance and paleogeography of the Nama Group (Ediacaran to early Palaeozoic, Namibia): Petrography, geochemistry and U-Pb detrital zircon geochronology. *Precambrian Research, 187,* 15–32.

Blatt, H., Middleton, G., & Murray, R. (1972). Origin of sedimentary rocks. Prentice Hall, New Jersey. Herron, M.M., 1988. Geochemical classification of terrigenous sands and shales from core log data. *Journal of Sedimentary Petrology, 58,* 820–829.

Bokhorst, M. P., Beets, C. J., Markovic, S. B., Gerasimenko, N. P., Matviishina, Z. N., & Frechen, M. (2009). Pedo-chemical climate proxies in Late Pleistocene Serbian-Ukranian loess sequences. *Quaternary International, 198,* 113–123.

Buggle, B., Glaser, B., Hambach, U., Gerasimenko, N., & Markovic, S. (2011). An evaluation of geochemical weathering indices in loess-paleosol studies. *Quaternary International, 240,* 12–21.

Caracciolo, L., Von Eynatten, H., Tolosana-Delgado, R., Critelli, S., Manetti, P., & Marchev, P. (2012). Petrological, geochemical and statistical analysis of Eocence-Oligocene sandstones of the Western Thrace Basin, Greece and Bulgaria. *Journal of Sedimentary Research, 82,* 482–498.

Chen, J., An, Z., & Head, J. (1999). Variation of the Rb/Sr ratios in the loess-paleosol sequences of Central China during the last 130,000 years and their implications for monsoon paleoclimatology. *Quaternary Research, 51,* 215–219.

Cox, R., Lowe, D. R., & Cullers, R. (1995). The influence of sediment recycling and basement composition on evolution of mudrock chemistry in the southwestern United States. *Geochimica et Cosmochimica Acta, 59* (14), 2919–2940.

Craigie, N. W. (2015a). Applications of chemostratigraphy in Cretaceous sediments encountered in the North Central Rub'al-Khali Basin, Saudi Arabia. *Journal of African Earth Sciences, 104,* 27–42.

Craigie, N. W. (2015b). Applications of chemostratigraphy in Middle Jurassic unconventional reservoirs in eastern Saudi Arabia. *GeoArabia, 20*(2), 79–110.

Craigie, N. W., Breuer, P., & Khidir, A. (2016a). Chemostratigraphy and biostratigraphy of Devonian, carboniferous and Permian sediments encountered in eastern Saudi Arabia: An integrated approach to reservoir correlation. *Marine and Petroleum Geology, 72,* 156–178.

Craigie, N. W., & Rees, A. J. (2016). Chemostratigraphy of glaciomarine sediments in the Sarah Formation, northwest Saudi Arabia. *Journal of African Earth Sciences, 117,* 263–284.

Craigie, N. W., Rees, A., MacPherson, K., & Berman, S. (2016b). Chemostratigraphy of the Ordovician Sarah Formation, North-West Saudi Arabia: An integrated approach to reservoir correlation. *Marine and Petroleum Geology, 77,* 1056–1080.

Ding, Z. L., Sun, J. M., Yang, S. L., & Liu, T. S. (2001). Geochemistry of the Pliocene red clay formation in the Chinese Loess Plateau and implications for its origin, source provenance and paleoclimatic change. *Geochimica et Cosmochimica Acta, 65*(6), 901–913.

Elderfield, H., & Greaves, M. J. (1982). The rare earth elements in seawater. *Nature, 296,* 214–219.

Ellis, D. V., & Singer, M. (2007). *Well logging for earth scientists: Dordrecht* (p. 692p). The Netherlands: Springer Science and Business Media B.V.

Englund, J. O., & Jorgensen, P. (1973). A chemical classification system for argillacoues sediments and factors affecting their composition. *Geologiska Föreningen i Stockholm Förhandlingar, 95,* 72–80.

Fedo, C. M., Nesbitt, H. W., & Young, G. M. (1995). Unraveling the effects of potassium metasomatism in sedimentary rocks and paleosols, with implications for paleoweathering conditions and provenance. *Geology, 63,* 921–924.

Garzanti, E., Padoan, M., Andò, S., Resentini, A., Vezzoli, G., & Lustrino, M. (2013). Weathering and relative durability of detrital minerals in equatorial climate: Sand petrology and geochemistry in the East African Rift. *The Journal of Geology, 121,* 547–580.

Garzanti, E., & Resentini, A. (2016). Provenance control on chemical indices of weathering (Taiwan river sands). *Sedimentary Geology, 336,* 81–95.

Ghosh, S., Sarker, S., & Ghosh, P. (2012). Petrography and major element geochemistry of the PermoTriassic sandstones, central India: implications for provenance in an intracratonic pull-apart basin. *Journal of Asian Earth Sciences, 43,* 207–240.

Harnois, L. (1988). The CIW index: A new chemical index of weathering. *Sedimentary Geology, 55,* 319–322.

Haskin, L. A., Helmke, P. A., Paster, T. P., & Allen, R. O. (1971). Rare earths in meteoric, terrestrial, and lunar matter. In A. Brunfelt, E. Steinnes (Eds.), *Activation analysis in geochemistry and cosmochemistry. Proceedings of NATO conference on activation analysis in geochemistry* (pp. 201–218). Oslo: Universitesforlaget.

Haskin, L. A., Wildeman, T. R., & Haskin, M. A. (1968). An accurate procedure for the determination of the rare earths by neutron activation. *Journal of radioanalytical chemistry, 1,* 337–348.

Herron, M. M. (1988). Geochemical classification of terrigenous sands and shales from core or log data. *Journal of Sedimentary Petrology, 58,* 820–829.

Hildred, G., Ratcliffe, A., Schmidt, K. (2011). *Application of inorganic whole-rock geochemistry to shale resource plays: An example from the Eagle Ford shale* (pp. 31–38). Texas., Houston: Geological

Society Northsiders Luncheon Meeting, Tuesday, April 19, 2011, Houston Geological Society Bulletin.

Holmes, N., Atkin, D., Mahdi, S., & Ayress, M. (2015). Integrated biostratigraphy and chemical stratigraphy in the development of a reservoir-scale stratigraphic framework for the Sea Lion Field area, North Falkland Basin. *Petroleum Geoscience, 21,* 171–182.

Hurst, A., & Morton, A. (2014). Provenance models: The role of sandstone mineral-chemical stratigraphy. In R. A. Scott, H. R. Smyth, A. C. Morton & N. Richardson (Eds.) (Vol. 386, pp. 7–26). Geological Society Special Publications.

Kocsis, L., Gheerbrant, E., Mouflih, M., Cappetta, H., Ulianov, A., Chiaradia, M., & Bardet, N. (2016). Gradual changes in upwelled seawater conditions (redox, pH) from the late Cretaceous through early Paleogene at the northwest coast of Africa: Negative Ce anomaly trend recorded in fossil bio-apatite. *Chemical Geology, 421,* 44–54.

Liu, Y. G., Miah, M. R. U., & Schmitt, R. A. (1987). Cerium: A chemical tracer for paleo-oceanic redox conditions. *Geochimica et Cosmochimica Acta, 52,* 1361–1371.

Lowey, G. W. (2015). Element/aluminum ratios in chemostratigraphy: A dubious normalization resulting in spurious correlations. *Geoconvention* 2015, New Horizons 4 pp.

Madhavaraju, J., Hussain, S. M., Ugeswan, J., Nagarajan, R., Ramasamy, S., & Mahalakshmi, P. (2015). Paleo-redox conditions of the Albian-Danian carbonate rocks of the Cauvery Basin, south India: Implications for chemostratigraphy. In M. Ramkumar (Ed.), Chemostratigraphy—Concepts, techniques and application (pp. 247–271). Amsterdam: Elsevier.

Mongelli, G., Sinisi, R., Mameli, P., & Oggiano, G. (2015). Ce anomalies and trace element distribution in Sardinian lithiophorite-rich Mn concentration. *Journal of Geochemical Exploration, 153,* 88–96.

Nesbitt, H. W., & Young, G. M. (1982). Early Proterozoic climates and plate motions inferred from major element chemistry of lutites. *Nature, 299,* 715–717.

North, C. P., Hole, M. J., & Jones, D. G. (2005). Geochemical correlation in deltaic successions: A reality check. *Geological Society of America Bulletin, 117*(5/6), 620–632.

Pearce, J. A., Harris, N. B. W., & Tindle, A. G. (1984). Trace elements discrimination diagrams for the tectonic interpretation of granitic rocks. *Journal of Petrology, 25,* 956–983.

Pearce, T. J., Martin, J. H., Cooper, D., & Wray, D. S. (2010). Chemostratigraphy of upper carboniferous (Pennsylvanian) sequences from the southern North Sea (United Kingdom). In K. T. Ratcliffe, & B. A. Zaitlin (Eds.), *Modern alternative stratigraphic techniques; Theory and case histories* (pp. 109–129). SEPM Special Publication No. 94.

Pearce, T. J., McLean, D., Wright, D. K., Jeans, C. J., & Means, E. W. (2005). Stratigraphy of the upper carboniferous schooner formation, southern North Sea: Chemostratigraphy, mineralogy, palynology and Sm-Nd isotope analysis. In J. D. Collinson, D. W. Evans, D. W. Holliday, N. S. Jones (Eds.), *Carboniferous hydrocarbon geology: The southern North Sea and surrounding onshore areas* (Vol. 7, pp. 165–182). Yorkshire Geological Society, Occasional Publications series.

Pearce, T. J., Wray, D. S., Ratcliffe, K. T., Wright, D. K., & Moscariello, A. (2005). Chemostratigraphy of the upper carboniferous schooner formation, southern North Sea. In J. D. Collinson, D. J. Evans, D. W. Holliday, & M. S. Jones (Eds.), *Carboniferous hydrocarbon geology: The southern north sea and surrounding onshore areas* (Vol. 7, pp. 147–164). Yorkshire Geological Society, Occasional Publications series.

Pettijohn, F. J., Potter, P. E., & Siever, R. (1972). *Sand and sandstones*. New York: Springer.

Potter, P. E. (1978). Petrology and chemistry of modern big river sands. *Journal of Geology, 86,* 423–449.

Ramkumar, M. (2015). Toward standardization of terminologies and recognition of chemostratigraphy as a formal stratigraphic method. In M. Ramkumar (Ed.), *Chemostratigraphy—Concepts, techniques, and applications* (Chap. 1, pp. 1–22). Amsterdam: Elsevier.

Ratcliffe, K. T., Morton, A. C., & Ritcey, D. H. (2007). Whole-rock geochemistry and heavy mineral analysis as petroleum exploration tools in the Bowser and Sustut basins, British Columbia, Canada. *Bulletin of Canadian Petroleum Geology, 55,* 320–333.

Ratcliffe, K. T., Wilson, A., Payenberg, T., Rittersbacher, A., Hildred, G. V., & Flint, S. S. (2015). Ground trothing chemostratigraphic correlations in fluvial systems. *American Association of Petroleum Geologists Bulletin, 99,* 155–180.

Ratcliffe, K. T., Wright, A. M., Haalsworth, C., Morton, A. C., Zaitlin, B. A., Potocki, D., et al. (2004). An example of alternative correlation techniques in a low accommodation setting, non-marine hydrocarbon system: The (lower Cretaceous) Mannville Basdal Quartz succession of southern Alberta. *American Association of Petroleum Geologists Bulletin, 88,* 1419–1432.

Retallack, G. J. (1997). *A colour guide to paleosols* (p. 175). Chichester, England: Wiley.

Rollinson, H. (1993). *Using geochemical data: Evaluation, presentation, interpretation*. Pearson Prentice Hall.

Roser, B. P., & Korsch, R. J. (1986). Determination of tectonic setting of sandstone-mudstone suites using SiO_2 content and K_2O/Na_2O ratio. *Journal of Geology, 94,* 635–650.

Roser, B. P., & Korsch, R. J. (1988). Provenance signatures of sandstone-mudstone suites determined using discriminant function analysis of major-element data. *Cemical Geology, 67,* 119–139.

Sahraeyan, M., Seif, H., Haddad, E. E., Mohammadzadeh, N. (2015). Sedimentology and geochemistry of the Larte Miocence-Pliocence succession in the Fars interior (SW Iran): Implications on depositional and tectonic setting, provenance and paleoweathering in the Zagross Basin (Chap. 5). In M. Ramkumar (Ed.), *Chemostratigraphy: Concepts, techniques, and applications* (pp. 103–126). Amsterdam: Elsevier.

Scott M. McLennan (1993). Weathering and Global Denudation. *The Journal of Geology 101*(2), 295–303.

Tan, H., Ma, H., Zhang, X., Lu, H., & Wang, J. (2006). Typical geochemical elements in loess deposits in the Northeastern Tibetan Plateau and its paleoclimatic implications. *Acta Geologica Sinica, 80,* 110–117.

Tang, M., McDonough, W. F., & Ash, R. A. (2017). Europium and strontium anomalies in the MORB source mantle. *Geochimica et Cosmochiimica Acta, 197,* 132–141.

Thompson, A., Amistadi, M. K., Chadwick, O. A., & Chorover, J. (2013). Fractionation of yttrium and holmium during basaltic soil weathering. *Geochemical et Cosmochimica Acta, 119,* 18–30.

Tobia, F. H., Mustafa, B. H. (2016). Geochemistry and mineralogy of the al-rich shale from Baluti Formation, Iraqi Kurdistan region: Implications for weathering and provenance. *Arabian Journal of Geosciences, 9* (20), 1–23

Tostevin, R., Shields, G. A., Tarbuck, G. M., He, T., Clarckson M. O., & Wood, R. (2016). Effective use of cerium anomalies as a redox proxy in carbonate-dominated marine settings. *Chemical Geology, 438*, 146–162.

Tribovillard, N., Algeo, T. J., Baudin, F., & Riboulleau, A. (2012). Analysis of marine environmental conditions based on molybdenum-uranium covariation—Applications to Mesozoic paleoceanography. *Chemical Geology, 324,* 46–58.

Verma, S. P., & Armstrong-Altrin, J. S. (2013). New multi-dimensional diagrams for tectonic discrimination of siliciclastic sediments and their application to Precambrian basins. *Chemical Geology, 355,* 117–133.

Wilde, P., Quinby-Hunt, M. S., & Erdtmann, B. D. (1996). The whole-rock cerium anomaly: A potential indicator of eustatic sea-level changes in shales of the anoxic facies. *Sedimentary Geology, 101,* 43–53.

Wronkiewicz, D. J., & Condie, K. C. (1987). Geochemistry of Archean shales from the Witwatersrand Supergroup, South Africa: Source area weathering and provenance. *Geochemica et Cosmochimica Acta, 51,* 2401–2416.

Yassin, M. A., & Abdullatif, O. (2017). Chemostratigraphic and sedimentological evolution of the Wajid Group (Wajid Sandstone): An outcrop analog study from the Cambrian to Permian, SW Saudi Arabia. *Journal of African Earth Sciences, 126,* 159–175.

Zaid, S. D. M. (2012). Provenance, diagenesis, tectonic setting and geochemistry of Rudies sandstone (Lower Miocene), Warda Field, Gulf of Suez, Egypt. *Journal of African Earth Sciences, 66–67,* 56–71.

Application of Chemostratigraphy in Clastic, Carbonate and Unconventional Reservoirs

5

Abstract

Chemostratigraphy is most commonly applied to clastic sediments, particularly where biostratigraphic control is lacking. An example of such a study is in the Berkine Basin, Algeria, where it was almost impossible to use lithostratigraphy and sedimentology alone to correlate the Triassic TAGI Formation. More recently, however, the technique has been used in conjunction with other correlation tools to provide higher levels of resolution and correlation confidence. This is certainly true of a study performed on Devonian, Carboniferous and Permian sediments encountered in eastern Saudi Arabia, where both chemostratigraphy and biostratigraphy were employed. Good biostratigraphic control is largely absent in the glaciogenic Sarah Formation of NW Saudi Arabia, where chemostratigraphy, sedimentology, borehole image and seismic data were employed as part of a multidisciplinary approach to reservoir correlation. Chemostratigraphy is less commonly applied to carbonate sediments, but can be used on these lithologies with an equal degree of success. When utilizing the technique on carbonates, the correlation scheme may either relate to changes in the distribution and chemistry of the carbonate fraction, or to the provenance of detrital components (e.g. heavy minerals). A more recent development in chemostratigraphy involves the use of technique on unconventional (source rock) reservoirs where there are normally two objectives, the first being to produce a correlation scheme. A secondary aim is to apply the inorganic geochemical data to recognize organic rich zones, changes in base level and redox, and unconventional seals/cap rocks.

N. Craigie, *Principles of Elemental Chemostratigraphy*, Advances in Oil and Gas Exploration & Production, https://doi.org/10.1007/978-3-319-71216-1_5

5.1 Introduction

Previous chapters in this book have concentrated on sample preparation procedures, analytical techniques, geochemistry and mineralogy, methodologies and chemostratigraphic rationale. The present chapter focuses on specific case studies where chemostratigraphy has been utilized as a reservoir correlation tool. Thousands of papers have been written on chemostratigraphy over the last 15 years, and it would be completely impractical to discuss each one in depth. Instead the author has provided a summary of a limited number of published case studies, where chemozones have been clearly defined, and where the technique has added considerable value to the development of a correlation scheme. The intention is not to restate every aspect of these studies, but simply to provide a summary of their salient points.

Chemostratigraphy studies can be divided into the following three broad categories:

(a) Clastic sediments.
(b) Carbonate sediments.
(c) Unconventional (source rock) reservoirs.

A selection of case studies from these groups are presented in the following paragraphs.

5.2 Clastic Sediments

Clastic sediments comprise sandstones and mudrocks and may be deposited in a range of depositional environments. Chemostratigraphy is often performed on continental clastic sediments as these are typically deposited under oxic conditions, not conducive to the preservation of bioclasts and palynomorphs that could be used to propose a biostratigraphic correlation. Arguably more projects have been completed on aeolian, fluvial and fluviodeltaic sediments than on any other depositional system, and the reader is referred to the excellent works of Pearce et al. (1999, 2005), Ratcliffe et al. (2006), Craigie et al. (2016a, b), Craigie and Polo (2017) for more information on the applications of the technique in continental environments. Lacustrine sediments are considered less important in terms of their potential to produce hydrocarbon bearing reservoirs and this may explain the lower number of studies performed on the deposits of these systems. One exception is the work of Craigie (1998) on Middle Devonian lacustrine sediments encountered in the Orcadian Basin, NE Scotland.

Many studies have been completed on deep water turbidite systems (e.g. Pearce and Jarvis 1991; Holmes et al. 2015; Craigie et al. 2016b), where there is often very good recovery of palynomorphs and/or microfossils. Because the sediments are often deposited rapidly, however, biostratigraphic resolution may be limited and chemostratigraphy may be one of the few techniques that can be employed for reservoir correlation. Shallow marine conditions cover a wide range of depositional environments including, but not limited to, deltaic, lagoonal, upper shoreface, middle shoreface and lower shoreface settings. Biostratigraphic recovery may be limited where these sediments were deposited under high energy and/or oxic conditions, whilst resolution may be restricted where deposition was rapid.

The aforementioned paragraphs may give the reader the impression that chemostratigraphy is generally used where biostratigraphic control is limited. This statement would certainly apply to many studies completed prior to 2007 (e.g. Pearce and Jarvis 1991; Pearce et al. 1999; Ratcliffe et al. 2006) but the technique is increasingly used in conjunction with biostratigraphy as part of a fully integrated approach to reservoir correlation (e.g. Craigie et al. 2016a). In addition to this, many studies now involve chemostratigraphy in conjunction with wireline log/lithostratigraphy, reservoir sedimentology and even seismic data (e.g. Holmes et al. 2015; Craigie et al. 2016a, b). By employing such a multidisciplinary approach to correlation, it is possible to add confidence to the placement of boundaries and to improve the levels of resolution. In some instances one or more of these datasets may be unavailable, but it is always advisable to utilize all available chemostratigraphic, biostratigraphic, sedimentological, mineralogical, lithostratigraphic, wireline log and seismic data for reservoir correlation purposes. Adopting such multidisciplinary thinking, however, demands that the chemostratigrapher is sufficiently knowledgeable in these subjects to provide scientific scrutiny of both chemostratigraphic and non-chemostratigraphic data.

5.2.1 Case Study of Triassic Fluvial Sediments, Berkine Basin, Algeria

The Triassic Argilo-Gréseux Inférior Formation (TAG-I) comprises prolific hydrocarbon-bearing sandstone reservoirs forming a SW-NE trending fluvial-dominated depositional system, extending eastwards from Algeria into southern Tunisia. This formation occurs immediately above shallow marine Carboniferous sediments, with the regionally defined Hercynian unconformity marking the boundary between the two stratigraphic intervals. The TAG-I is succeeded by estuarine to shallow marine sediments of the Triassic Carbonate interval. Detailed information on the sedimentology of the TAG-I Formation is provided by Turner et al. (2001) and Sabaou et al. (2005). The following chemostratigraphy study was completed by Ratcliffe et al. (2006) and the following paragraphs represent a summary of their findings. This project involved the study of wells in Blocks 405a and 402, Eastern Algeria. A lack of diagnostic fauna or flora from wellbore samples and lateral changes in facies between the two study area resulted in difficulties associated with correlating fluvial sandstones both within and between the blocks. It is also noteworthy that the distance between Blocks 405a and 402 is around 200 km, adding a further complication to the correlation of lithostratigraphic units on such a large subregional scale.

By comparing geochemical, petrographic, XRD and heavy mineral data, Ratcliffe et al. (2006) established the mineralogical affinities of key/index elements with a high degree of confidence. The following key element ratios (and their mineralogical associations) were used to identify chemozones:

K/Al: K feldspar and illite.
Zr/Nb: zircon versus rutile.
Ti/K: anatase versus K feldspar and illite.
Rb/K: illte and mica versus K feldspar.

Ratcliffe et al. (2006) identified four chemozones labelled CP10, CP20, CP30 and CP40 in ascending stratigraphic order, with CP10 characterised by high Zr/Nb and Ti/K ratios, but low K/Al. This is succeeded by zone CP20 which is geochemically similar to CP10 but generally yields slightly higher K/Al values. The sandstones of CP30 produce higher K/Al ratios than in both CP10 and CP20, with petrographic data revealing such increases in this ratio from CP10 to CP30 to be related to an upward increase in K feldspar. The mudrocks show a similar upward increase in K/Al but, unlike the sandstones, this increase is related to higher concentrations of illite. Zone CP40 occurs at the top of the TAG-I, with sandstones of this interval generally defined by the highest values of K/Al. The K/Al ratios of the CP30 and CP40 mudrocks are very similar, but the top of the latter is clearly marked by elevated Mg/Al, thought to be linked to the presence of a dolocrete. Another notable feature of CP40 mudrocks is an upward decrease in Rb/K ratios at the top of this zone. The characterisation of chemozones is shown for wells MLN-1 and BSFN-1 in Figs. 5.1 and 5.2 respectively.

The correlation of chemozones between the study wells is summarized in Fig. 5.3. Zones CP10, CP30 and CP40 are correlative between the fourteen wells in spite of the c. 200 km distance between the ones drilled in Blocks 405a and 402. By contrast, zone CP20 is identified in all of the Block 402 wells, but only in two of the ones located in Block 405a. This unit is likely to be absent as a result of erosion or non-deposition in wells MLW-1, MLN-1, MLN-3, MLN-4, MLC-1 and MLC-2.

The study of Ratcliffe et al. (2006) was ground breaking as it represented one of the first chemostratigraphy projects to be undertaken on a large subregional scale. Chemozones were clearly defined and the mineralogical affinities of key

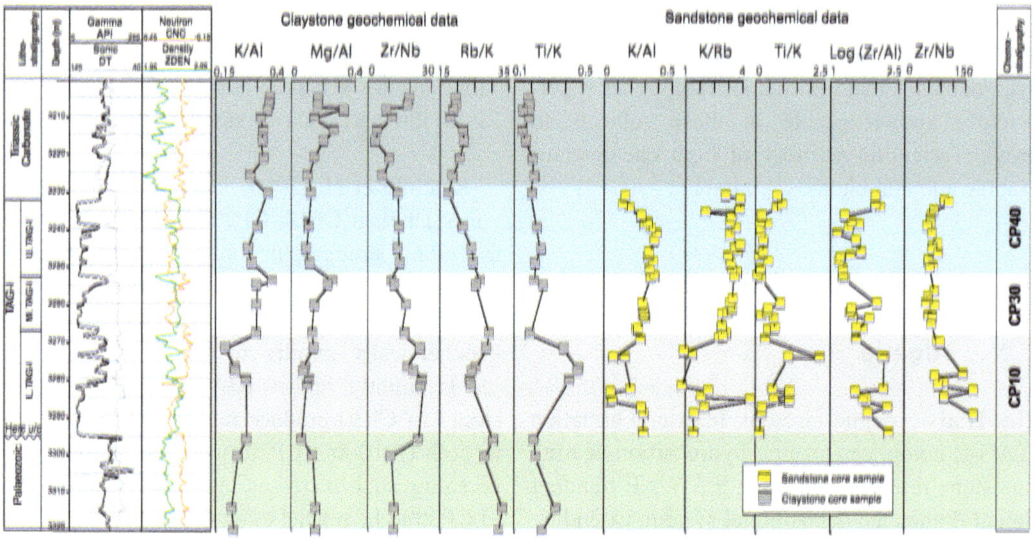

Fig. 5.1 Key elemental ratio profiles plotted for claystone and sandstone core samples in well MLN-1, Block 405a. All depths are log depths in feet. Here u/c refers to the Hercynian unconformity (modified, after Ratcliffe et al. 2006)

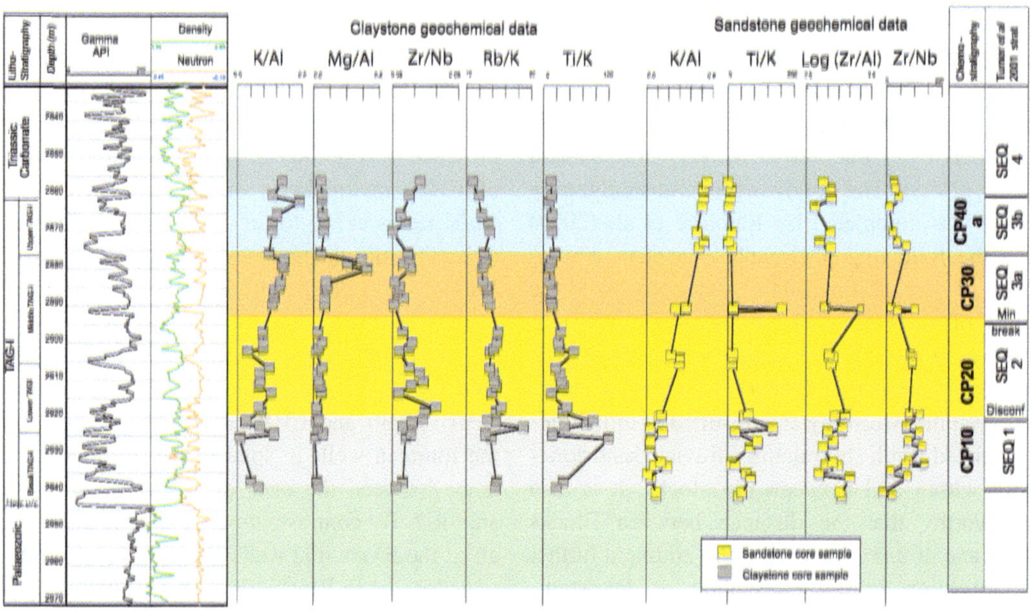

Fig. 5.2 Key elemental ratio profiles plotted for claystone and sandstone core samples in well BSFN-1, Block 402. All depths are log depths in feet. Here u/c = the Hercynian unconformity; Discon = disconformity; Min Break = mineralogical break (modified, after Ratcliffe et al. 2006)

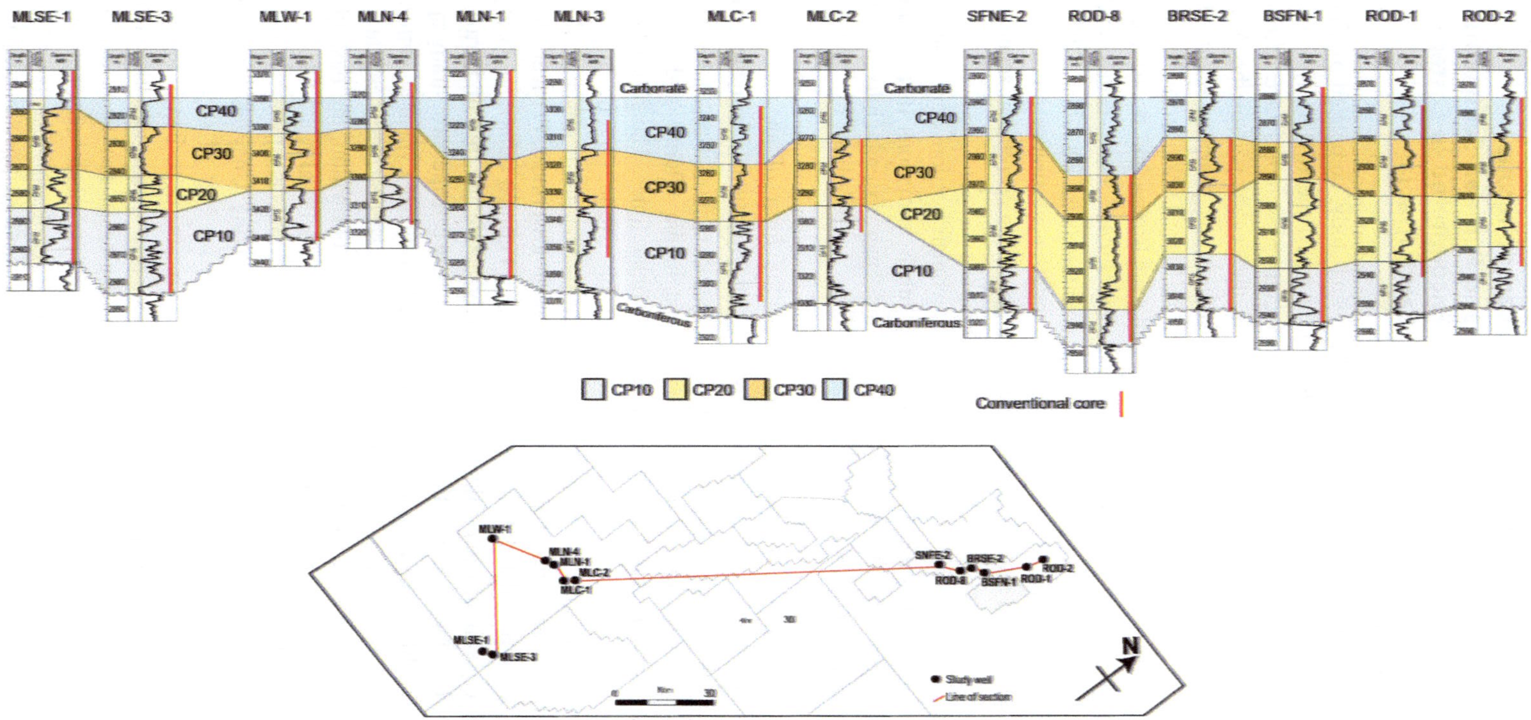

Fig. 5.3 Summary of chemostratigraphic correlation of selected wells in Blocks 402 and 405a (modified, after Ratcliffe et al. 2006)

elements were established by comparing geochemical data with heavy mineral, XRD and petrographic data. The main criticism is that no cut off values were used to distinguish the chemozones of each well. For example, zone CP10 produces higher Zr/Nb than in CP30, but the precise values/ranges of this ratio are not defined for the two zones. It is also noteworthy that the horizontal scales of profiles plotted for key element ratios are different in wells MLN-1 and BSFN-1 (Figs. 5.1 and 5.2). The Zr/Nb scale ranges from 0 to 30 in the mudrock profile of well MLN-1 but extends from 0.05 to 0.08 in BSFN-1. Similarly, Ti/K profiles plotted for mudrocks in wells MLN-1 and BSFN-1 are in the ranges 0.1–0.5 and 0–150 respectively. This makes a direct comparison of the values of key ratios difficult between the two wells. The values of Ti/K, for instance, are higher in the mudrocks of CP10 than in CP30 in both wells, but are an order of magnitude higher in BSFN-1 (generally 50–150) than in MLN-1 (0.4–0.5). These differences may be explained by variations in source/provenance, hydrodynamic sorting or depositional environment over such a vast scale, though little mention of these inconsistences are made by Ratcliffe et al. (2006). Without having access to the entire dataset from all of the study wells, it is difficult to pass further comment on the differences in key element/ratio values between individual wells, or on the validity of making a correlation between the two sets of wells. It is postulated that the chemozones may yield very similar values of key ratios in closely spaced wells in each Block, meaning that there is probably a robust correlation between wells drilled in Blocks 402 and 405a, whilst the correlation between the two areas may be more questionable. In spite of these uncertainties, this was one of the first published studies to use geochemical and mineralogical data to establish element:mineral links and was, almost certainly, one of the first to apply the technique to more than 10 wells and on a subregional scale. Another impressive aspect of this study is that data for sandstone and mudrock samples were interpreted separately- this does not always happen in chemostratigraphy projects performed on clastic sediments.

5.2.2 Case Study of Devonian, Carboniferous and Permian Sediments Encountered in Eastern Saudi Arabia

The following study, completed by Craigie et al. (2016a), involved the ICP analysis of 195 core and 655 cuttings samples taken from 6 wells located in the central part of the Arabian Gulf (Saudi Arabia), and penetrate sediments ranging in age from Early Devonian to Permian. These are mainly sandstones and occasional mudrocks deposited in fluvial, fluvio-deltaic and shallow marine environments. The aim of the study was to utilize both chemostratigraphy and biostratigraphy to produce a correlation scheme for these sediments.

The Devonian section comprise the Tawil, Jauf and Jubah formations in ascending order, though the latter extends into the lowermost Carboniferous. According to Al-Hajri et al. (1999), the Devonian of the Middle East generally takes the form of an overall coarsening upwards, progradational sequence along the passive margin of northeastern Gondwana that commenced in the Silurian and culminated in Upper Devonian times. The Jubah Formation is unconformably overlain by the Berwath Formation, which is then succeeded by the Pennsylvanian-lowermost Permian Juwail Formation. The Permian Unayzah Group occurs immediately above the Juwail Formation, with the contact between the two stratigraphic intervals marked by the Hercynian unconformity. The top of the study sections are characterised by the Permian Khuff Formation, the base of which is represented by the Basal Khuff Clastics Member. The base of this member forms an unconformity with the underlying Unayzah Group and comprises an 'erosive lag' of sandstones conglomerates and breccias deposited in shallow marine environment in response to a rise in sea level. Further increases in sea level resulted in the precipitation of carbonate sediments immediately above the Basal Khuff Clastics Member. For more detailed information on the sedimentology and stratigraphy of the Palaeozoic sediments of

Saudi Arabia, the reader is referred to the work of Sharland et al. (2001) and references therein.

Unlike the previous study of Ratcliffe et al. (2006), no mineralogical data were available in this project, so the mineralogical affinities of elements were established by employing eigenvector analysis and binary diagrams. The following key elements (and their mineralogical significance) were used for chemostratigraphic purposes:

Cr/P: Cr-bearing heavy minerals (e.g. chrome spinel) versus P-bearing heavy minerals (e.g. apatite, monazite and/or biogenic phosphate).
Th/Nb: Th-bearing heavy minerals versus Nb-bearing heavy minerals (i.e. rutile, anatase, sphene, titanomagnetite, magnetite and/or ilmenite).
Th/Ta: Th-bearing heavy minerals versus Ta-bearing heavy minerals (i.e. rutile, anatase, sphene, titanomagnetite, magnetite and/or ilmenite).
Zr/Th: zircon versus Th-bearing heavy minerals.
Zr/P: zircon versus P-bearing heavy minerals (e.g. apatite, monazite and/or biogenic phosphate).
Nb/Ti: Nb-bearing heavy minerals versus Ti-bearing heavy minerals.
Nb/P: Nb-bearing heavy minerals versus P-bearing heavy minerals (e.g. apatite, monazite and/or biogenic phosphate).
Al/(Mg+K+Na): weathering index.

With the exception of the Al/(Mg+K+Na) weathering index, variations in these parameters are largely related to changes in source/provenance, as the elements Th, Cr, Nb, Zr and Ta are highly immobile, being almost exclusively concentrated in heavy minerals. The only possible exception to this rule may be P as, in the absence of XRD, petrographic and heavy mineral data, it is not possible to determine whether this element is linked with heavy minerals (e.g. xenotime, apatite, monazite) or biogenic phosphate. The latter, however, is less plausible based on the statistical and graphical evaluation of the dataset. Consequently, it is considered more likely that the observed variations in Cr/P, Zr/P and Nb/P are related to changes in provenance. Craigie et al. (2016a) concluded that even if some of the P is associated with biogenic phosphate, it is highly unlikely that this material would have been 'in-situ' in depositional environments dominated by oxic fluvial, fluvio-deltaic and shallow marine environments, but may have been transported by these processes prior to deposition. This led to the conclusion that variations in Cr/P, Zr/P and Nb/P most likely reflect changes in source/provenance, rather than than depositional environment.

In many chemostratigraphy studies, the Al/(Mg+Ca+K+Na) ratio or other similar parameters involving Ca, are used to model the intensity of subaerial weathering, with the assumption that Ca is concentrated in the siliciclastic fraction (see Chap. 4 for more details). In the study of Craigie et al. (2016a), however, it is apparent that Ca is, at least in part, concentrated in carbonate minerals and/or Ca-bearing drilling additives. For this reason, the Al/(Mg + K + Na) ratio was employed to identify such zones of weathering, particularly those associated with unconformities.

The principal geochemical characteristics of each chemozone identified in the sandstone dataset are summarized in Fig. 5.4, with the correlation of chemozones present in two study wells illustrated in Fig. 5.5. Profiles of Si/Al are included for both wells, but this parameter was not used to define chemozone boundaries. The Si/Al ratio is used for quality control purposes as values of this ratio normally relate to grain size, with higher Si/Al linked with a higher quartz/clay mineral ratio. Any ratio or element showing strong positive or negative associations with Si/Al should not be used for chemostratigraphic purposes as it is probably associated with changes in bulk lithology/mineralogy, rather than source/provenance. This correlation panel shows zones labelled CS1, CS2 and CS3 in ascending stratigraphic order, with the former producing the lowest values of Cr/P. The differentiation of the other two zones is based on variations in Th/Nb, as CS3 is defined by higher values of this ratio (Fig. 5.4). A threefold subdivision is proposed for CS1, with subzones CS1-1 and CS1-3

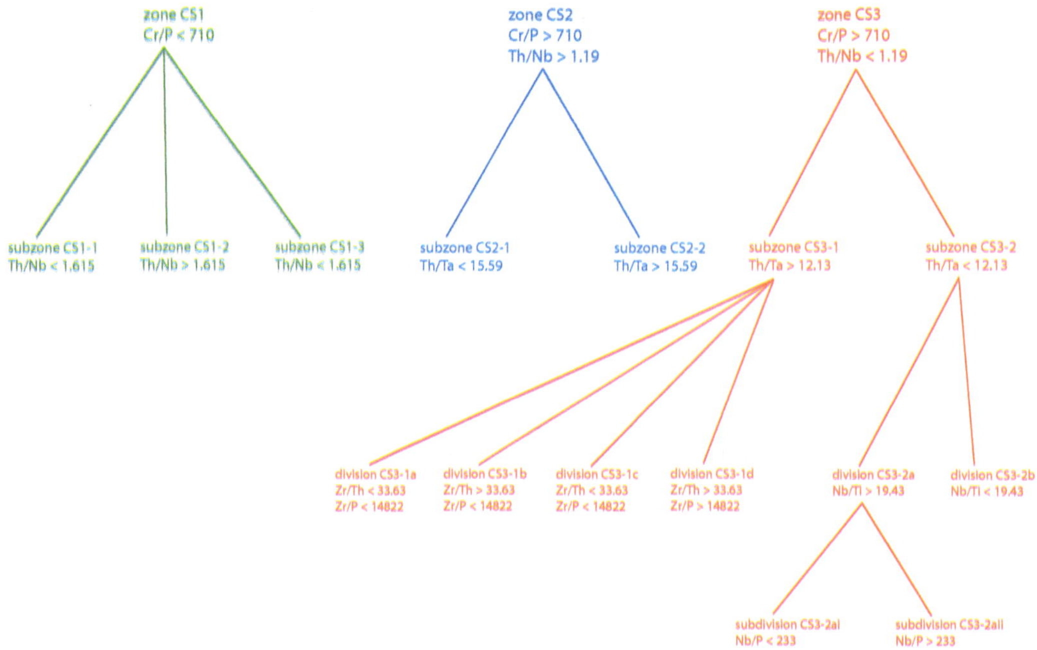

Fig. 5.4 Principal geochemical characteristics of chemozones recognized in the sandstone dataset (after Craigie et al. 2016a)

occurring at the base and top of this zone respectively, and yielding lower Th/Nb ratios than in the intervening subzone CS1-2. Subzone CS2-1 is noted at the base of zone CS2 and produces lower Th/Ta values than in the overlying subzone CS2-2. Of the two zone CS3 subzones, CS3-1 occurs at the base and is characterized by higher Th/Ta ratios than in the overlying subzone CS3-2. Four divisions are identified in subzone CS3-1, their differentiation being based on changes in Zr/Th and Zr/P. Division CS3-2a occurs at the base of subzone CS3-2 and produces higher Nb/Ti ratios than in the overlying division CS3-2b. Two further subdivisions are recognized in division CS3-2a, with CS3-2ai noted at the base and yielding lower Nb/P values than in the overlying subdivision CS3-2aii.

Figure 5.5 shows that most chemozones are correlative between the two study wells, with the absence of CS3-1a in Well 2 probably explained by the existence of a sample gap. Conversely, CS3-1d is considered missing in the same section as the high Zr/Th and Zr/P ratios used to define this chemozone in Well 5 do not occur at the top of CS3-1 in Well 2. Chemozone CS3-2b is identified at the top of the study interval of Well 5 and may exist in the unsampled section immediately above CS3-2aii in Well 2.

Figure 5.5 also illustrates that Al/(Mg+K+Na) ratios are generally highest towards the top of the Jubah Formation of both wells which may be explained by a period of exposure and subaerial weathering prior to the deposition of the overlying Berwath Formation. The existence of a 'known' unconformity between the two formations may provide further proof of this. Values of the same parameter are generally low in the Berwath Formation and Unayzah Group, but increase toward the top of the latter stratigraphic interval. This is not surprising as the top of the Unayzah Group represents an unconformity recognized on a regional scale throughout Saudi Arabia and other parts of the Middle East (Sharland et al. 2001).

A summary of the geochemical parameters used to identify chemozones in the mudrock dataset is presented in Fig. 5.6. This scheme is

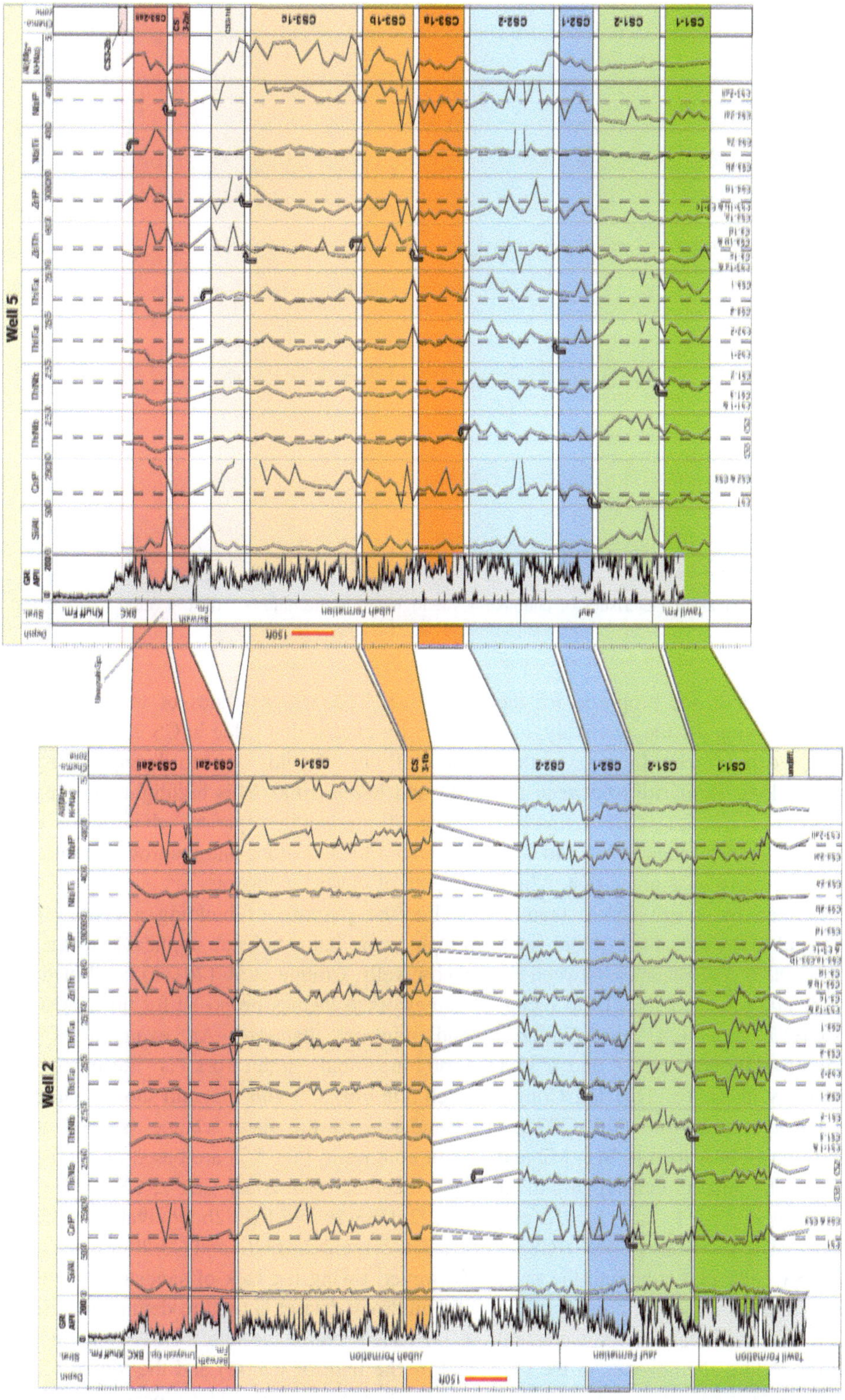

Fig. 5.5 Chemostratigraphic correlation between wells 2 and 5 based on geochemical data acquired from sandstone samples. All depths are log depths in feet (after Craigie et al. 2016a)

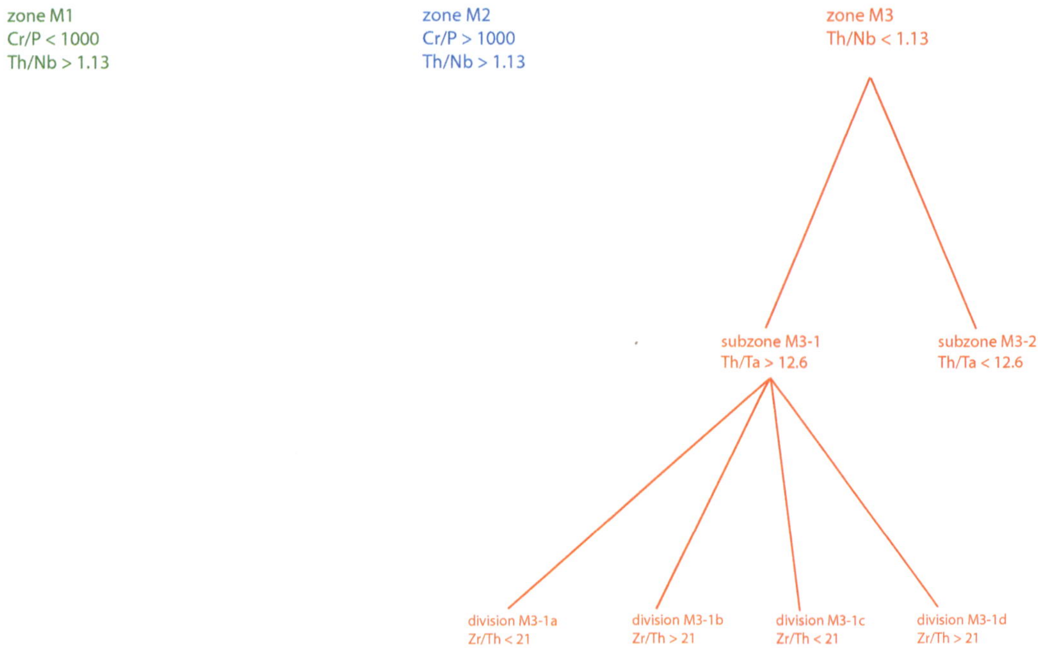

Fig. 5.6 Principal geochemical characteristics of chemozones recognized in the mudrock dataset (after Craigie et al. 2016a)

far less detailed than the one proposed for the sandstone dataset, owing to the relatively low number of analysed mudrocks. In spite of this, there are some striking similarities between the two frameworks. For example Cr/P ratios are higher in zones M2 and CS2 than in M1 and CS3 respectively. Furthermore, it is noted that, whilst CS3 is characterized by lower Th/Nb than in CS2, M3 also yields lower values of the same ratio than in M2. Other consistencies between the two schemes include the use of Th/Ta to recognize two subzones in zones CS3 and M3, and a fourfold subdivision of both M3-1 and CS3-1 based on variations in Th/Ta. This led Craigie et al. (2016a) to suggest that zones CS1, CS2 and CS3 are roughly time-equivalent to M1, M2 and M3 respectively. The authors concluded similar associations between CS3-1 and M3-1, and CS3-2 and M3-2. The divisions of M3-1/CS3-1 were also probably deposited at around the same time. In spite of these similarities, however, it is clear that different 'cut off' values of key element ratios are used to define chemozones in the two datasets. For example, zone CS-1 yields Cr/P of less than 710, whilst values of the same ratio generally exceed this value in CS-2. By contrast, the Cr/P cut off value used to distinguish zones M1 and M2 is 1000. This highlights the importance of producing separate schemes for data generated for sandstone and mudrock samples on studies of clastic sediments.

For the purpose of building a chemostratigraphic correlation scheme applicable to the six study wells, Craigie et al. (2016a) combined the frameworks/schemes proposed for the sandstone and mudrock datasets. An example of the 'combined/final' framework is illustrated for Well 3 in Fig. 5.7. This diagram shows that most boundaries of the final scheme (second from right column in Fig. 5.7) are based on the sandstone scheme, with zones CS1-2, CS1-3, CS2-1, CS2-2, CS3-1b and CS3-2b, equivalent to C1-2, C1-3, C2-1, C2-2, C3-1b and C3-2b respectively. However, owing the lack of sandstone samples, the mudrock dataset is used to define the base of C1-1 (occurring at the base of M1 immediately below the base of CS1-1) and the top of C3-1c (equivalent to top of M3-1c occurring above

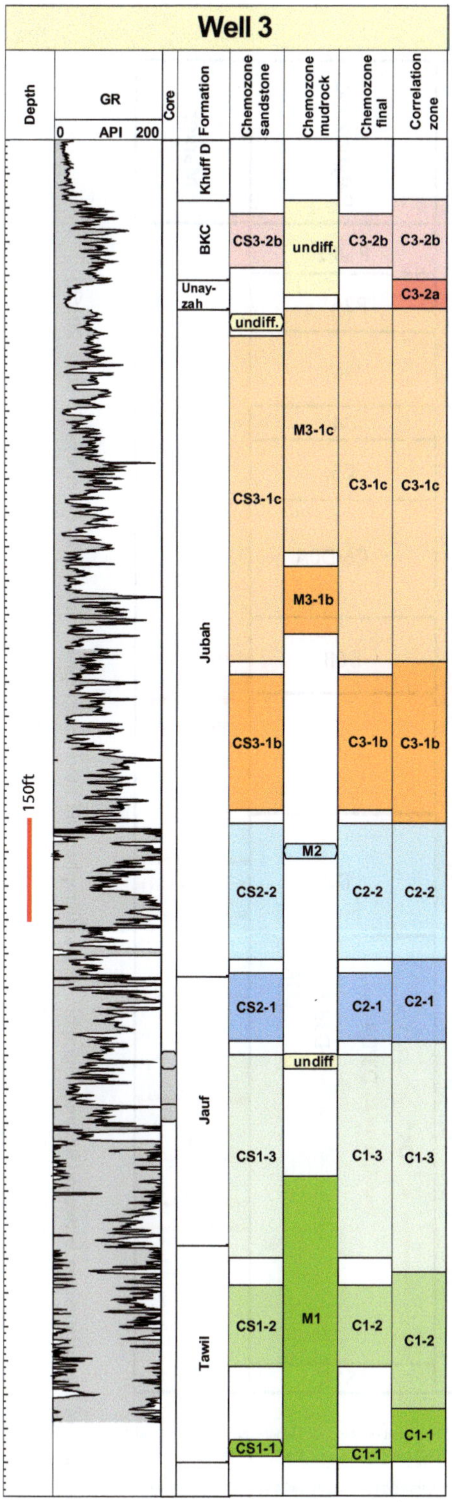

Fig. 5.7 Placement of chemostratigraphic boundaries in Well 3. See text for further explanation (modified, after Craigie et al. 2016a)

CS3-1c). The only inconsistency in this well is that subzone M3-1b occurs towards the base of CS3-1c, rather than within C3-1b. This may be explained by the low number of mudrock samples used to define M3-1b. Consequently, the sandstone dataset is considered more reliable, with the top of C3-1b and the base of C3-1c related entirely to the interpretation of geochemical data acquired from this lithology.

In most studies, the 'gaps' between chemozones should be honoured with no attempt made to interpret the 'true' depth of chemozone boundaries between these gaps. There are exceptions to this rule where the sample coverage is good and the 'gaps' between chemozones small, and where there are very obvious lithostratigraphic or biostratigraphic boundaries within these 'gaps'. This certainly holds true for Well 3 and other wells in the study of Craigie et al. (2016a). The 'final' zonation for Well 3 (with the gaps between sample points closed) is shown in the left hand column of Fig. 5.7. Subzone C3-2a was not identified in this well, but corresponds to the Unayzah Group in other wells. Given that this group is defined by a relatively low and 'blocky' GR response in Well 3 (as it is in the other study wells), it is thought that this subzone exists in the unsampled interval between C3-1c and C3-2b. Furthermore palynozone Cm, a diagnostic biostratigraphic marker for the Unayzah Group, has also been identified in this interval.

It is beyond the scope of this chapter to discuss the biostratigraphic scheme used in the study, but details are provided by Craigie et al. (2016a) and references therein. Figure 5.8 shows a generalized comparison between the sandstone, mudrock and combined schemes, and the biostratigraphic zonation based on palynological data. The correlation of chemozones and palynozones between the six study wells is presented in Fig. 5.9. Lithostratigraphic markers are included in Figs. 5.8 and 5.9, their placement being based on very obvious trends in the GR log. Most are correlative between at least two wells, if not three, but, with the single exception of marker 2 in Well 1, these cannot be identified with any confidence in Wells 1 and 3.

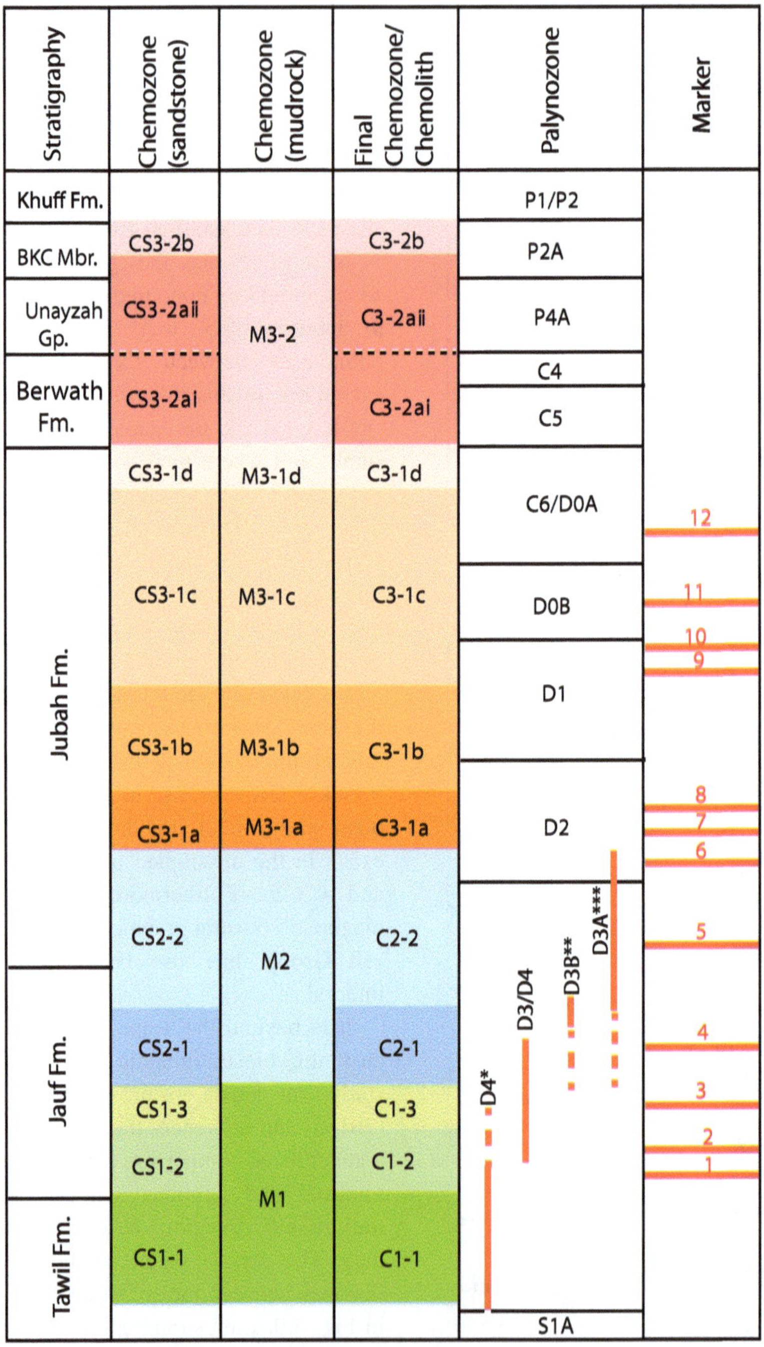

Notes
* In general D4 encompasses chemozone C1-1 and the basal part of C1-2. The exception is in well 3 where this biozone extends from the base of C1-1 to the middle part of C1-3. It is also noted that D4 transcends the C1-2:C1-3 boundary in well 6.

** Biozone D3B is normally associated with chemozones C2-1 and C2-2. A notable exception is in well 6 where this biozone is noted towards the center of C1-3. In well well 3, this biozone occurs at the C1-3: C2-1 boundary.

*** Biozone D3A is generally associated with chemozone C2-2 but transcends the C1-3:C2-1 and C2-1:C2-1 boundaries in wells well 6 and well 3 respectively.

Fig. 5.8 Summary of the comparison between chemostratigraphic, biostratigraphic and lithostratigraphic frameworks (after Craigie et al. 2016a)

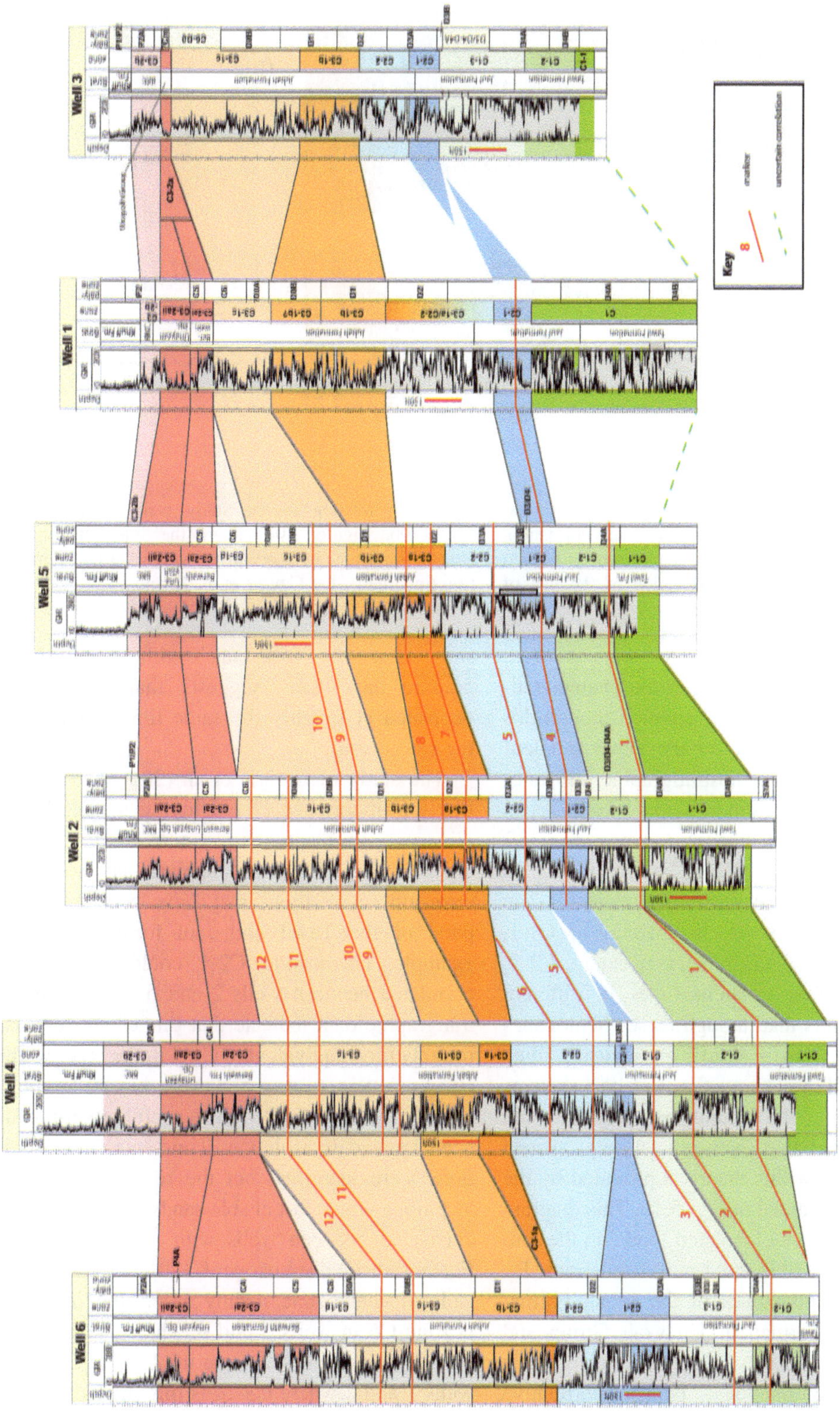

Fig. 5.9 Summary of the chemostratigraphic, lithostratigraphic and biostratigraphic correlation schemes in Wells 1-6 (after Craigie et al. 2016a)

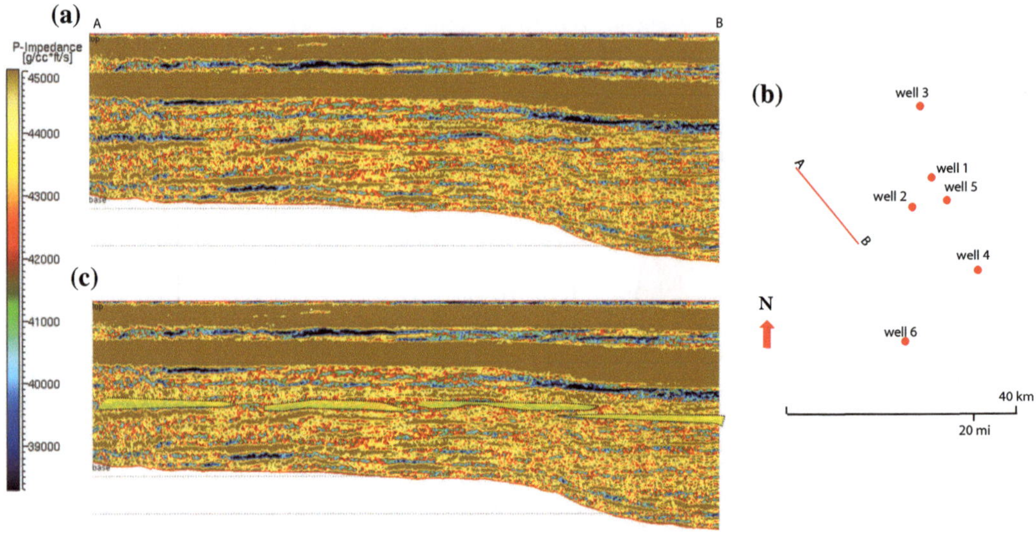

Fig. 5.10 Interpretation of the distribution and correlation of sandstone beds within the Jauf Formation. Figure 5.20a shows a seismic impedance section filtered to seismic frequency, with the location of the line of section in relation to the study wells illustrated in Fig. 5.10b. Figure 5.10c shows the interpretation of isolated sandstone beds occurring within the Jauf Formation (modified, after Craigie et al. 2016a)

Figures 5.8 and 5.9 show that there is a reasonably close 'match' between the chemostratigraphc and biostratigraphic schemes, though this is definitely not the case for the Tawil and Jauf formations, where chemostratigraphic subzones occur at different times in individual wells. As an obvious example of this is Palynozone D3B which transcends the C2-1:C2-2 boundary in Wells 2 and 5, but exists within subzone C1-3 in Well 6. The lowermost chemozones of C1-1, C1-2, C1-3, C2-1 and, to a lesser extent, C2-2, were deposited at different times in different wells. This infers that, though these chemozones were derived from a similar provenance, they were deposited by separate fluvio-deltaic systems at different times throughout the study area and are only correlative between closely spaced wells. Support for this theory is provided by the seismic data presented in Fig. 5.10. This diagram shows the existence of isolated 'pods' in the Jauf and, to a lesser extent the Tawil, formations. By contrast, the chemozones of the Jubah Formation, Berwath Formation, Unayzah Group and Basal Khuff Clastics Member display a 'layer-cake' correlation, being correlative between the study sections of most wells and showing a very close match with the biostratigraphic data. This suggests that they were deposited at roughly the same time throughout the study area.

The study of Craigie et al. (2016a) exemplified the importance of employing a multidisciplinary approach to correlation. Had chemostratigraphy been applied in isolation, erroneous correlations may have been made in parts of the Tawil and Jauf formations. For example, chemozone C2-1 could have been correlated between wells 2 and 6, but the biostratigraphic data indicate that this chemozone is roughly time equivalent to C1-3 in the latter well. By a similar token, however, biostratigraphy would have been limited in its usefulness if it had been applied in isolation. Time-equivalent sections were identified but the resolution of the technique would not enable sandstone beds to be correlated between wells with any confidence. For instance, chemozone C3-1d represents an isolated chemozone mainly associated in palynozone C6 in wells 5 and 6. The same palynozone extends from C3-1c to the base of C3-2 in wells 1 and 2 where C3-1d is missing as a result of erosion/non-deposition (Fig. 5.9). In the absence of chemostratigraphic data, the existence and lateral extent of C3-1d could not have been determined.

The integration of seismic data is rare in chemostratigraphy studies, mainly owing to the general lack of expertise that chemostratigraphers have in the interpretation of this data and the rarity of geophysicists possessing a sufficient knowledge of chemostratigraphy. The seismic data did not provide a sufficient level of resolution to correlate individual sandstone beds with any confidence but did provide additional evidence for the existence of isolated sandstone 'pods' in the Jauf and Tawil formations, and more laterally widespread sandstones above this. Without using the chemostratigraphic, biostratigraphic and seismic data as part of a multidisciplinary approach to reservoir correlation, these techniques would have been limited in their usefulness.

5.2.3 Case Study from the Permo-Carboniferous Unayzah Group, Central Saudi Arabia

Craigie and Polo (2017) used a combination of chemostratigraphy, sedimentology and, to a limited extent, biostratigraphy, to propose a correlation scheme for the Unayzah Group and overlying Basal Khuff Clastics Member encountered in six wells in central Saudi Arabia. A total of 186 core and 863 cutting samples were analysed, most of which were derived from sandstones. Some mudrock samples were also taken, as were carbonate samples from the Khuff Formation occurring immediately above the Basal Khuff Clastics Member.

The Permo-Carboniferous Unayzah Group forms significant sandstone reservoirs in Saudi Arabia and is divided into the lowermost Juwail, and uppermost Nuayyim, formations. The Juwail Formation is further divided into the Ghazal and Jawb members, with the former occurring at the base and dominated by sediments deposited by glacial processes. The overlying Jawb Member mainly comprises fluvial outwash and lacustrine sediments associated with the terminal melting of the ice sheets. A twofold subdivision also exists in the Nuayyim Formation, with the members labelled Wudayhi and Tinat in ascending order. The Wudayhi Member consists of red siltstones and silty sandstones deposited in ephemeral (playa) lakes in an arid continental setting. This interval also contains occasional coarser-grained sandstones showing fluvial and aeolian affinities. The Tinat Member is identified at the top of the Nuayyim Formation and is characterized by a fluvial dominated succession of sandstones with subordinate mudrocks. Aeolian and occasional shallow-marine sediments have also been recognized in this member, which is truncated at the top by the pre-Khuff unconformity.

The Basal Khuff Clastics Member succeeds the Unayzah Group and consists of a thin transgressive lag followed by a regressive pulse of offshore marine siltstones, storm-influenced shoreface sandstones and coastal plain sediments of mixed clastic lithologies. These sediments are overlain by limestones and dolomites of the Khuff Formation.

For further reading on the sedimentology and stratigraphy of the Unayzah Group and Basal Khuff Clastics Member, the reader is referred to the works of Wender et al. (1998), Al-Husseini (2004), Melvin et al. (2005) and Price et al. (2008).

In the study of Craigie and Polo (2017), the mineralogical affinities of elements were established using statistical and graphical techniques. The chemostratigraphic scheme was based on variations in Nb/U, Zr/Nb, Zr/Ti, Zr/Th, Ta/U, Zr/Ta and Ca. With the exception of Ca (concentrated in carbonate minerals, particularly calcite), these parameters reflect changes in provenance. In addition to this, the intensity of subaerial weathering was measured using the Al/(K+Ca+Mg+Na) and Y/Ho ratios.

Details of the sandstone scheme are summarized in Fig. 5.11 and consists of a hierarchical order of three zones, five subzones and nine divisions. The zones are labelled C1, C2 and C3 in ascending stratigraphic order and are defined by progressively higher values of Nb/U. It is also noted that Zr/Nb ratios are generally lowest in C3.

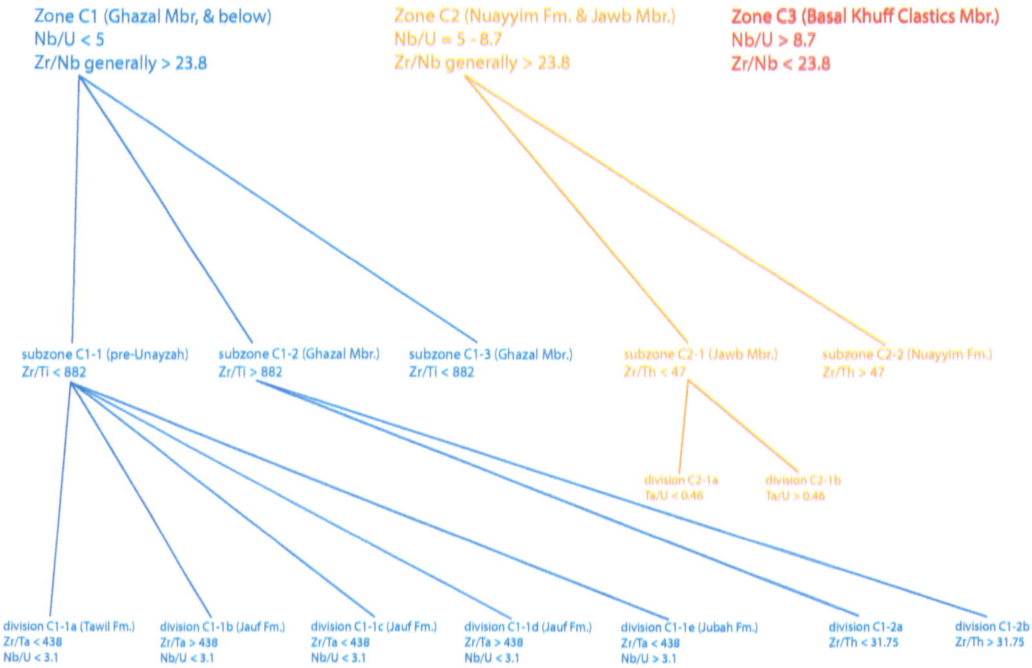

Fig. 5.11 Summary of the principal geochemical characteristics of chemozones identified in the sandstone dataset (after Craigie and Polo 2017)

Of the three C1 subzones, C1-1 and C1-3 occur at the base and top of the zone respectively, and yield lower Zr/Ti values than in the intermediate subzone C1-2. A broad twofold subdivision exists for zone C2, with C2-1 identified at the base and producing lower ratios of Zr/Th than in the overlying subzone C2-2. Five divisions occur in subzone C1-1, their differentiation being related to changes in Zr/Ta and Nb/U. Of the two subzone C1-2 divisions, C1-2a occurs at the base and is characterized by lower Zr/Th values than in the overlying division C1-2b. The divisions of subzone C2-1 are labelled C2-1a and C2-1b in ascending stratigraphic order, with the latter yielding the highest Ta/U. Figure 5.12 illustrates the correlation of chemozones between two of the study wells. Although chemozones C2-1a, C2-2 and C3 are correlative between the two study wells, this does not hold true for the subzones/divisions of zone C1, or for division C2-1a. Subzone C1-1 (and associated divisions) are absent in Well 5, exemplified by relatively high Zr/Ti values (> 882) of C1 in this study section. This subzone may exist, however, in the unsampled interval immediately below C1-2a in this well. Subzones C1-2a and C1-2b are identified in Well 5, with the latter defined by subtlety, though consistently, higher values of Zr/Th. It is not possible to subdivide subzone C1-2 with any confidence in well 4 as Zr/Th ratios are more variable. The reasons for this inconsistency are unclear, but may relate to somewhat subtle difference in provenance or even depositional environment between the locations of the two study wells. Division C2-1a generally produces lower Ta/U values than in C2-1b in Well 4. By contrast, Ta/U ratios are higher than 0.46 in C2-1 in Well 5, inferring the absence of division C2-1a, probably as a result of erosion/non-deposition on a local scale.

The principal geochemical characteristics of each chemozone recognized in the mudrock scheme are summarized in Fig. 5.13. The zones are labelled M1, M2, M3 and M4 in ascending stratigraphic order, with M1 characterized by lower Nb/U values than in M2. The zone M3 samples yield lower Zr/Nb than in M1 and M2, whilst zone M4 is defined by the highest

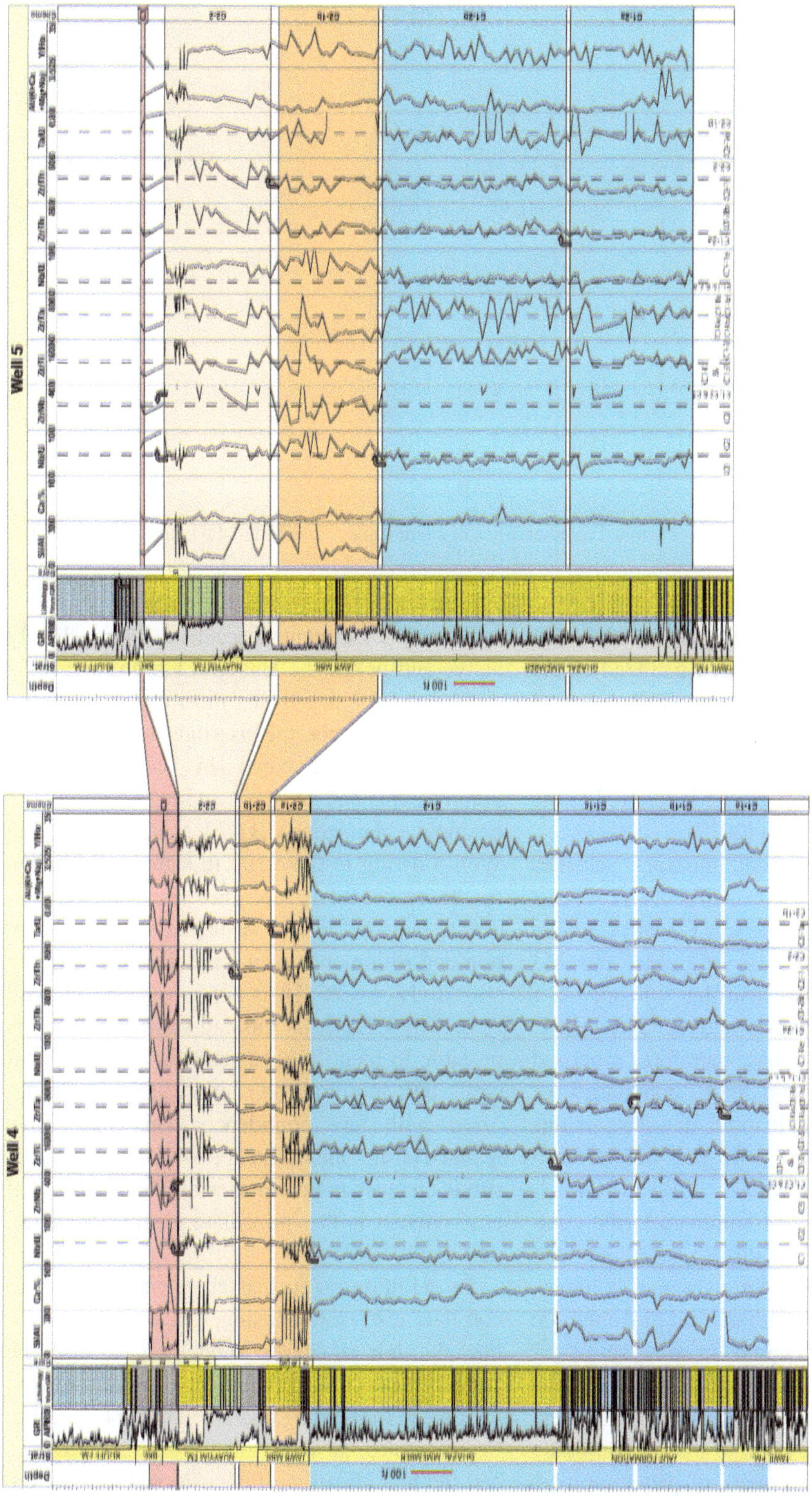

Fig. 5.12 Chemostratigraphic correlation proposed for wells 4 and 5 based on geochemical data acquired from sandstone core and cuttings samples. All depths are log depths in feet (after Craigie and Polo 2017)

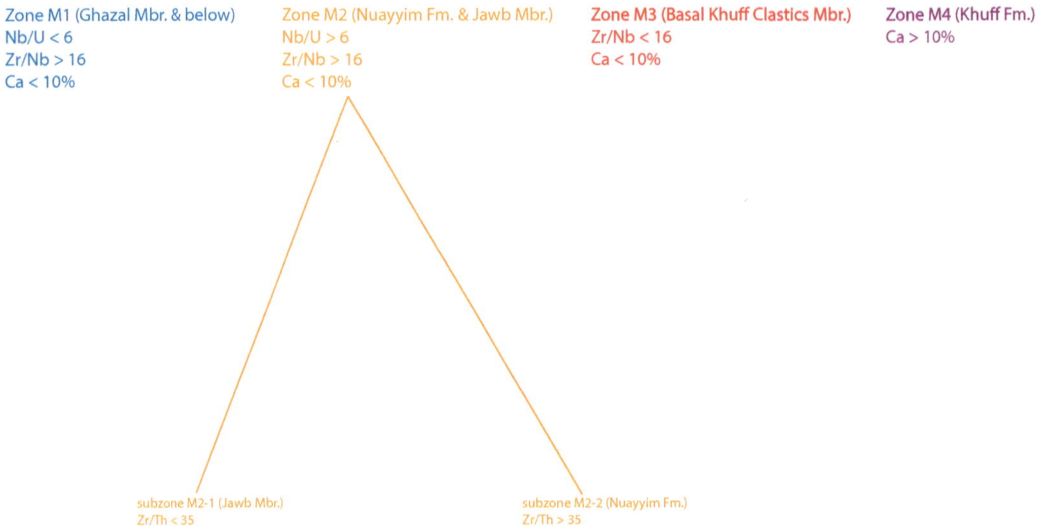

Fig. 5.13 Summary of the principal geochemical characteristics of chemozones identified in the mudrock dataset (after Craigie and Polo 2017)

concentrations of Ca. No figures of key element/ratio profiles have been included in the present work to illustrate the definition of chemozones for the mudrock dataset, owing to the low number of analysed mudrocks. For further details, however, the reader is referred to the study of Craigie and Polo (2017).

When the sandstone and mudrock schemes were compared by Craigie and Polo (2017), it became apparent that zones C1, C2 and C3 were roughly time equivalent to M1, M2 and M3 respectively. It was also clear that subzone C2-1 could be related to M2-1, and C2-2 to M2-2. A 'combined' scheme was, therefore, proposed to honour the top and base boundaries of chemozones identified in both datasets. In this scheme the suffix CB was put before each chemozone (e.g. CB1, CB2, CB3 m CB2-1, CB1-1a etc.). Figure 5.14 summarizes the comparison between the mudrock, sandstone and combined frameworks. Stratigraphic columns, including formations/members, and biozones, are also included in the diagram. In some studies, it is possible to use biostratigraphy to correlate the formations and members of the Unayzah Group, but this was not achievable in the present study owing to the limited recovery of palynomorphs. Palynology did prove useful, however, in age-dating sediments occurring immediately above and below the Unayzah Group.

Figure 5.15 shows that most chemozones are correlative between the six study wells, with the absence of chemozones in particular wells mainly explained by the sampling strategy. For example, chemozone CB1-2 was not identified in Well 3 owing to the c.100 ft. sample gap below the base of CB2-1b. In other instances, chemozones may be absent as a result of erosion/non-deposition on a local scale. This may explain why division CB2-1b is not recognized in Well 5, or why CB2-1a is missing from Well 2. It is possible to identify zones and subzones in most wells but, in some, divisions cannot by recognized with any confidence. This may have resulted from the existence of local and very subtle changes in provenance and/or depositional environment in the immediate vicinity of particular wells.

In addition to using chemostratigraphy to correlate reservoir sandstones, it was also possible, through stochastic modelling, to use the technique in conjunction with lithostratigraphy, to place formation and member boundaries. Sedimentological data, in the form of core logs and the identification of depositional facies were used to recognise formations/members in each

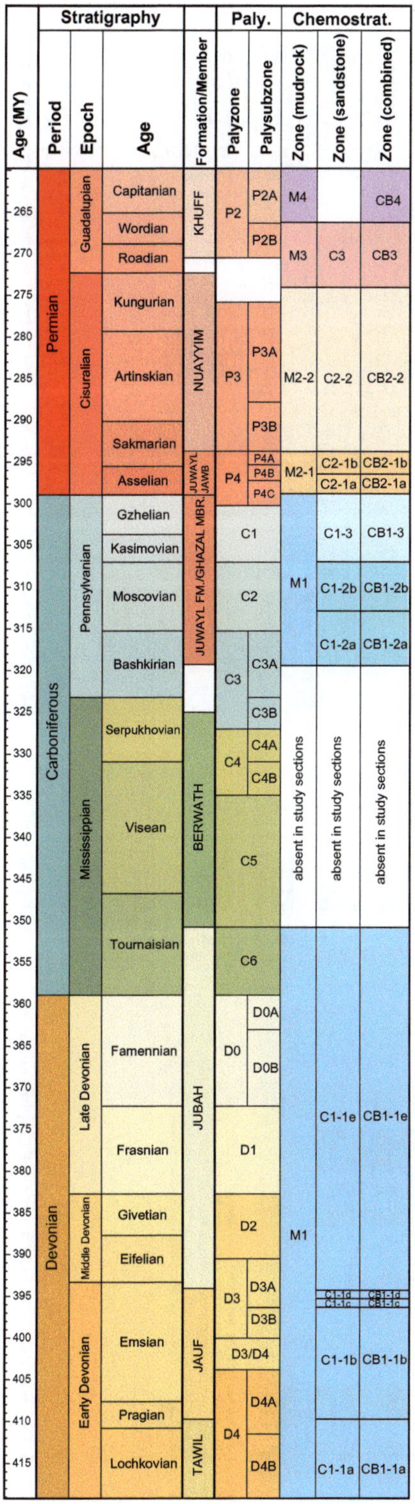

Fig. 5.14 Generalized comparison between lithostratigraphic, biostratigraphic and chemostratigraphic frameworks (after Craigie and Polo 2017)

core. Unfortunately, the relatively low coverage of core in each well, meant that it was not possible to use this data in isolation for correlation. This was possible, however, when the results of the sedimentological and chemostratigraphy study were compared. The cored sections of wells 2, 3, 4, and 5 revealed the following associations:

Subzone CB2-1 = within the Jawb Member.
Subzone CB2-2 = within Nuayyim Formation.
Zone CB 3 = within Basal Khuff Clastics Member.
Zone C4 = Khuff Formation carbonate immediately above the Basal Khuff Clastics Member.

Biostratigraphic control was considered poor over the study intervals but the following observations were made when a comparison was drawn between the, albeit limited, palynological data, and chemostratigraphic scheme. In the sections below the Unayzah Group (Ghazal Member) in Wells 3 and 4, palynozone D4B was used to identify the Tawil Formation which is associated with chemozone CB1-1a in the latter well (zone CB1 cannot be differentiated in Well 3 owing to the low number of analysed samples). Figure 5.15 also illustrates an association between Palynozone D3A (Jauf Formation) and division CB1-1c in Well 1. In well 5, sparse recovery of palynomorphs were identified towards the base of division CB2-1b. Along with the aforementioned palynological data, this provides further confirmation of the link between this chemozone and the Jawb Member. Palynozone P2A (indicator of Basal Khuff Clastics Member) occurs within zone CB3 in Well 4, whilst P1 and P2 are recognized in the overlying Khuff Formation carbonates and chemozone CB4 in Well 5.

By utilizing the aforementioned associations between particular chemozones and specific formations/members, it was possible to use chemostratigraphy and trends in the GR log to place stratigraphic boundaries. For example, the boundary between the Jawb Member and Nuayyim Formation was placed to coincide with a distinct upward decrease in GR, occurring close

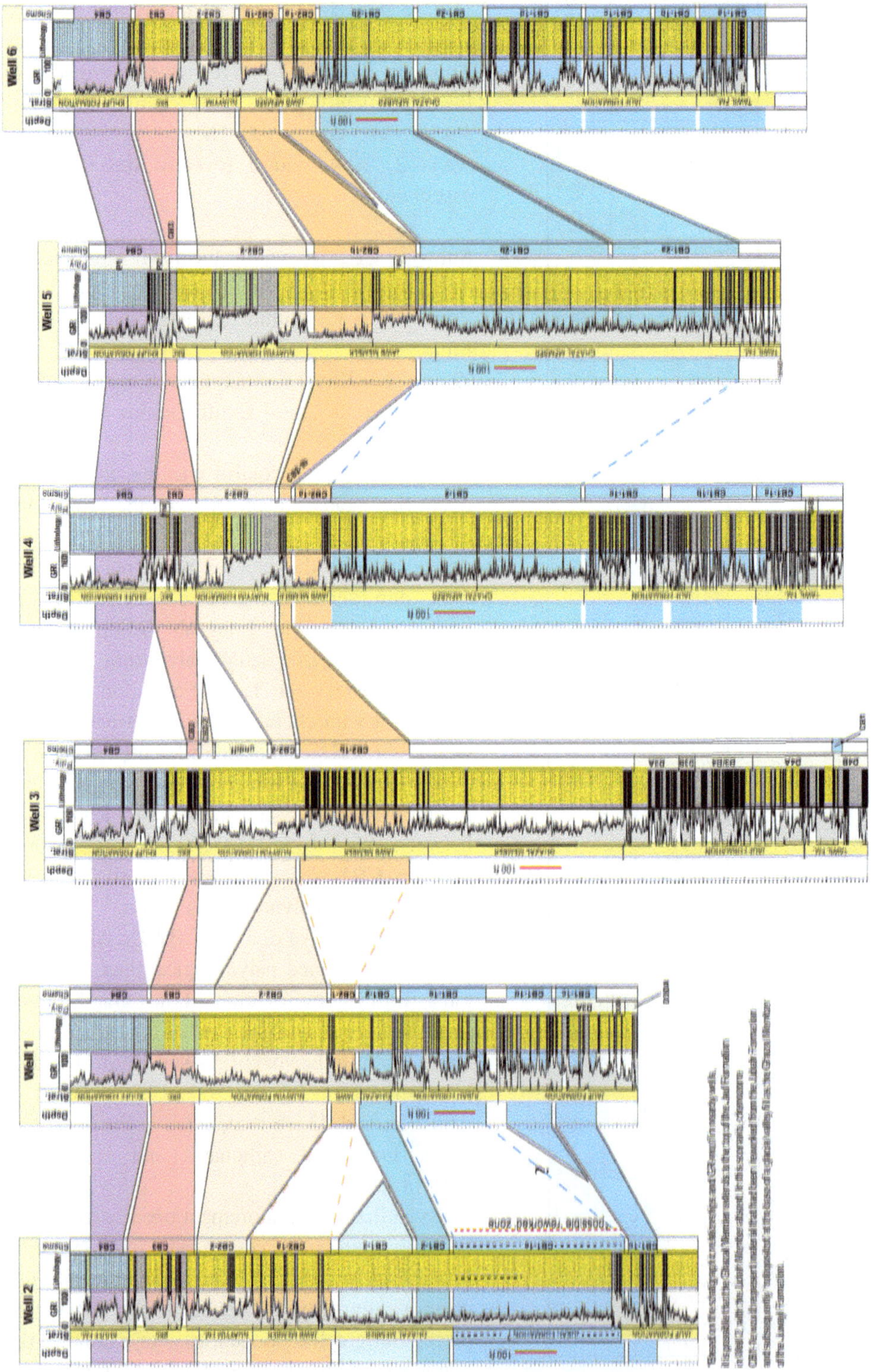

Fig. 5.15 Summary of the chemostratigraphic correlation scheme proposed for wells 1-6, with biostratigraphic data and stratigraphic boundaries also provided. All depths are log depths in feet (after Craigie and Polo 2017)

to the top of chemozone CB2-1. Similarly an upward decrease in GR was used to define the Ghazal:Jawb Member boundary at depths very close to the top of zone CB1. The Jawf Formation is represented by a 'ratty' wireline log response, and is thought to encompass chemozones CB1-1b, CB1-1c and CB1-1d. The only major source of uncertainty is found towards the base of Well 2, where it is unclear whether the Jubah Formation or the Ghazal Member succeeds the Jauf Formation.

In summary, chemostratigraphy was used in conjunction with reservoir sedimentology, biostratigraphy and trends in the GR log to place formation and member boundaries. In addition to this, it was possible to employ chemostratitgraphy to provide a higher level of resolution. For example, three chemozones were identified in the Ghazal Member, whilst two occur in the Jawb Member. Relating to this, the chemostratigraphic study showed that particular section were missing in certain wells as a result of erosion/non-deposition (e.g. absence of CB2-1a in Well 5)—something that could not have been realized without using this technique. It is beyond the scope of this chapter to discuss the sedimentological or biostratigraphical data of this study in detail, but the reader is referred to the study of Craigie and Polo (2017) for more information.

5.2.4 Case Study of the Ordovician Sarah Formation, North West Saudi Arabia

The following study, completed by Craigie et al. (2016b), utilized chemostratigraphy, reservoir sedimentology, borehole image and seismic data to provide a correlation scheme for the Ordovician Sarah Formation, encountered in 17 wells in northwestern Saudi Arabia. For brevity, the details of facies analysis and the chemostratigraphic schemes proposed for sandstone and mudrock lithologies are not covered in much detail in the following paragraphs. It is noteworthy, however, that the definition of chemozones was based solely on elemental ratios reflecting changes in the distribution of heavy minerals, and hence, sedimentary provenance.

In Ordovician times, the Arabian Plate occupied an intra-cratonic location at relatively high latitudes (c. 45° S), with a passive margin interpreted to have been present between this part of the Gondwana margin and the Paleo-Tethys (Sharland et al. 2001). The plate drifted into lower southern latitudes of less than 40° S during this time, and an ice sheet covered much of the African platform and Western Arabia (Vaslet 1989, 1990; Eyles 1993; Sutcliffe et al. 2000). The Sarah Formation of NW Saudi Arabia comprises sandstone dominated sediments deposited by outwash turbidite channels and fans emanating from the terminal parts of glaciers in response to glacial retreat (Craigie et al. 2016b). The pressurized release of water during glacial maximum or early glacial retreat, resulted in the cutting of deep incisions into the underlying Qasim Formation, eroding a network of paleovaleys (Melvin 2009). Large quantities of sand were deposited in the form of pro-glacial ribbon channels and fans at the terminus of tunnel valleys (Rees 2015).

Figure 5.16a shows a map of the study area of Craigie et al. (2016b), whilst Fig. 5.16b illustrates a two way time basemap for this region showing deeper basin development towards the NE, with thinning in more proximal southwesterly locations. The basal part of the Sarah Formation of many wells is dominated by sandstones deposited in a series of randomly oriented turbidite channels, the locations of which have been mapped using seismic data. The sites of the study wells in relation to these channels are presented in Fig. 5.16c. Some wells penetrate these lowermost Sarah Formation channels (e.g. wells 2 and 4), whilst others are located in interchannel areas (e.g. wells 5 and 17). Figure 5.17 shows the thickest development of the Sarah Formation occurring in wells that penetrate turbidite channels such as wells 2, 4, and 6. The sedimentological data show these to be dominated by climbing dune cross bedded channel fill sandstones (facies association FA2 of

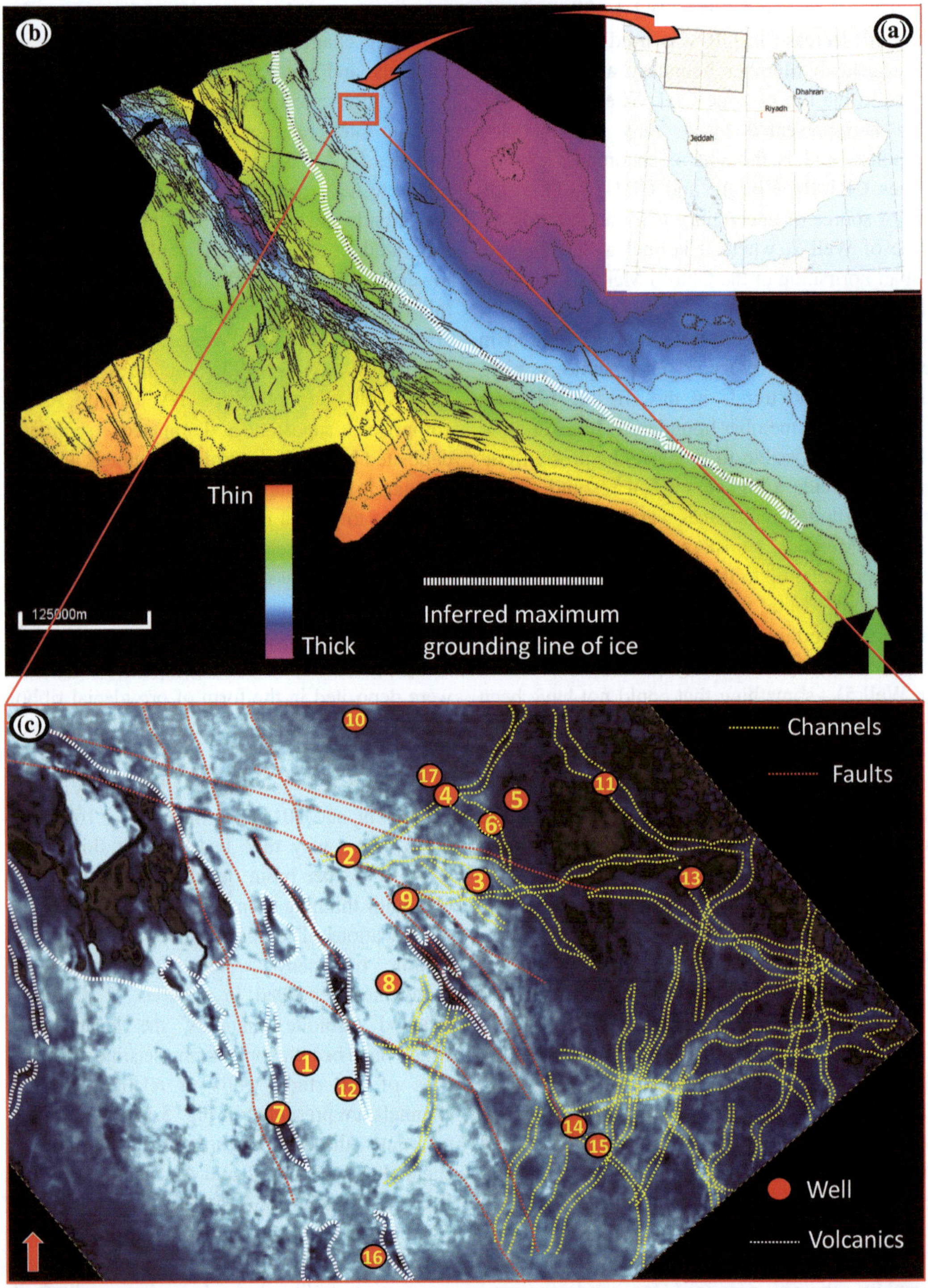

Fig. 5.16 Figure **a** shows a map of the study area, whilst **b** illustrates a two-way time basemap. **c** Shows the location of study wells in relation to channels identified using seismic data (after Craigie et al. (2016b)

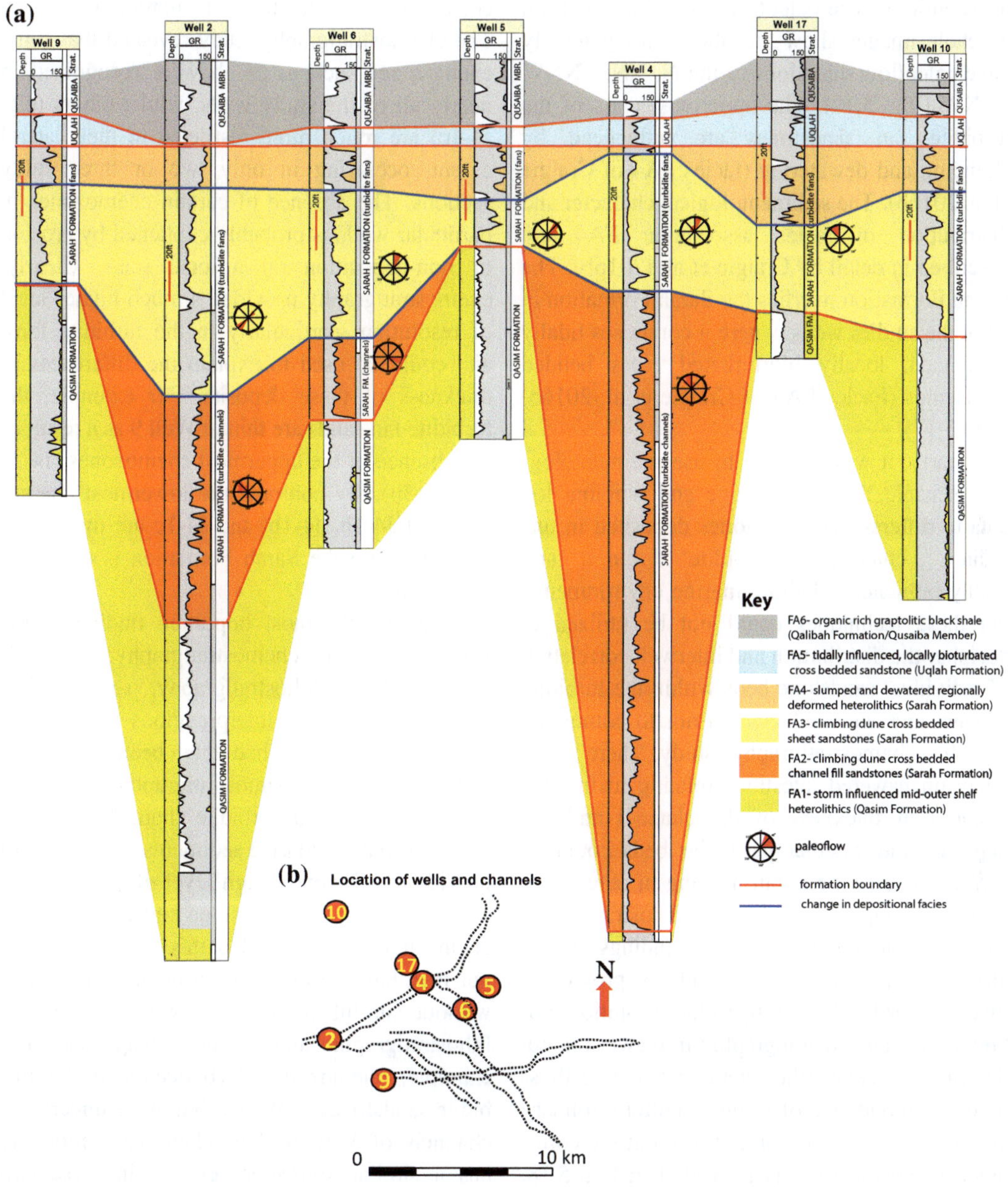

Fig. 5.17 **a** Shows the correlation scheme proposed for a selection of study wells prior to the chemostratigraphy study, whilst **b** illustrates the location of the wells in relation to turbidite channels at the base of the Sarah Formation (modified, after Craigie et al. 2016b)

Craigie et al. 2016b). The interpretations of borehole image data indicate the orientation of the channels to be entirely random, a fact that is supported by the seismic data (Figs. 5.16 and 5.17). Another feature of the channel sandstones is that their GR response shows a general overall upward fining pattern (Fig. 5.17). The channel sandstones are succeeded by climbing dune cross bedded sheet sandstones deposited by turbidite fans (facies association FA3 of Craigie et al. 2016b). In inter-channel wells, such as wells 15, 10 and 17, these account for the Sarah Formation in its entirety. The paleoflow data obtained from borehole images show that these fans generally have paleoflow directions to the N, NNE, NNW or NW (Fig. 5.17). The uppermost parts of the turbidite fan sandstones are influenced by slumping and dewatering (facies FA4 of Craigie et al. 2016b). The sedimentological character and distribution of facies association FA4 are described in detail by Craigie et al. (2016b). The Ulqah Formation overlies the Sarah Formation in all of the studies wells, where it comprises tidally influenced, locally biorturbated cross bedded sandstones (Facies FA5 of Craigie et al. 2016b) (Fig. 5.17).

Clearly, it was possible to use sedimentological, seismic, borehole image and wireline log data to differentiate sandstones deposited in turbidite channel, turbidite fan and tidally-influenced shallow marine environments, but these techniques could not be utilized to decipher the distribution and interwell correlation of individual sandstone beds within the turbidite channel and fan sandstones. Prior to commencing the chemostratigraphy study, there were many unanswered questions relating to the differences in thickness of the channel and fan deposits, and whether particular beds could be present in some wells but missing in others.

The principal aim of the chemostratigraphy study was to complement the findings of the aforementioned techniques and to provide a much higher level of stratigraphic resolution. For brevity, the chemostratigraphic framework is not described in detail (the intention being to illustrate the importance of using a multidisciplinary approach to reservoir correlation), but the correlation of chemozones is presented in Fig. 5.18. The turbidite channels are represented by a thick development of sandstones but the seismic data indicates that these are not laterally widespread. Chemozone J3 occurs at the base of these channel sandstones in wells 2 and 3 but, as these wells penetrate different channels, J3 cannot be correlated between these wells. The channel sandstones of Well 6 comprise chemozones J4-1a and J4-1b but again, as these channel sands are isolated in their distribution, they cannot be correlated on wider field or subregional scales. The chemostratigraphy study revealed that some chemozones such as J4-3b were recognized in nearly all of the study wells, whilst others (e.g. J4-1c) are much more restricted in their lateral extent, occurring in only two or three study sections. The absence of certain chemozones in particular wells is probably explained by erosion or non-deposition on a local scale. Clearly, chemostratigraphy provided a much higher level of resolution, particularly in the turbidite fans, and could be used to explain the differences in thickness of these deposits. For example, the turbidite fan sands are thin in Well 9 as a result of the absence of the uppermost chemozones (J4-3a and J4-3b). By contrast, the lowermost chemozones of J4-1b, J4-1b′ and J4-1c are missing in Well 5 where the Sarah Formation is also relatively thin.

Arguably the most important finding of the study was that chemostratigraphy, reservoir sedimentology, lithostratigraphy, seismic data and borehole image interpretations would have been of limited use had they been applied in isolation. The non-chemostratigraphic techniques were used to define turbidite channels, turbidite fans and shallow marine sediments, but could not provide a sufficiently high level of resolution to correlate individual sandstone beds. Conversely, chemostratigraphy enabled this to be achieved, but may have resulted in erroneous correlations without careful comparisson with the seismic, lithostratigraphic and borehole image data. For example, chemozone J4-1b occurs in the turbidite sandstones of Well 9, but in the underlying channels of Well 6. This chemozone probably has a similar source/provenance in these two localities, but was deposited at an earlier time in the latter well and cannot be correlated between the two. Similarly, chemozone J3 is defined using the same geochemical criteria in wells 2 and 3 but occurs in separate channels in each well and cannot be correlated. It is hoped that this study provides a very good example of why a multidisciplinary approach should be adopted when attempting to correlate reservoir sandstones.

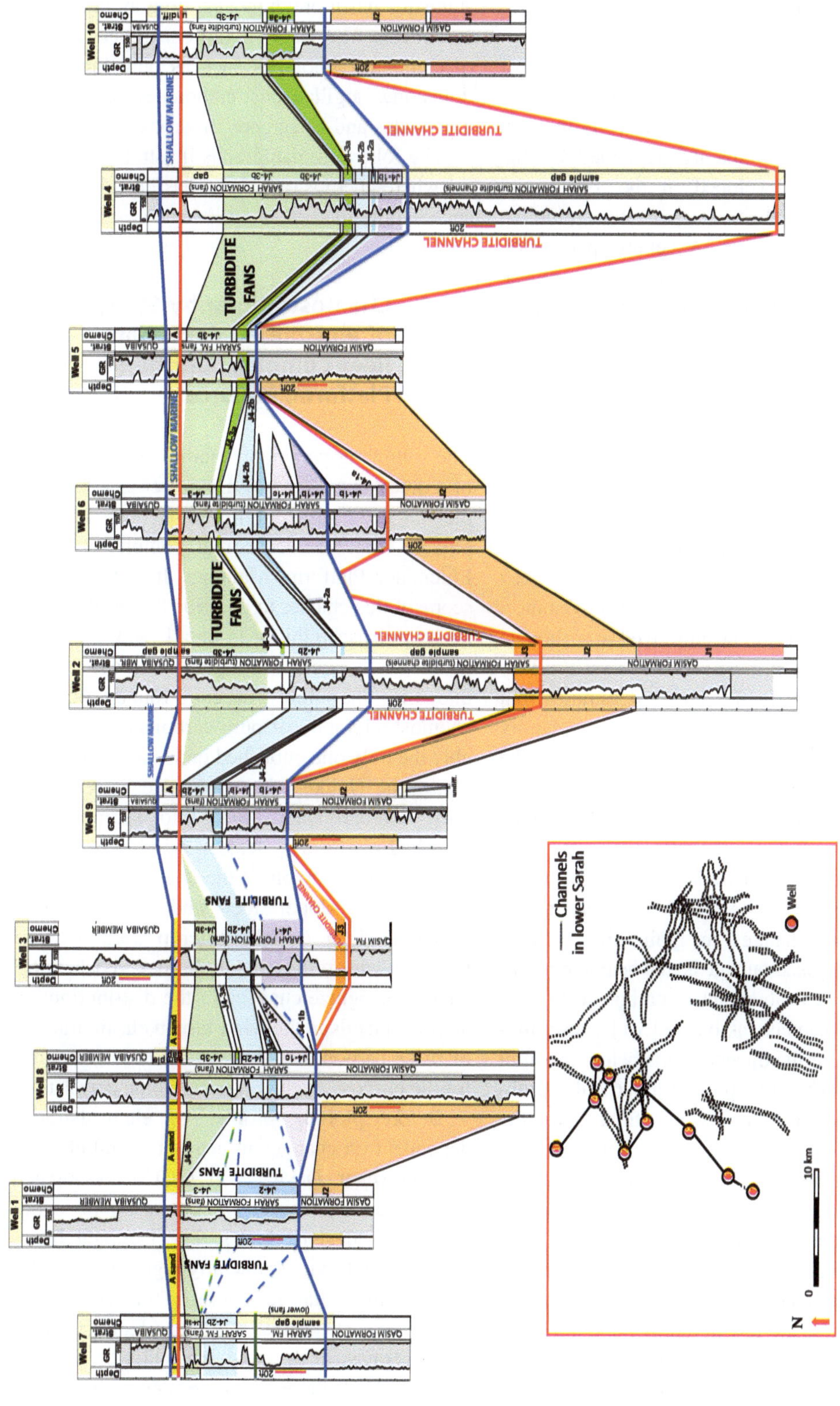

Fig. 5.18 Summary of the chemostratigraphic correlation scheme proposed for ten of the study wells. See text for further explanation. All depths are log depths in feet (after Craigie et al. 2016b)

5.3 Carbonate Reservoirs

One unfortunate aspect of the study of chemostratigraphy is that far less projects have been completed on carbonate reservoirs than on clastic ones. This is mainly explained by the technique being developed in the 1970s and 1980s in the North Sea where the principal reservoirs are sandstones. A general perception developed that chemostratigraphy could only be used in clastic sediments. Studies have, however, been completed on carbonate reservoirs (e.g. Jorgensen 1986; Davies et al. 2013, Craigie 2015a, b; Vishnevskaya et al. 2015; Eltom et al. 2017), providing definitive proof that chemostratigraphy can be used to correlate both types of reservoir. The approach is slightly different to that employed for clastic sediments, where the main objectives are to define chemozone boundaries based on trends in key elements/ratios relating to changes in provenance. This approach can also be employed in carbonate reservoirs, provided there is a sufficient quantity of clastic material in the analysed samples, but other factors, such as changes in the type, quantity and chemistry of the carbonate fraction should be considered when proposing a correlation scheme.

A significant advantage of using chemostratigraphy on carbonate sediments is that variations in elements/ratios linked with the heavy mineral fraction are less likely to be influenced by changes in grain size than in sandstones. A disadvantage is that heavy minerals are normally less abundant in carbonate lithologies and the elements associated with these minerals may occur in concentrations below the limit of detection. This may present a particular problem where XRF analysis is utilized, as the detection limits of most trace elements and REE are much higher than those of ICP. A further challenge when using XRF is that it is often very difficult to achieve good calibration methods for carbonate lithologies. Standard reference materials can be purchased for limestones and dolostones, but these are few in number and normally have very 'pure' compositions. It is advisable that published standard reference materials are used in conjunction with non-published/internal standards when building a calibration method for carbonates. The latter should encompass a range of carbonate lithotypes such as pure limestone, dolostone, argillaceous carbonates, arenaceous carbonates and calcareous mudrocks.

The following paragraphs illustrate case studies where chemostratigraphy has been used very successfully to correlate carbonate reservoirs.

5.3.1 Chemostratigraphy Study of an Isolated Carbonate Reef Complex, Llucmajor Platform, Majorca

The following paragraphs summarize the results of a study completed by Davies et al. (2013) on the magnetic susceptibility and chemostratigraphy of Upper Miocene reef complexes of the Llucmajor Platform, Mallorca. It is beyond the scope of this chapter to describe the results of the magnetic susceptibility in detail, with the focus being on the chemostratigraphy study. This is one of the few papers found on the chemostratigraphy of pure carbonates and probably the only one published on isolated carbonate platforms. Unlike other carbonate systems, the only non-carbonate input into sediments of detached platform systems is wind-blown material. The study involved the analysis of samples taken from slope, reef and lagoon facies encountered in 6 field outcrop sections. The intention was to the use whole rock geochemical data to recognize changes in the distributions of heavy minerals in the wind-blown clastic fraction of these carbonates, and to use these variations for the purpose of chemostratigraphic correlation.

Before plotting profiles of elements/ratios dependent on changes in heavy mineral chemistries, however, Davies et al. (2013) investigated the differences between the chemistry of lagoonal, reef and slope facies. The lagoonal facies were shown to yield generally higher average Si, Al, Fe and K than reef core or fore reef facies, irrespective of the fact that these facies were of different ages in the field area (Fig. 5.19). This suggests that there is some degree of facies control on the terrigenous-related elemental data,

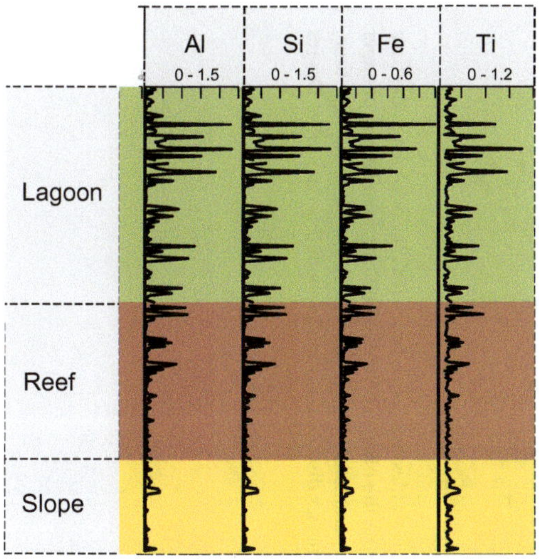

Fig. 5.19 Selected major element values recorded in reef, lagoon and slope environments (modified, after Davies et al. 2013)

with much higher non-carbonate inputs recorded by the lagoonal facies. For this reason, Davies et al. (2013) treated these facies separately when producing a chemostratigraphic framework.

Profiles are displayed for the lagoonal facies on the right side of Fig. 5.20, with samples from field locality PNC being stratigraphically older and plotting below the ones from section SGW. Samples from the former interval clearly yield higher Cr/Nb and Cr/Zr ratios than in the latter. Figure 5.20 also shows profiles plotted for the reef core facies, the oldest sediments being at the base (PC section), working through to the youngest at the top (CB section). There are upward step increases and decreases in Cr and Cr/Zr going from the PC to the PNC sections, and from PNC to CO. It is also noted that P concentrations are lowest in PC. There are similarities between the values of key elements/ratios plotted for the CO and CC sections, though the latter generally yields slightly lower levels of P. The CB section is the youngest and normally produces the lowest Cr/Zr ratios.

A significant finding of the study was that, though the lagoonal and core roof facies have different abundances of detrital heavy minerals, and Cr/Zr profiles are plotted at a different scale for each, the trends in this ratio are very similar, showing that the stratigraphically oldest sediments (lagoonal facies of PNC, reef core facies of PC, PNC and CO) are generally defined by higher Cr/Zr values than in the younger ones (SGW of lagoonal facies, CC and CB sections of the core reef facies).

5.3.2 Chemostratigraphy of Cretaceous Sediments Encountered in North Central Rub' al-Khali Basin, Saudi Arabia

The following paragraphs summarize the study of Craigie (2015a) on the chemostratigraphy of Cretaceous carbonate sediments encountered in 11 wells in the North Central Rub' al-Khali Basin, Saudi Arabia. The objective was to propose a chemostratigraphic scheme for these wells and to compare this with a previous lithostratigraphic scheme. The following key elements and ratios were used for chemostratigraphic purposes.

Ca: carbonate minerals.
Al: clay minerals.
Zr: detrital zircon.
P/Y: biogenic phosphate and/or phosphatic heavy minerals versus Y-bearing heavy mineral.
Zr/P: zircon versus biogenic phosphate and/or phosphatic heavy minerals.
Zr/Y: zircon versus Y-bearing heavy minerals.
Zr/Nb: zircon versus Nb-bearing heavy minerals.
Zr: zircon.
Mo: pyrite/anoxia association.

The mineralogical affinities of these key elements and others were established using statistical and graphical techniques. The elements Ca and Al are concentrated in calcite and clay minerals respectively, with variations in these mainly resulting from changes in depositional environment, reflecting changes in sea level. Rudistic limestones, forming in shallow marine environments, yield the highest levels of Ca, whilst calcareous mudrocks, deposited in lagoonal-marginal marine environments during a fall in sea level, produce the highest concentrations of

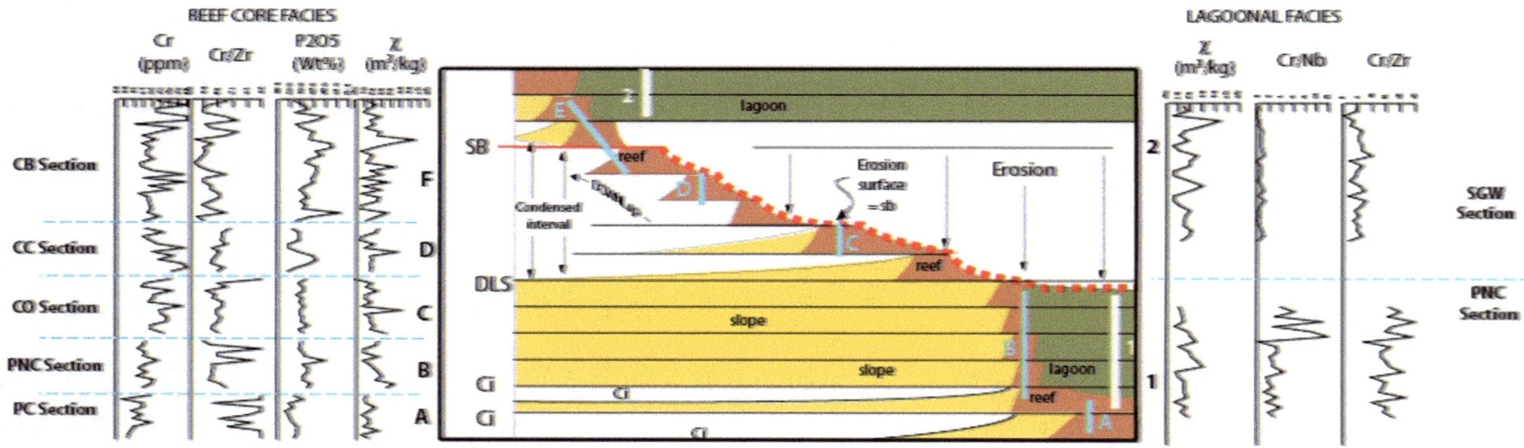

Fig. 5.20 Changes in elemental ratios through time. On the left, samples from reef core facies are arranged so that the oldest reef core facies (PC Section) are at the base and youngest (CB Section) are at the top. Similarly, samples from lagoonal facies are arranged on the right, such that the oldest (PNC Section) are beneath the younger ones from the SGW Section. Data for selected elements and ratios are arranged in such a manner to show how there parameters varied within a single facies through time. Magnetic susceptibility data are also presented in this figure but show no such secular change. For this reason, no further discussion is made about this data (after Davies et al. 2013)

Al. The top of the Raumaila Formation, occurring towards the centre of the study intervals, is the only exception to this general rule, where the deposition of calcareous mudrocks and argillaceous limestones occurred during a slight increase in water depth. The elements Zr, Y and Nb are highly 'stable' and associated with heavy minerals, so variations in Zr, Zr/Nb and Zr/Y model changes in the provenance of non-carbonate minerals. It is thought that heavy minerals were sourced from the Arabian Shield to the West and transported to shallow marine waters by fluvial systems. In the absence of heavy mineral and petrographic data, it was impossible for Craigie (2015a) to determine whether P could be linked with biogenic phosphate, phosphatic heavy minerals or both. Consequently, the P/Y and Zr/P ratios may have be influenced by changes in depositional conditions and/or provenance. However, as variations in P/Y and Zr/P mainly relate to trends in Y and Zr respectively, variations in these two ratios are considered largely to reflect changes in provenance. Elevated concentrations of Mo were used to identify beds deposited under anoxic paleoenvironments.

Figure 5.21 shows the recognition and correlation of chemozones in two of the study wells. The zones were labelled by Craigie (2015a) as C1, C2, C3, C4 and C5 in ascending stratigraphic order, with C1, C3 and C5 generally yielding higher Al and lower Ca than in C2 and C4. Differentiation of zones C2 and C4 is more difficult, though the latter yields lower P/Y values in three of the 11 wells (wells A, B and C). It is not possible to distinguish zones C1, C3 and C5 based on geochemical data alone. In addition to the aforementioned variations in Ca and Al, concentrations of Zr have also been utilized to define zone boundaries, with C1, C3 and C5 characterized by the highest levels of this element. Variations in Zr often represent changes in provenance, but this is unlikely to hold true in the present study where there is a close association between the trends plotted for Al and Zr, inferring that the distribution of this element simply reflects the proportion of detrital material and is highest where the amount of carbonate (and Ca) is lowest, and vice versa.

The definition of chemostratigraphic subzones are based on variations in Zr/P, Zr/Y, Zr/Nb, P/Y and Zr/P, reflecting changes in provenance (Fig. 5.21). It is noteworthy that the subzones of the wells located in the western part of the study area (labelled wells A, B, C, D and E) are defined using different criteria than those of wells 1, 4, 5 and 6 in the East. Wells 2 and 3 are thought to have been located furthest from the hinterland (and the source of clastic sediments), hence the reason that that it is not possible to subdivide zone C2 in these wells. The subzones of zone C4 are also different in the eastern and western wells, suggesting separate clastic inputs for these sets of wells, though evidence of the mixing of sediment sources is apparent in wells D and E. Figure 5.22 summarizes the principal geochemical characteristics of the subzone C2 and C4 subzones, with the correlation of these chemozones and others presented in Figs. 5.23 and 5.24. The directions of sediment transport and generalized paleogeographic reconstructions are shown for zones C2 and C4 in Figs. 5.25 and 5.26 respectively. The precise directions of paleoflow are difficult to determine, but it is postulated that detrital non-carbonate minerals were sourced from easterly flowing fluvial systems in the eastern wells and from ones flowing to the NW in the western wells. The subzones of zones C1 and C5 are defined using the same geochemical criteria in all of the study wells, so it is thought that these were deposited at the same time, with detrital material supplied by a single feeder system.

Chemostratigraphic divisions were only identified by Craigie (2015a) in subzones C1-3 and C5-1. Division C1-3a occurs at the base of subzone C1-3 and produces higher concentrations of Mo than in the overlying division C1-3b. Of the two subzone C5-1 divisions, C5-1a is noted at the base and is defined by lower Zr/Nb ratios than in C5-1b (Fig. 5.21).

A second aim of the study was to compare the chemostrigraphic scheme with the placement of stratigraphic boundaries based largely on lithostratigraphic criteria. Both sets of boundaries are plotted in Figs. 5.23 and 5.24, with a summary of the comparison between the two schemes presented in Fig. 5.27. Clearly, it is possible to use

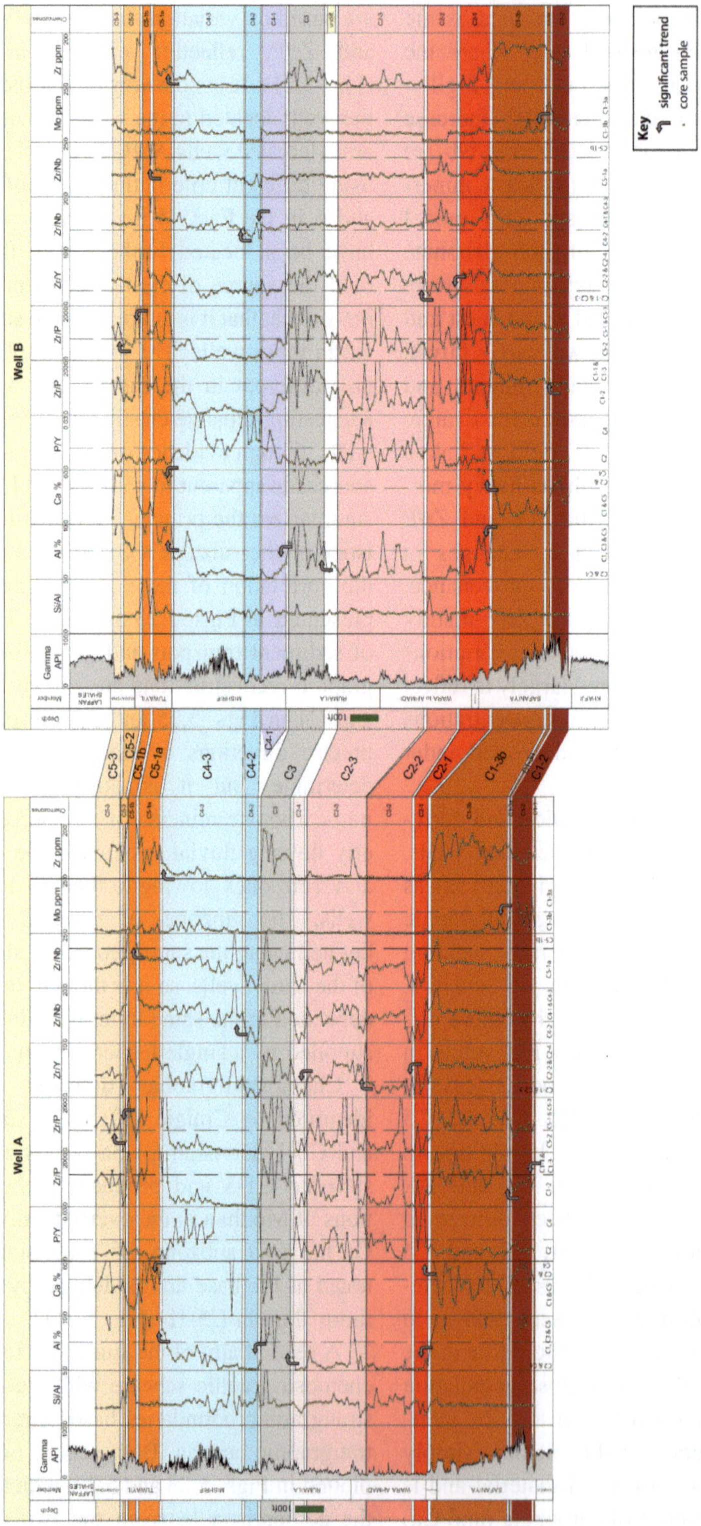

Fig. 5.21 Chemostratigraphic correlation of Cretaceous sediments encountered in Wells A and B. All depths are log depths in feet (after Craigie, 2015a)

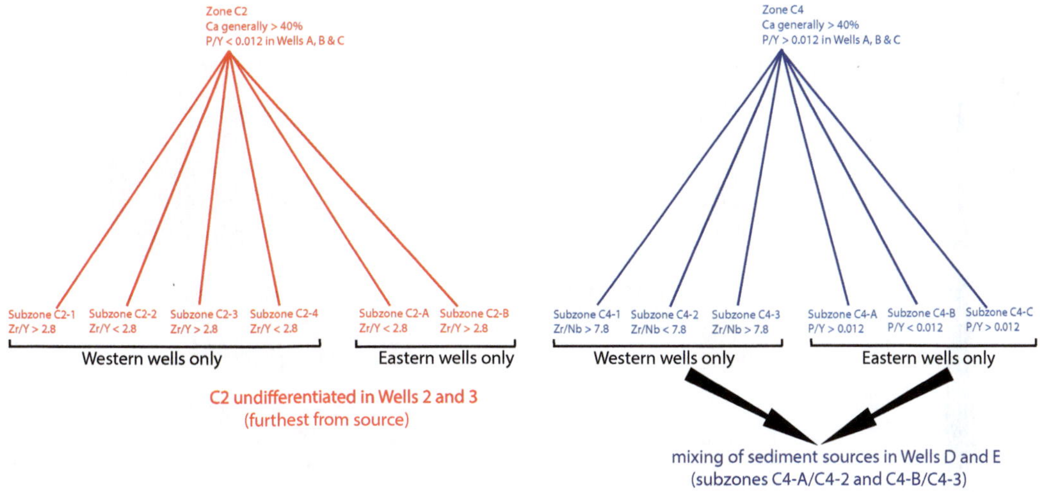

Fig. 5.22 Principal geochemical characteristics of zones C2, C4 and associated subzones. Note the differences in the way subzones are defined in the Eastern and Western wells (after Craigie, 2015a)

chemostratigraphy to define some lithostratigraphic boundaries with a high degree of certainty. For example the C1-1:C1-2, C2:C3 and C4:C5 boundaries approximately coincide with the tops of the Khafji, Rumaila and Mishrif members respectively. In other instances, there are mismatches between the two schemes. The base of zone, C2, for instance, occurs in the basal part of the Maddud Member in wells C, D E, 2, 3 and 4, but in the upper part of the underlying Safaniya Member in wells B, 5 and 6. The base of C2 is placed in the uppermost Mauddud Member in Well A. Another difference mentioned by Craigie (2015a) is that zone C3 is generally linked with the uppermost Rumaila, but extends into the lower part of the Mishrif Formation in wells C, 1 and 3. The reasons for these inconsistencies and others may relate to difficulties associated with the placement of lithostratigraphic boundaries, particularly where core data are unavailable. It is entirely possible that the placement of at least some of the boundaries may be revised in light of the chemostratigraphic study.

Craigie (2015a) also notes that a higher level of stratigraphic resolution was possible using chemostratigraphy than by employing the existing lithostratigraphic framework. For example, chemozones C1-2, C1-3a and C1-3b occur in the Safaniya Member, whilst the C2-1:C2-2, C2-2:C2-3 and C2A:C2B boundaries are present in the Wara to Ahmadi members. In addition to this, it is also possible to use chemostratigraphy to subdivide the Raumaila and Mishrif Members, and the lowermost Aruma Formation.

In summary, significantly less work has been completed on the chemostratigraphy of carbonate sediments and this is reflected by the relatively low number of publications on this subject. However, it is hoped that the works of Davies et al. (2013), Craigie (2015a) and others demonstrate the applicability of the technique in this lithology. It is also hoped that more studies will be published on carbonate reservoirs in the future as the popularity of the technique grows.

5.3.3 Case Studies on Dolostones

Despite their abundance in the rock record, very few chemostratigraphy and inorganic geochemical studies have been performed on dolostones—even less than on limestones. Two notable exceptions are the studies of Shukla (1988) and Holh et al. (2017) which are discussed in the following paragraphs.

Shukla (1988) produced a chemostratigraphic scheme for the Silurian Interlake Formation

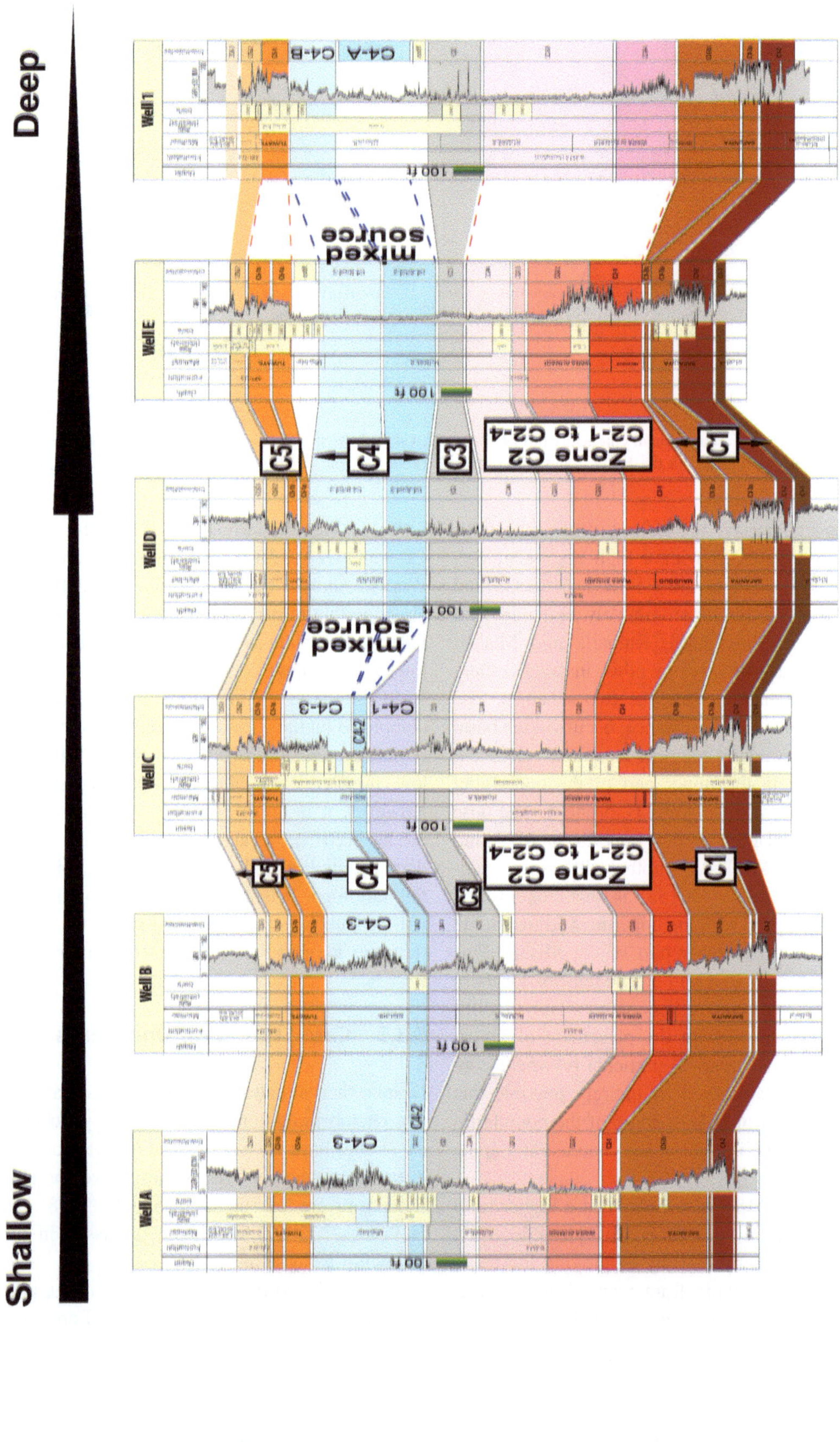

Fig. 5.23 Summary of chemostratigraphic correlation in wells A, B, C, D, E and 1. All depths are log depths in feet (after Craigie 2015a)

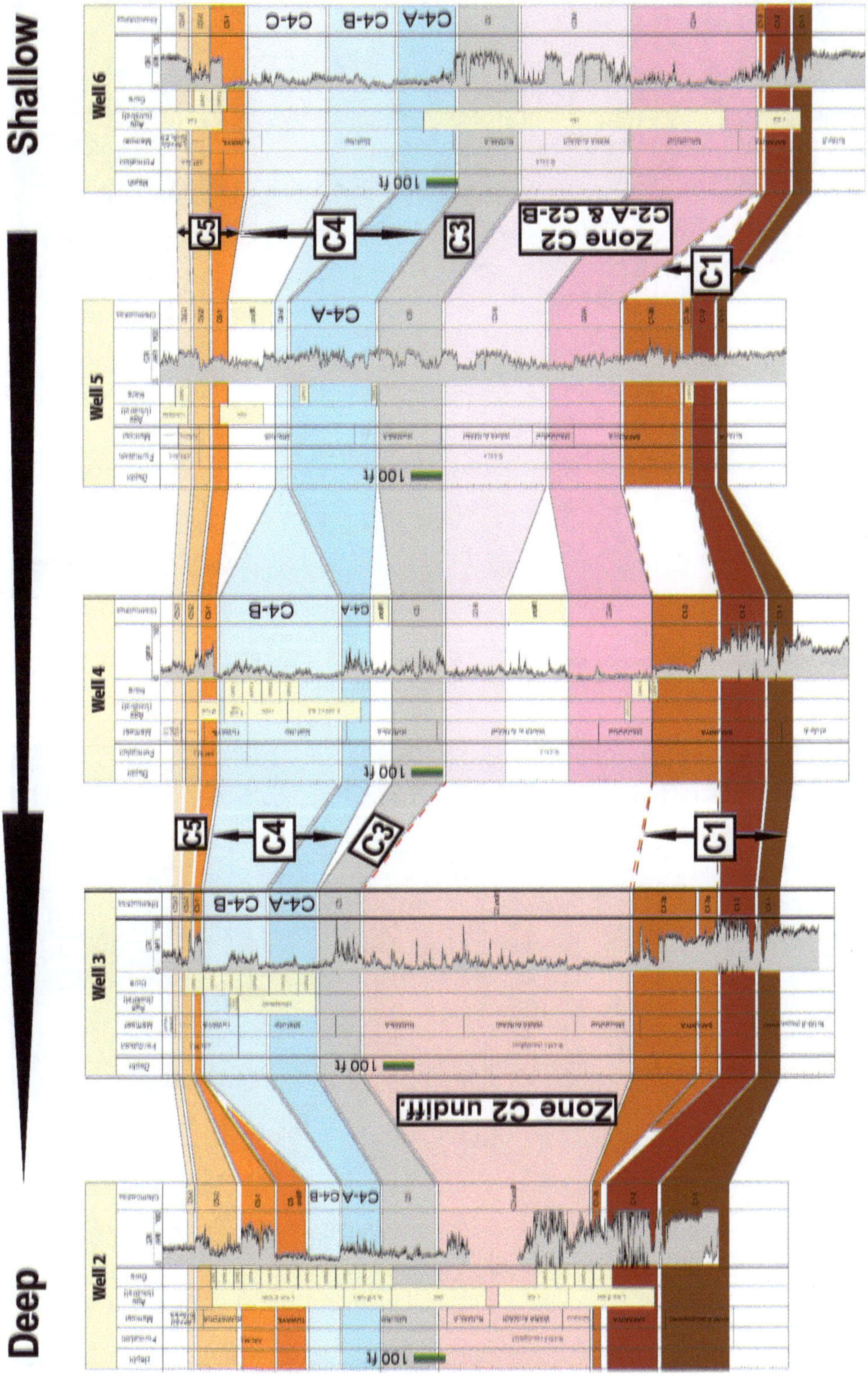

Fig. 5.24 Summary of chemostratigraphic correlation in wells 2, 3, 4, 5 and 6. All depths are log depths in feet (after Craigie 2015a)

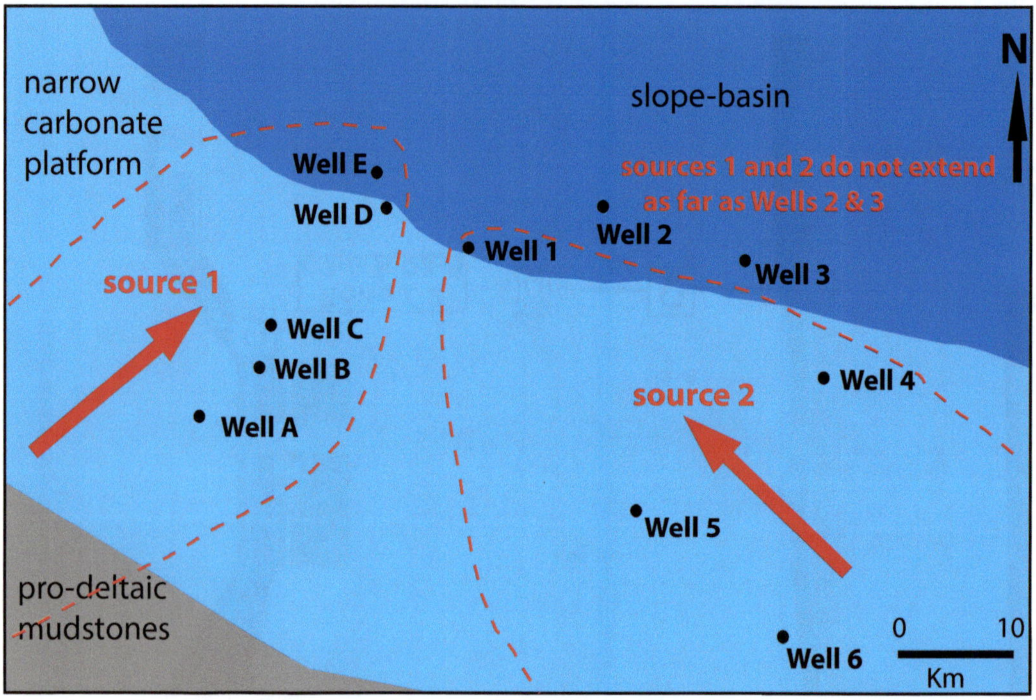

Fig. 5.25 Predicted directions of sediment transport and generalized paleogeographic reconstruction with respect to chemostratigraphic zone C2. Interpretation of depositional environments is based on unpublished work completed by Dr. Chris Reid (Saudi Aramco Geologist) (after Craigie, 2015a)

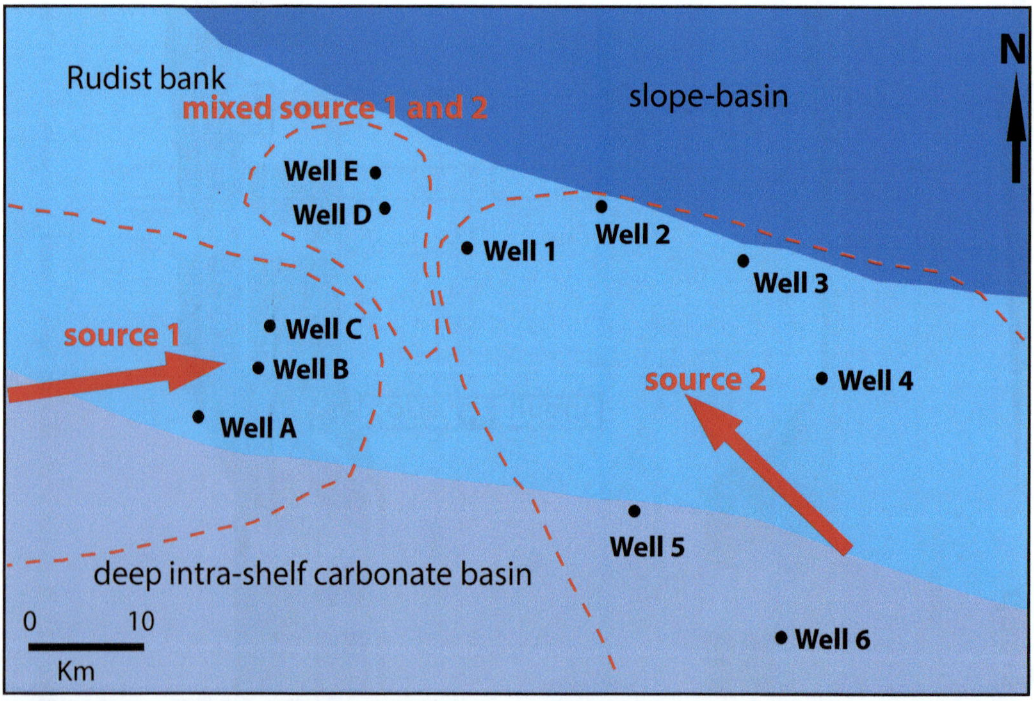

Fig. 5.26 Predicted directions of sediment transport and generalized paleogeographic reconstruction with respect to chemostratigraphic zone C4. Interpretation of depositional environments is based on unpublished work completed by Dr. Chris Reid (Saudi Aramco Geologist) (after Craigie, 2015a)

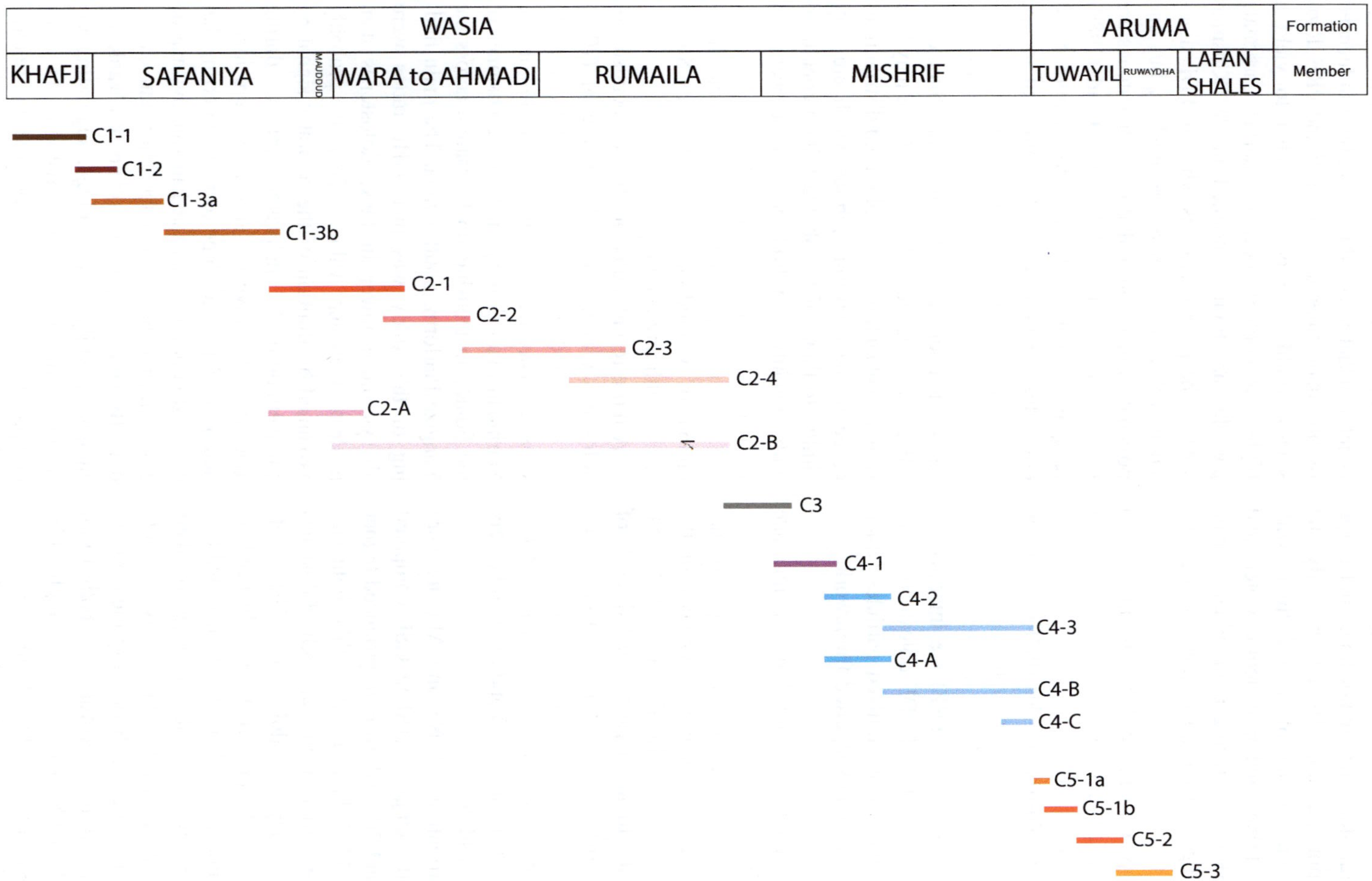

Fig. 5.27 Generalized comparison between the chemostratigraphic and the existing lithostratigraphic schemes (after Craigie 2015a)

encountered in North Dakota. The Interlake Formation comprises a 366 m-thick dolostone unit representing dolomitization of subtidally deposited lime grainstones and minor lime mudstones. Dolomitisation was considered to be early and caused by the mixing of marine fluids with hypersaline brines and/or meteoric water, and hypersaline brines occurring in supratidal settings. The following three dolostone fabrics were noted:

1. Grain-supported fabric (GSF) resulting from replacement of lime grainstones or packstones.
2. Pervasive-dolostone fabrics (PDF) taking the form of crystalline carbonates lacking evidence of depositional textures and fabrics.
3. Particle relic fabrics (PRF) are intermediate between the GSF and PDF dolostones, containing relics of dolomitized particles which are not in a grain-supported framework.

Burial dolomites are noted in the Interlake Formation but are rare.

A significant finding of the study was that these three fabrics showed differences in the concentrations of Sr, Mn, Fe, Na and B. GSF were generally found to yield elevated levels of Mn and Na but low Fe, Sr and B. Conversely, PDF are were characterised by high values of Fe, Sr and B, but low Mn and Na. The PRF produced elevated concentrations of B and Sr but relatively low Mn, Fe and Na.

The distributions of Na and Mn are very similar, both being elevated in GSF compared with PDF and PRF, whilst PDF produced higher levels than PRF. In addition to this, both elements were enriched in clay-rich dolostones compared to clay-poor dolostones. These patterns infer that Na and Mn were influenced by precursor minerals. Shukla (1988) suggested that both elements were probably released by their precursor minerals and became locally available for some time before being exported in an open-system setting. According to both Lippmann (1973) and Shukla (1988), it is highly like that the mechanism leading from ionized dissolved species to an ionic crystal would be:

$$Ca + Mg + 2CO_3^{2-} \rightarrow CaMg(CO_3)_2 \text{ (Solid)}$$

Thus, as with Ca and Mg, Na and Mn also become briefly available as ionic species.

In contrast to Na and Mn, Shukla (1988) concluded that Sr could not have been influenced by precursor mineralogy as GSF and PRF dolostones would have been expected to yield a higher Sr content, because marine carbonate particles are highly enriched in Sr (Kinsman 1969). The precise reasons why the PDF dolostones are enriched in Sr are unclear, but one possible explanation is that these dolostones are finer grained. Veizer (1983) and Land (1980) recognised that early (penecontemporaneous) dolomites are enriched in Sr compared to late (diagenetic) dolomites. Shukla (1988) also notes the significance of dolomite fabric because penecontemporaneous dolomites are finer crystalline than diagenetic dolomites. There may be a similar explanation for the observed distributions of Fe in the dolostone fabrics which are very similar to those of Sr. Shukla (1988) concluded with certainty that both Sr and Fe were not influenced by precursor mineralogy.

In contrast to Sr, Fe, Na and Mn, the distributions of B could not be interpreted by Shukla (1988) with any certainty.

A more recent study on the geochemistry of dolostones was completed by Holh et al. (2017) who evaluated temporal and special variations of seawater chemistry in the aftermath of the Marinoan glaciation by analysing dolostones of the Ediacaran Doushantuo Formation across the Yangtze Platform, South China. The main finding of this study was that Y/Ho ratios were always much lower in these dolostones than modern seawater values. This is probably explained by dilution of the seawater signal by continent-derived meltwater influx during glaciation. Negative Ce anomalies were identified in platform and slope sections, inferring that oxidized conditions persisted in shallow marine environments shortly after the Marinoan glaciation. By contrast, positive or no resolvable Ce anomalies typify basin settings, suggesting the existence of anoxic paleoenvironments. This interpretation is further supported by relative enrichments of redox sensitive trace metals, including V, Mo and U in basinal settings.

The studies of both Shukla (1988) and Holh et al. (2017) were very useful, but focused on relating variations in the geochemistry of dolostones to paleoenvironment and diagenesis, rather than using the data for chemostratigraphic correlation purposes. The author has completed several studies on dolostones occurring in Western Canada and the Middle East region but these were proprietary studies of a highly confidential nature and, as such, the details cannot be discussed in the present publication. It is hoped that more geochemical studies, and chemostratigraphy projects in particular, will be published on dolostones in the future.

5.4 Unconventional (Source Rock) Reservoirs

There are many different types of unconventional hydrocarbon reservoirs including 'tight' sands, 'tight' shales, tar sands, oil shales, shale oil reservoirs and others. Many chemostratigraphy studies have been performed on 'tight' sandstone reservoirs (e.g. Craigie et al. 2016b) where they are treated (in chemostratigraphic terms) in the same way as conventional sandstone reservoirs. The following discussion is related entirely to shale and limestone reservoirs which, in conventional hydrocarbon plays, would be considered as source rocks. The approach in these reservoirs may include, but not be restricted to the following objectives:

(a) Use chemostratigraphy to define chemozone that can be correlated between wells (i.e. the objective of most studies performed on conventional reservoirs).
(b) Use inorganic geochemical data to define zones of high TOC (Total Organic Carbon).
(c) Use inorganic geochemical data to identify potential seal/cap rocks, most of which have high clay contents.
(d) Use inorganic geochemical data to model mineralogy and to identify zones that are more and less prone to hydraulic fracture.

The elements Al, U, Mn, Mo, Zn, Ni, Cu, Co and Zn are considered most useful in identifying brittle zones of high TOC. The elements Mo, Zn, Ni, Cu, Co and V are most closely associated with the development of anoxia, where the preservation of organic matter is usually highest. The strong link between such elements and TOC does not always hold true, however, as was demonstrated by Craigie (2015b). The element U may be influenced by anoxia but also occurs in high abundances in organic matter accumulating in both anoxic and oxic sediments. Concentrations of Mn are usually highest in suboxic environments and low where conditions were oxic and anoxic. Some authors (e.g. Tribovillard et al. 2006) make a clear distinction between anoxic and euxinic paleoenvironments. However, as it is often very difficult to distinguish the two based on inorganic geochemical data alone, no such distinctions are made in case studies described in the present chapter and the word 'anoxic' is used as a general term to described paleoenvironments lacking oxygen.

In addition of having high levels of TOC, the best reservoirs also have low clay contents (typically $< 15\%$) as the presence of clay minerals may inhibit the propagation of man-induced hydraulic fractures and the subsequent 'flow' of hydrocarbons. Consequently, the best reservoirs should be characterised by relatively low levels of Al. Of the clay minerals, smectite is the least favourable as it is prone to swelling. For this reason, unconventional reservoirs are rarely exploited where concentrations of this mineral exceed 4%. Using geochemical data to model the abundances or TOC and specific clay minerals can only be done through stochastic modelling, where mineralogical and TOC data are already available. Under such circumstances, it may then be possible, for example, to relate a particular abundance of Mo to a level of TOC. The abundance of particular clay minerals may also be modelled using such a predictive methodology. The mineralogical affinities of elements, and techniques used to established element:mineral links, are discussed in detail in Chap. 3 of this book.

The study of shales is arguably one of the fastest growing areas of research in the field of chemostratigraphy and the following two case studies have been carefully chosen as the author considers these to represent the most effective use the chemostratigraphy in this type of reservoir.

5.4.1 Chemostratigraphy of Middle Jurassic Unconventional Reservoirs in Eastern Saudi Arabia

The study of Craigie (2015b) involved the ICP analysis of 1032 core and cutting samples acquired from five wells penetrating Middle and Upper Jurassic carbonate sediments, which were found to be organic rich, yielding TOC contents exceeding 15% in places. The objectives of the study were to produce a chemostratigraphic correlation scheme, and to utilize the inorganic geochemical data to recognize organic-rich zones and seals.

The geochemical ratios used to define chemozones were as follows: La/Lu, Cr/Th, Cr/Ta, Zr/Cr, Ti/U, Zr/Yb, Nb/Cr and Zr/Nb. Figure 5.28 shows the correlation of chemozones between two closely spaced wells. Based on the statistical and graphical analysis of the data it was thought that these ratios related to variations in heavy mineral abundances, and hence provenance. The zones were labelled C0, C1, C2, C3 and C4 in ascending stratigraphic order, with C0 producing lower Nb/Cr than in C1, whilst C1 is characterized by lower Zr/Nb than in C2. The zone C3 samples yield higher values of Cr/Th and Cr/Ta, but lower Zr/Cr, than those assigned to C2 and C4. Zone C3 was also shown to yield lower Nb/Cr than in C1 and C2. The principal geochemical characteristics of the subzones and divisions are presented in Tables 5.1 and 5.2.

Figure 5.29 illustrates the chemostratigraphic correlation between the five study wells. Zone C0 has only been recognized in wells 1 and 2 but probably exists below the study intervals of the other wells. The other zones are correlative between all five wells, where they are identified with a high degree of confidence. The correlation is less robust with respect to the subzones, with only C4-1 and C4-2 occurring in all of the wells. The absence of other zones in particular wells (e.g. absence of C2-2 in Well 4) is likely to be explained by localised erosion/non-deposition. The zone C3 subzones are labelled C3-1, C3-2 and C3-3 in ascending stratigraphic order in wells 1 and 2, but it is not possible to further subdivide this zone in wells 3 and 4. In addition to this, a completely different subdivision of C3 is proposed for Well 5 (I.e. labelled C3-A and C3-B), where variations in La/Lu (as opposed to Zr/Yb in wells 1 and 2) are used to characterize two subzones. Differences in the way the C3 subzones are defined in Well 5 compared with wells 1 and 2 may be explained by local variations in provenance. The fact that C3 cannot be further subdivided in wells 3 and 4, suggests that there were no significant changes in provenance during the deposition of this zone in the vicinity of these wells. Another inconsistency in the correlation scheme is that subzone C4-0 is only identified in Well 4. Given the large distances (> 60 km) between most study wells, except for wells 1 and 2, it is perhaps unsurprising that some subzones are absent in particular wells. A localized variation in provenance was probably responsible for the deposition of subzone C4-0 in the immediate vicinity of Well 4.

The correlation of chemostratigraphic subzones and divisions, is strongest between wells 1 and 2. All of the subzones, together with the divisions of C2-2, C3-2 and C4-1, are correlative between these wells. The robustness and resolution of this correlation is explained by these wells being drilled within 1 km of each other. Conversely, the other wells are located at more than 80 km from each other (and from wells 1 and 2), where the subtle differences in provenance, resulting in the recognition of high resolution zone and subzone boundaries, did not exist over such a vast area.

A secondary aim of Craigie (2015b) was to use the geochemical data to recognize potential unconventional reservoirs and seals/cap rocks. TOC data were provided in most wells but, where such data were absent, the Schmoker ratio (see Craigie 2015b for details) was used estimate this parameter. The best unconventional reservoirs are defined by TOC values exceeding 4% and low concentrations of clay minerals,

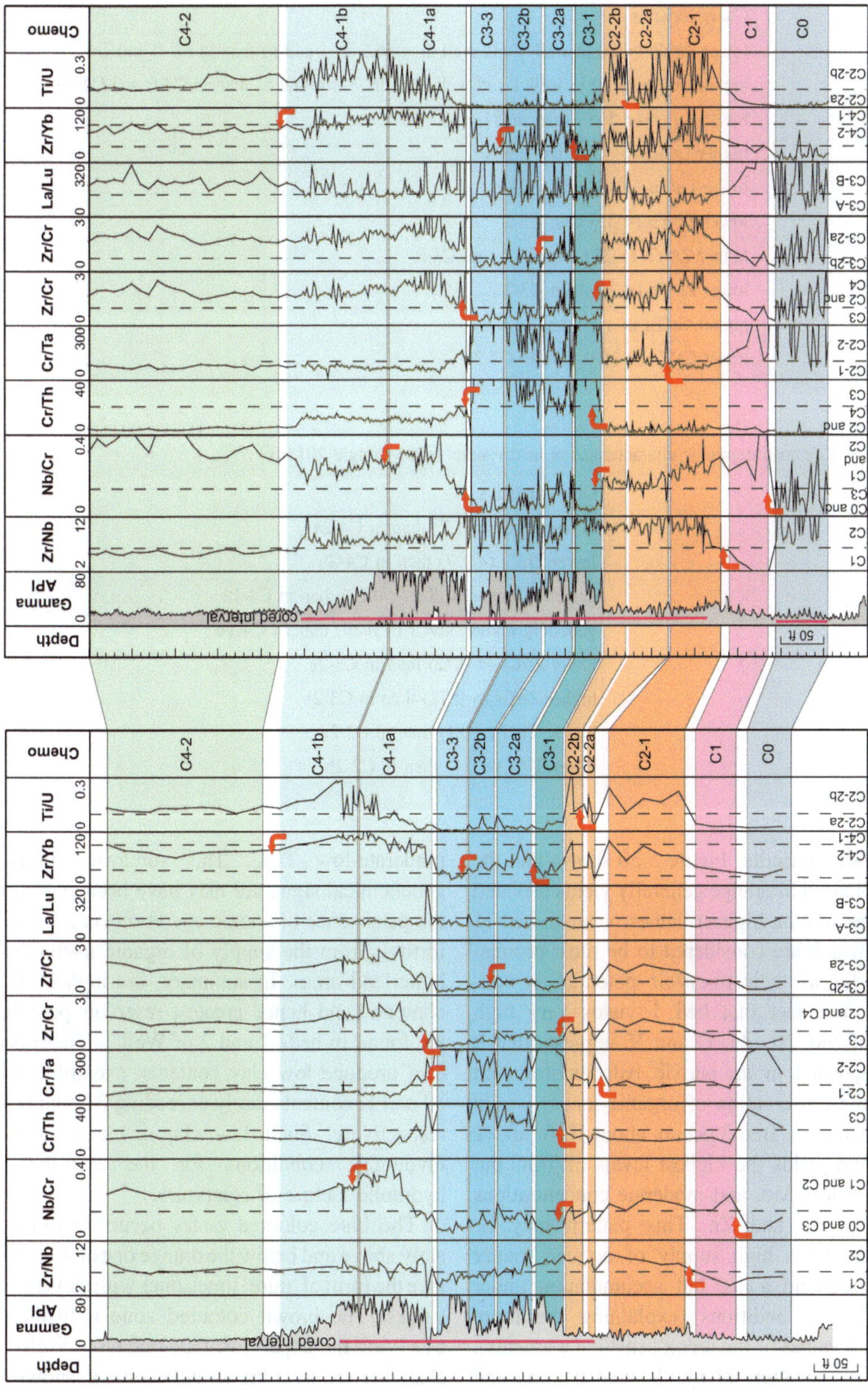

Fig. 5.28 Correlation of chemozones in Wells 1 and 2 based on variations in key element ratios. All depths are log depths in feet (after Craigie 2015b)

Table 5.1 Chemostratigraphic characterisation of subzones (after Craigie 2015b)

Sub-zone	Characteristics
C4-2	lower Zr/Yb (< 80 in well 1 and well 2, < 56.9 in 4 and 5) than in C4-0 and C4-2
C4-1	higher Zr/Yb (> 80 in well 1 and well 2, > 56.9 in 4 and 5) than in C4-0 and C4-2
C4-0	lower Zr/Yb (< 56.9) than in C4-1
C3-B	higher La/Lu (> 132) than C3-A
C3-A	lower La/Lu (< 132) than C3-B
C3-3	lower Zr/Yb (< 35) than C3-2
C3-2	higher Zr/Yb (> 35) than C3-1 and C3-3
C3-1	lower Zr/Yb (< 35) than C3-2
C2-2	higher Cr/Ta (> 103) than in C2-1
C2-1	lower Cr/Ta (< 103) than in C2-2

Table 5.2 Chemostratigraphic characterisation of divisions (after Craigie 2015b)

Divisions	Characteristics
C4-2b	higher Zr/Cr (> 1.7) than in C4-2a
C4-2a	lower Zr/Cr (< 1.7) than in C4-2b
C4-1b	generally lower Nb/Cr (< 0.29) than in C4-1a
C4-1a	generally higher Nb/Cr (> 0.29) than in C4-1b
C3-2b	lower Zr/Cr (< 0.72) than in C3-2a
C3-2a	higher Zr/Cr (> 0.72) than in C3-2b
C2-2b	higher Ti/U (> 0.1) than in C2-2a
C2-2a	lower Ti/U (< 0.1) than in C2-2b

particularly smectite. Figure 5.30 shows that the zone coloured in orange generally yields elevated TOC, though the highest values occur in beds 1 and 2, which are considered to be most promising as far as their reservoir potential is concerned. The fact that bed 1 yields very high levels of Mo, Ni, Zn, U and V suggests that it was deposited in an anoxic paleoenvironment where the preservation of organic matter would have been high. Bed 2 occurs about 10 ft. above bed 1 and yields the highest levels of TOC but relatively low Mo, and moderate concentrations of V, U, Ni and Zn. This pattern may be explained by a high supply of organic matter from hinterland areas, but accumulation under oxic-suboxic conditions (explaining the more moderate values of the redox sensitive elements). Bed 3 is characterised by elevated gamma and high levels of Mo, Ni, U, Zn and V but moderate-low TOC. This somewhat unusual geochemical signature may have been caused by the persistence of anoxic conditions in a location furthest from the supply of organic matter from hinterland areas. In summary, the highest TOC contents, and hence greatest reservoir potential, are found in beds 1 and 2 of Well 1. These beds also produce low clay contents, exemplified by Al and K contents rarely exceeding 3.5 and 0.5% respectively. Such low clay contents provide favourable conditions for the man-induced hydraulic facture of reservoirs.

The blue coloured zones occurring immediately above and below the orange one on Fig. 5.30 take the form of 'pure' limestones with elevated Ca values. The brown coloured zone immediately overlying the uppermost blue one on in Well 1 is characterised by a higher clay content, shown by Al concentrations in the range 5–8%, and K of 1–

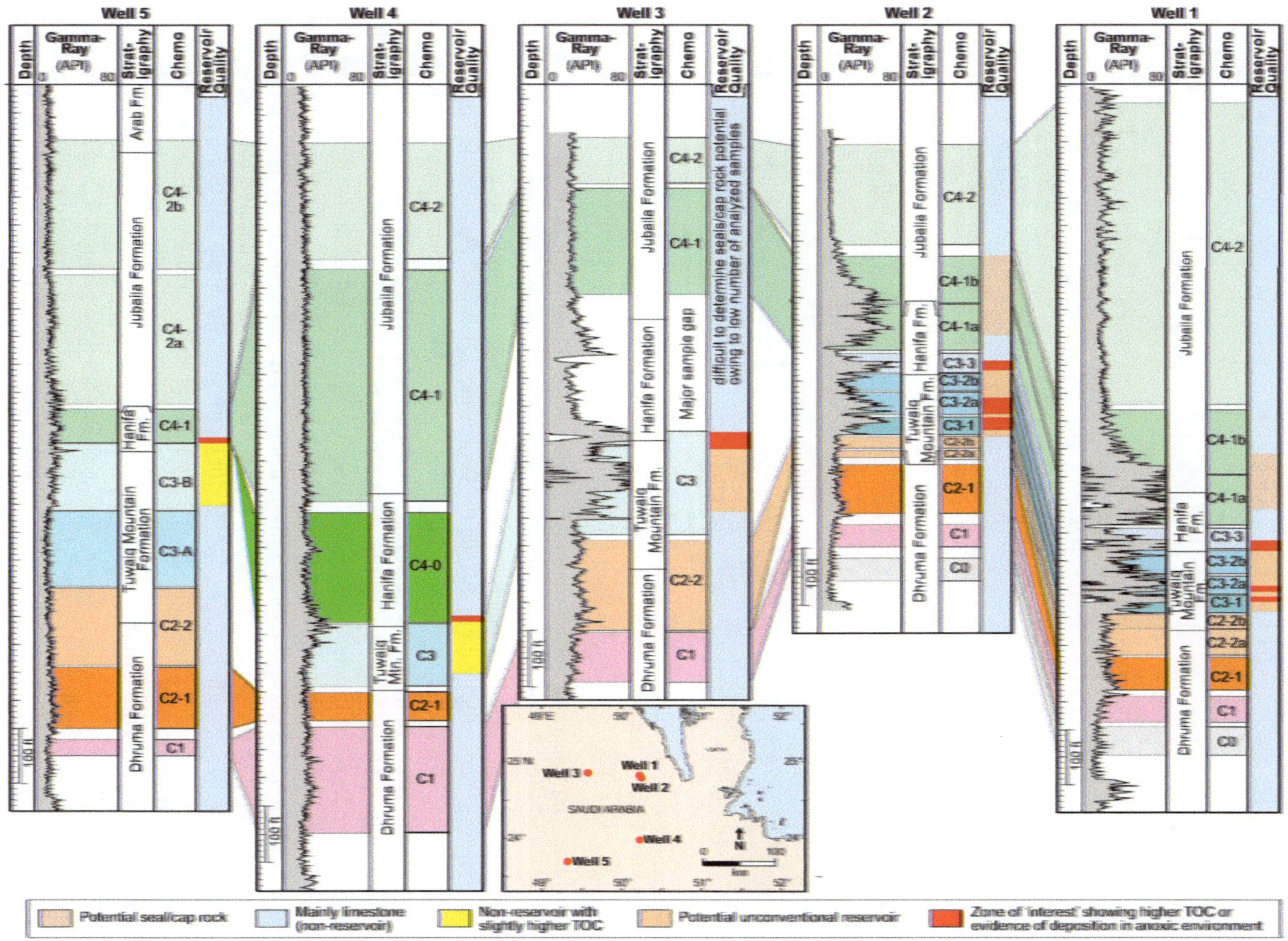

Fig. 5.29 Summary of the correlation of chemozones identified in wells 1-5. Potential unconventional reservoirs and seals/cap rocks are also shown in this diagram. See text for further explanation. All depths are log depths in feet (after Craigie 2015b)

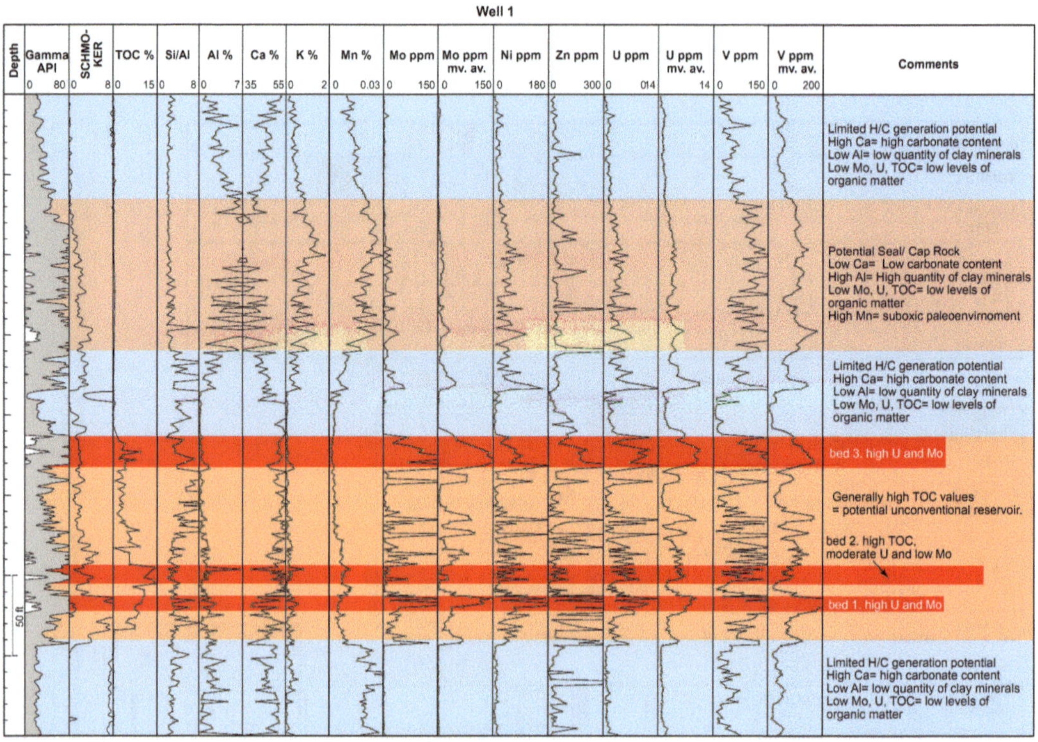

Fig. 5.30 Recognition of potential unconventional reservoirs and seals/cap rocks in Well 1 using geochemical data. All depths are log depths in feet (after Craigie 2015b)

2%. This may represent an unconventional seal/cap rock, as this bed contains a much higher proportion of clay minerals- man-induced hydraulic fractures may stop within this bed without penetrating permeable water bearing horizons immediately above it. The fact that this bed also yields high values of Mn may infer that it is was deposited under suboxic, rather than oxic or anoxic conditions. The unconventional reservoir and seal/cap rock properties of the five study wells are summarized in Fig. 5.29.

5.4.2 Chemostratigraphy of the Haynesville Shale, Eastern Texas and Northwestern Louisiana

Sano (2013) completed a chemostratigraphy study of the Upper Jurassic Haynesville Shale, a calcareous and locally organic-rich mudrock representing a prominent shale gas play in North America. A total of 1421 core samples were analysed from 10 wells located in the east Texas/west Louisiana Haynesville/Lower Bossier, USA. Most elements were analysed by ICP-OES and ICP-MS, though data for S were obtained by XRF. Part of the study involved the use of geochemical data to differentiate the Haynesville Shales from the formations immediately above and below it, though a further aim was to produce a more detailed zonation scheme for the Haynesville Shale and to use geochemical data to predict variations in paleoredox and organic matter within this interval.

The mineralogical affinities of elements were established by comparing XRD and geochemical data acquired from the same samples and by employing principal component analysis and binary diagrams. The following key elements and ratios were employed for chemostratigraphic purposes:

Al: clay minerals.
Si: detrital quartz.
Th: heavy minerals, some of which are probably associated with clay grade material.
Ca: calcite.
Na: plagioclase.
Mg: chlorite and dolomite.
Fe/Mg: reflects the proportion of pyrite.
U, S and EFV (EFV = $(V/Al)_{sample}/(V/Al)_{average\ shale}$): these parameters are generally highest where high levels of TOC are recorded.

Figure 5.31 shows the definition of chemostratigraphic packages in Well Watson-4. This illustrates that Package 1, equivalent to the Smackover Formation, is defined by very high Ca/Al ratios, explained by the lithologies of this interval being limestones. The Gilmer Lime interval, or chemostratigraphic package 2, yields Ca/Al ratios that are intermediate between the high levels of Package 1 and the significantly lower ones of Package 3. This interval represents backstepping carbonate facies and contains dark

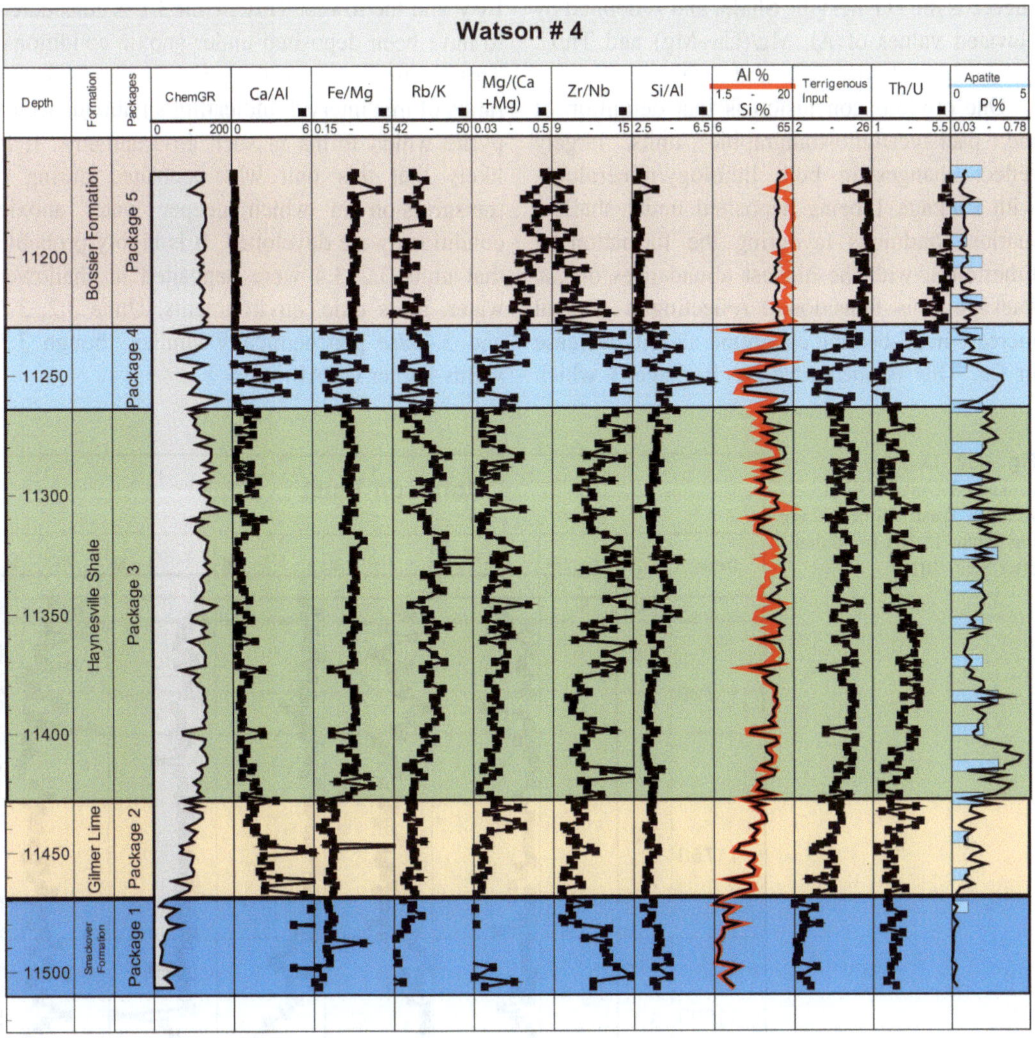

Fig. 5.31 Key elements and ratios plotted for Well Watson 4. Note that ChemGR is derived from K, Th and U concentrations of the samples, using an algorithm to convert those values to API values. All depths are log depths in feet (modified, after Sano 2013)

grey to brown micritic limestone and thin beds of dark grey shale. Package 3 encompasses most of the Haynesville Shale and produces lower Ca/Al, and generally higher Fe/Mg and Rb/K ratios, than in packages 1 and 2. Sedimentological data show this interval to be a black, organic rich calcareous mudrock that was deposited under arid climatic conditions in a restricted intrashelf basin on the evolving Gulf Coast passive margin. Package 4 occurs at the top of the Haynesville shale and is characterized by generally higher Ca/Al ratios than in Package 3, but lower Fe/Mg. Figure 5.31 shows that the Bossier Formation (Package 4) succeeds the Haynesville Shale, and is defined by elevated values of Al, Mg/(Ca+Mg) and Th/U, combined with low Si/Al and Zr/Nb.

One obvious conclusion is that definition of the packages/lithostratigraphic units largely reflect changes in bulk lithology/mineralogy, with Package 1 being deposited under shallow marine conditions favouring the formation of limestones, with the highest abundances of Ca. Package 2 is transitional, reflecting a gradual increase in Al-bearing clay minerals and decrease in Ca. This is succeeded by Package 3 which yields higher clay and lower carbonate contents, shown by higher Al and lower Ca respectively. The higher Ca/Al ratios of Package 4 reflects an increase in the proportion of carbonate minerals. Package 5 is a mudrock, defined by higher Al, but lower Ca/Al and Si/Al. The fact that this interval also produces lower Zr/Nb ratios than in the other packages may suggest that it was sourced from a different provenance.

Sano (2013) subdivided Package 3 into four units labelled 3.1, 3.2, 3.3 and 3.4 in ascending stratigraphic order, with the definition of these presented in Fig. 5.32. With the highest values of EFV and the lowest Th/U, Unit 3.1 is considered to have been deposited under anoxic conditions. This is supported by the generally high Fe/Mg ratios of this interval, indicating an abundance of pyrite which forms in such environments. It is likely that this unit was deposited during a transgression in which deeper water anoxic conditions were developed. It is highly probable that units 3.2–3.4 were deposited in shallower water, more oxic, environments. Units 3.2, 3.3 and 3.4 are geochemically similar, though 3.3 yields higher Ca/Al ratios.

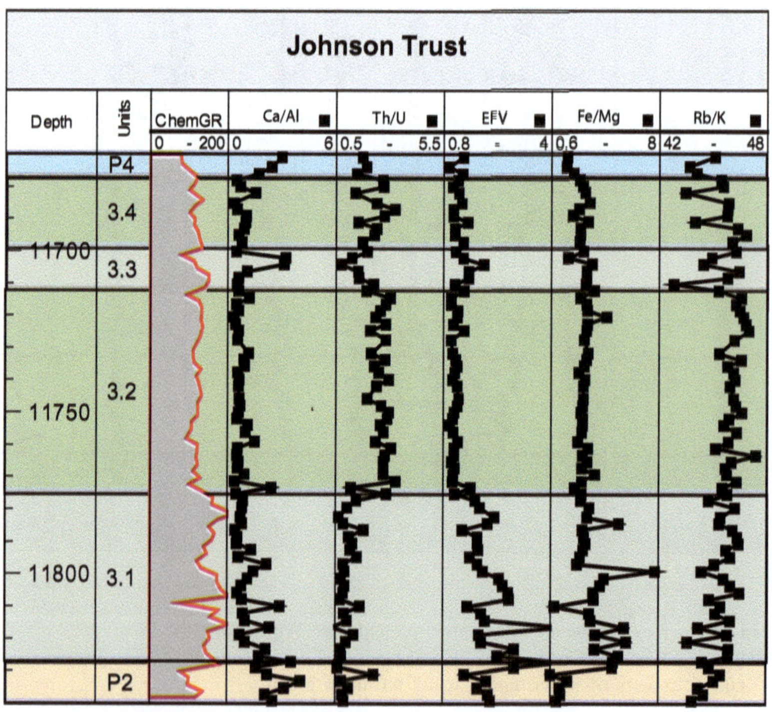

Fig. 5.32 Definition of package 3 Units in Well Johnson Trust. All depths are log depths in feet (modified, after Sano 2013)

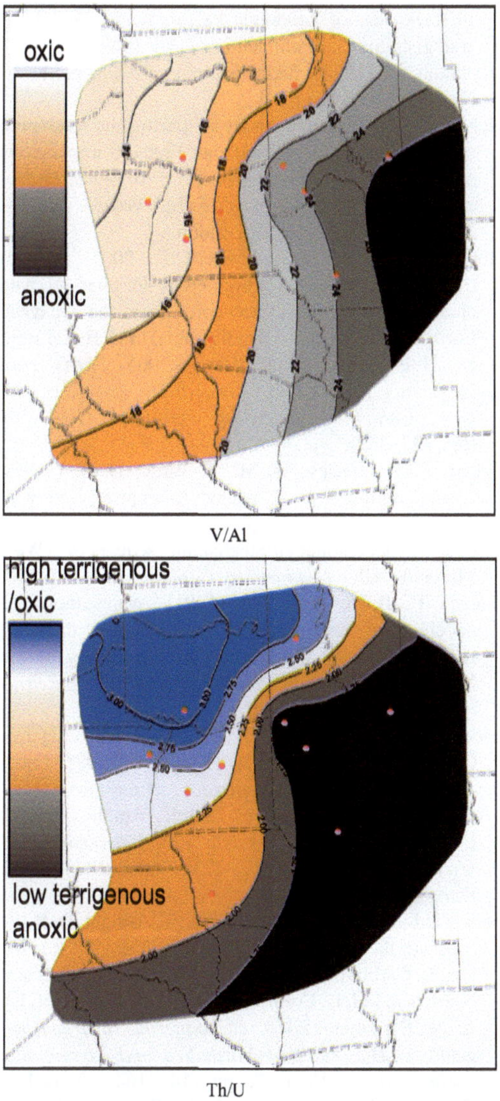

Fig. 5.33 Isochemical maps showing average V/Al and Th/U ratios throughout the basin during the depostion of the Haynesville Shale interval (Package 3). Red dots illustrate the location of wells in this study. The values between study wells are interpolated based on the values at each well point (modified, after Sano 2013)

According to Sano (2013), an exact relationship does not exist between the redox sensitive parameters (i.e. EFV, S, Th/U) and TOC. The reasons for this are unclear but may relate to minor variations in paleoenvironment and/or the supply of organic matter. In spite of this, a broad relationship does exist between these elements/ratios and TOC, which led Sano (2013) to produce isochemical maps for Package 3, based on values of V/Al and Th/U. Figure 5.33 shows that the highest ratios of V/Al, and lowest Th/U, occur to the SE of the study area, inferring the existence of sediments deposited in anoxic paleoenvironments. It is also significant that the highest values of TOC exist in this location, suggesting that the inorganic geochemical data could be used as a predictive tool to locate shales with the highest TOC contents, which may be the most favourable in terms of unconventional reservoir potential.

5.5 Concluding Remarks

Whilst none of the aforementioned studies have been described in the level of detail that they were in their respective publications, it is hoped that these will provide an indication of how elemental chemostratigraphy can been applied to clastic, carbonate and unconventional reservoirs. It is also hoped that the author has demonstrated the importance of employing a fully integrated approach to reservoir correlation. Chemostratigraphy may be applied on a stand-alone basis, but some studies (e.g. Holmes et al. 2015; Craigie et al. 2016a, b; Craigie and Polo 2017) clearly demonstrate that it often works more effectively when used in conjunction with other correlation techniques such as biostratigraphy, reservoir sedimentology, borehole images and seismic data.

References

Al-Hajri, S. A., Filatoff, J., Wender, L. E., & Norton, A. K. (1999). Stratigraphy and operational palynology of the Devonian system in Saudi Arabia. *GeoArabia, 4* (1), 53–68.

Al-Husseini, M. I. (2004). Pre-Unayzah unconformity, Saudi Arabia. In M. I. Al-Husseini (Ed.), *Carboniferous, Permian and early Triassic Arabian stratigraphy: GeoArabia special publication 3* (pp. 15–59). Bahrain: Gulf PetroLink.

Craigie, N. W. (1998). *Chemostratigraphy of middle Devonian lacustrine sediments* (Unpublished PhD thesis). N.E. Scotland: University of Aberdeen.

Craigie, N. W. (2015a). Applications of chemostratigraphy in Cretaceous sediments encountered in the North Central Rub' al-Khali Basin, Saidi Arabia. *Journal of African Earth Sciences, 104,* 27–42.

Craigie, N. W. (2015b). Applications of chemostratigraphy in middle Jurassic unconventional reservoirs in eastern Saudi Arabia. *GeoArabia, 20*(2), 79–110.

Craigie, N. W., Breuer, P., & Khidir, A. (2016a). Chemostratigraphy and biostratigraphy of Devonian, Carboniferous and Permian sediments encountered in eastern Saudi Arabia: an integrated approach to reservoir correlation. *Marine and Petroleum Geology, 72,* 156–178.

Craigie, N. W., Rees, A., MacPherson, K., & Berman, S. (2016b). Chemostratigraphy of the Ordovician Sarah formation, North West Saudi Arabia: An integrated approach to reservoir correlation. *Marine and Petroleum Geology, 77,* 1056–1080.

Craigie, N. W., & Polo, C. A. (2017). *Applications of Chemostratigraphy and sedimentology in a complex reservoir: A case study from the Permo-Carboniferous Unayzah group.* Central Saudi Arabia: Marine and Petroleum Geology (in prep.)

Davies, E. J., Ratcliffe, K. T., Montgomery, P., Pomar, L., Ellwood, B. B., & Wray, D. S. (2013). Magnetic susceptibility (X stratigraphy) and chemostratography applied to an isolated carbonate platform reef complex; Llucmajor Platform, Mallorca. SEPM Special Publication dedicated to the deposits, architecture and controls of carbonate margin, slope and basin systems.

Eltom, H., Abdullatif, O. M., Makkawi, M. H., & Eltoum, I. E. (2017). Rare earth element geochemistry of shallow carbonate outcropping strata in Saudi Arabia: Application for depositional environments prediction. *Sedimentary Geology, 348,* 51–68.

Eyles, N. (1993). Earth's glacial record and it's tectonic setting. *Earth Science Review, 35,* 1–248.

Holh, S. V., Becker, H., Jiang S. Y., Ling, H. F., Guo, Q., & Struck, U. (2017). Geochemistry of Ediacaran cap dolostones across the Yangtze Platform, South China: Implications for diagenetic modification and seawater chemistry in the aftermath of the Marinoan glaciation. *Journal of the Geological Society London* (in press).

Holmes, N., Atkin, D., Mahdi, S., & Ayress, M. (2015). Integrated biostratigraphy and chemical stratigraphy in the development of a reservoir-scale stratigraphic framework for the Sea Lion Field area, North Falkland Basin. *Petroleum Geoscience, 21,* 171–182.

Jorgensen, N. O. (1986). Chemostartigraphy of Upper Cretaceous chalk in the Danish sub-basin. *Bulletin of American Association of Petroleum Geologists, 70,* 309–317.

Kinsman, D. J. J. (1969). Interpretation of $Sr^{0}U^{20}D$ concentrations in carbonate minerals and rocks. *Journal of Sedimentary Petrology, 39,* 486–508.

Land, L. S. (1980). The isotopic and trace element geochemistry of dolomite: The state of the art. In: D. H. Zenger, J. B. Dunham & R. C. Ethington (Eds.), *Concepts and models of dolomitization* (29, pp. 87–110). Society of Economic Paleontologists and Mineralogists Special Publication.

Lippmann, F. (1973). *Sedimentary carbonate minerals.* New York: Springer.

Melvin, J. (2009). Heterogeneity in glaciogenic reservoirs: examples from the Ordovician and Permo-Carboniferous of Saudi-Arabia. In *Glaciogenic Reservoirs and Hydrocarbon Systems, Conference Abstract Volume* (pp. 37–38), Geological Society of London.

Melvin, J., Sprague, R. A., Heine, C. J. (2005). Diamictites to aeolianites: Carboniferous–Permian climate change seen in subsurface cores from the Unayzah Formation, east-central Saudi Arabia. In G. E. Reinson, D. Hills & L. Eliuk (Eds.), 2005 *CSPG core conference papers and extended abstracts CD: Calgary, Canadian Society of Petroleum Geologists* (pp. 237–282).

Pearce, T. L., Besley, B. M., & Wray, D. S. (1999). Chemostratigraphy: A method to improve interwell correlation in barren sequences—a case study using onshore Duckmantian/Stephanian sequences (West Midlands, UK). *Sedimentary Geology, 124,* 197–220.

Pearce, T. J., & Jarvis, I. (1991). Applications of geochemical data to modeling sediment dispersal patterns in distal turbidites: Late Quaternary of the Madeira abyssal plain. *Journal of Sedimentalr Petrology, 62,* 1112–1129.

Price R. J., Norton K. A., Melvin J. A., Filatoff, J., Heine, C, J., Sprague R. A., Al-Hajri, S. (2008). Saudi Aramco Permian-Carboniferous (Unayzah) stratigraphic nomenclature of Saudi Arabia. In M. I. Al-Husseini (Ed.), *Middle East petroleum geosciences conference* (p. 223). GEO'2008: Gulf PetroLink, Bahrain.

Pearce T. J., Wray, D. S., Ratcliffe, K. T., Wright, D. K., & Moscariella, A. (2005). Chemostratigraphy of the Upper Carboniferous Schooner Formation, southern North Sea. In J. D. Colinson, D. J. Evans, D. W. Holiday & N. S. Jones (Eds.), *carboniferous hydrocarbon geology: The southern North Sea and surrounding onshore areas* (Vol. 7, pp. 147–164). Yorkshire Geological Society, Occasional Publication Series.

Ratcliffe, K. T., Martin, J., Pearce, T. J., Hughes, A. D., Lawton, D. E., Wray, D. S., et al. (2006). A regional chemostratigraphically-defined correlation framework for the Late Triassic TAG-I Formation in Blocks 402 and 405a, Algeria. *Petroleum Geoscience, 12,* 3–12.

Rees, A. J. (2015). The Late Ordovician glacial record in the subsurface of NW Saudi Arabia. In *Clastic reservoirs of the Middle East-AAPG conference* 23rd–25th March, Kuwait.

Sabaou, N., Lawton, D. E., Turner, P., & Pilling, D. (2005). Floodplain deposits and soil classification: The prediction of channel sand distribution within the Triassic Argilo-Greseux Inferieur, Berkine Basin, Algeria. *Journal of Petroleum Geology, 28,* 3–20.

Sano, J. L., Ratcliffe K. T., & Spain D. (2013). Chemostratigraphy of the Haynesville Shale. In: U. Hammes & J. Gale (Eds.), *Geology of the Haynesville*

Gas Shale in East Texas and West Louisiana, USA. AAPG Memoir 105, pp. 137–154.

Sharland, P. R., Archer, R., Casey, D. M., Davies, R. B., Hall, S. H., Heward, A. P., Horbury, A. D., Simmons, M. D. (2001). *Arabian Plate sequence stratigraphy* (Vol. 2), GeoArabia Special Publication.

Shukla, V. (1988). *Sedimentology and geochemistry of a regional dolostone: Correlation of trace elements with dolomite fabrics*. SEPM Special Publication No. 43, pp. 145–157.

Sutcliffe, O. E., Dowdeswell, J. A., Whittington, R. L., Theron, J. N., & Craig, J. (2000). Calibrating the Late Ordovician glaciation and mass extinction by the eccentricity cycles of Earth's orbit. *Geology, 28,* 967–970.

Tribovillard, N., Algeo, T., Lyons, T. W., & Riboulleau, A. (2006). Trace elements as paleoredox and paleoproductivity proxies; an update 2006. *Chemical Geology, 232* (1–2), 12–32.

Turner, P., Pilling, D., Walker, D., Exton, J., Binnie, J., & Sabaou, N. (2001). Sequence stratigraphy and sedimentology of the Late Triassic TAG-1 (Blocks 401/402, Berkine Basin, Algeria). *Marine and Petroleum Geology, 18,* 959–981.

Vaslet, D. (1989). *Late Ordovician glacial deposits in Saudi Arabia: A lithostratigraphic revision of the Early Palaeozoic succession.* Saudi Arabian Deputy Ministry for Mineral Resources, professional papers, pp. 13–44.

Vaslet, D. (1990). Upper Ordovician glacial deposits in Saudi Arabia. *Episodes, 13,* 147–161.

Veizer, J. (1983). Chemical diagenesis of carbonates: Theory and application of trace element technique. In: M. A. Arthur, T. F. Anderson, I. R. Kaplan, J. Veizer & L. S. Land (Eds.), Stable isotopes in sedimentary geology. Society of Economic Paleontologists and Mineralogists Short Course No. 10, pp. 3-1 to 3-100.

Vishnevskaya, I., Letnikova, E., Pisareva, N., & Proshenkin, A. Chemostratigraphy of Neoproterozoic carbonate deposits of the Tuva-Mongolian and Dzabkhan continental blocks: constraints on the age, glaciation and sedimentation. In: M. Ramkumat (Ed.), 2015. *Chemostratigraphy: concepts, techniques and applications, Chapter 18*, pp. 451–487.

Wender, L. E., Bryant, J. W., Dickens, M. F., Neville, A. S., & Al-Moqbel, A. M. (1998). Paleozoic (pre-Khuff) hydrocarbon geology of the Ghawar area, eastern Saudi Arabia. *GeoArabia, 3,* 273–302.

Applications of Wellsite Chemostratigraphy

6

Abstract

Chemostratigraphy was first used at wellsite in the early 2000s but it is rather unfortunate that very little material has been published on the subject, with most articles taking the form of very short papers or conference abstracts. In spite of this, wellsite chemostratigraphy is becoming increasingly popular, as it can be used to provide lithology/bulk mineralogy predictions, stratigraphic control whilst drilling and to aid the placement of casing points, coring points and total depth (TD). Arguably, the most commonly used application, however, is to assist the geosteering/well trajectory monitoring of horizontal-subhorizontal wells drilled in structurally complex plays. Major advances in XRF technology over the last 15 years, particularly with respect to benchtop ED-XRF instruments, has resulted in the ability to generate high quality inorganic geochemical data in 'near real time'.

6.1 Introduction

The possibility of using chemostratigraphy at wellsite during the drilling of wells was first investigated in the late 1990s to early 2000s. Chemostrat Ltd. pioneered wellsite chemostratigraphy by forming an alliance agreement with Halliburton to deploy a LIBS (Laser Induced Breakdown Spectrometry) instrument (Laserstrat). Since that time, a number of companies (e.g. Baker Hughes, Schlumberger, RPS Ichron, Weatherford) have become involved in wellsite chemostratigraphy, mainly using portable XRF instruments such as the Spectro Xepos instrument. Some companies simply specialize in data generation, whilst others offer both data generation and interpretation. The principal applications of wellsite chemostratigraphy can be summarized as follows:

1. Lithology and bulk mineralogy determination
2. Stratigraphic control during drilling
3. Placement of casing points
4. Placement of coring points
5. Placement of total depth (TD)
6. Geosteering.

Each one of these applications are discussed in this chapter but consideration should also be given to the practicalities of transporting to, and

N. Craigie, *Principles of Elemental Chemostratigraphy*, Advances in Oil and Gas Exploration & Production, https://doi.org/10.1007/978-3-319-71216-1_6

using, analytical equipment at wellsite. Unfortunately, very few papers have been written on the subject of wellsite chemostratigraphy and most publications take the form of short articles and conference abstracts (e.g. Schmidt et al. 2010; Pilcher 2011; Hildred and Rice 2014) rather than 'full length' papers published in recognized journals. Consequently, very little information is available on the practicalities and challenges of generating/interpreting data in what is, invariably, a harsh high pressure environment. The reasons for this lack of material are unclear but may relate to confidentiality issues associated with publishing data on recently drilled wells. The consequence is that much of the information provided in Sects. 6.2, 6.3 and 6.4 is based on the author's personal experience of assisting in the development of wellsite chemostratigraphy in the early 2000s and of working at wellsite on numerous occasions from 2004 to 2012.

6.2 Practical Considerations of Wellsite Chemostratigraphy

One of the most important considerations at wellsite involves the packing and transportation of analytical equipment. XRF instruments are normally transported from manufacturers to clients in padded boxes designed for a single journey, but certainly not for multiple transportations to wellsite. Consequently, the type of box required for this purpose will probably have to be manufactured by other companies and should include extra padding to accommodate possible manhandling of equipment. Ideally, the box should be waterproof and provide protection against the harmful effects of sea air which can cause havoc to the internal workings of instruments. Other sensitive equipment such as microscopes and portable rock grinders should also be transported in protective boxes. Once packed in a minicontainer, the analytical and sample preparation equipment needs to be properly secured to ensure no or limited movement during transportation. Particular care should be taken if Helium or any other gasses are to be used. Ideally, gas cylinders should be transported in a vertical position, in a purpose built metallic cage that is clearly labelled. Alternatively, gas cylinders can be placed in a minicontainer, but should always be transported in the vertical position and be secured in such a way that they will not be displaced during transportation. Owing to the, albeit limited, risk of gas escape/explosion, it is recommended that gas cylinders are transported in separately to the analytical equipment. Probably the most important consideration should be to comply with health and safety regulations specific to the given country, region and rig. Minicontainers containing analytical equipment should always be labelled as 'fragile' to avoid breakages though, in the experience of the author, crane operators often pay little attention to such instructions.

Manual handling skills are also required to carry equipment and pack/empty minicontainers used for the transportation of equipment. Many portable XRF instruments weigh more than 65 kg, meaning that at least two individuals, and ideally as many as four (one person at each corner of the instrument or box), should be involved in its carry. Other equipment (e.g. box containing a microscope) can also be of a sufficient weight to warrant the use of two or more individuals in their carry. Particular care should be taken when lifting gas cylinders as the consequences of dropping such a cylinder may involve gas leakage and, in a worst case scenario, explosion. Whether the company/client insists on this or not, it is highly recommended that individuals involved in wellsite chemostratigraphy studies receive formal training in health and safety, manual handling and the use of gasses. Remember that the health and safety of individuals should always be a higher priority than economic considerations when working on wellsite chemostratigraphy studies.

Forward thinking is essential when planning the equipment to be taken to wellsite, and consideration should be given to the possibility of equipment malfunction, breakage and loss. For example, only two or three sieves may be used during a particular wellsite project, but it would be recommended to take at least six, to cover potential loss or breakage. In simplistic terms, try to take about twice the amount of equipment that is required, including duplicates of nearly every utensil. The number of gas cylinders to take will very much depend on the usage for a particular instrument, which should be monitored prior to deployment at wellsite. However, it is always better to overestimate usage and plan to take about double the number of cylinders required. It may well be impractical to transport two XRF instruments to wellsite, but a plan should exist to transport a second instrument to wellsite if required. Malfunction of XRF instruments may involve a simple fix such as a replacement of a fuse in a plug, to much more complicated repairs involving the changing of detectors or X-ray tubes. It is important for the wellsite chemostratigrapher to have the contact details of specialist engineers employed by the manufacturer. Some malfunctions may not be resolved at wellsite, but this is less likely to be the case for more simple repairs that may be undertaken by the analyst after holding a short conversation with a skilled engineer employed by the manufacturer. The same holds true for sample preparation equipment, such as portable rock grinders, hot plates, ovens and microscopes. The wellsite chemostratigrapher also needs to make himself/herself aware of transportation schedules in relation to the rig on which they are deployed, and to relay the necessary instructions to individuals working in the office/lab, should it be necessary to deploy additional equipment or personnel to wellsite. Such considerations may seem like common sense but it is not uncommon for wellsite chemostratigraphy studies to fail owing to complacency on the part of the analyst and/or a lack of sufficient planning.

The equipment to take to wellsite should include, but not necessarily be limited to the following items:

1. XRF instrument and related equipment (e.g. cables, sample cups etc.).
2. Helium cylinders.
3. Regulators for helium cylinders.
4. Microscope.
5. Picking trays (×4).
6. Small magnets (×4).
7. Hot plate and/or oven to dry samples.
8. Metallic trays and or metallic beakers.
9. Glass or plastic beakers.
10. Tweezers (×4).
11. 'Wet' sieve for washing samples (×4).
12. Dry sieves for sieving samples that have been washed and dried (×4 or more).
13. Paper and stationery.
14. Cloths.
15. Agate pestle and mortar and/or portable mechanical grinder.
16. Liquid detergent for washing samples.
17. First aid kit.
18. Oxygen monitor if Helium gas is to be used in unit.
19. Chairs for unit.
20. Rope.
21. Tool kit.
22. Necessary computers and software.
23. De-ionised water.

The type and set up the unit in which the processing/analysis is to take place is also important. In most offshore wellsite chemostratigraphy studies, A60 units are hired or bought by the service company to accommodate the necessary analytical and preparation equipment required to complete the given study. A particular consideration is that there is enough bench space for the XRF instrument, computer equipment, rock grinder and the microscope. Ideally, there should be a separate oven for sample drying, though a hotplate could be utilized for this purpose if an oven is unavailable. If the hotplate is to be used, it should be placed in a fume cupboard specifically designed with an extractor to remove harmful hydrocarbons or chemicals released during heating. Similarly, if an oven is used it must have an adequate extraction system. The A60 unit should also have a sink and comply with the necessary health and safety requirements of the given rig on which it is to be deployed. In some studies, there may be a lack of space on the rig to accommodate an A60 unit and the XRF and related equipment may have to operate in the mud logging unit or in another part of the rig (e.g. MWD cabin). These circumstances are hardly ideal but are not uncommon. It is important to ensure, before embarking on such a project, that the space and requirements are suitable. With respect to onshore wellsite chemostratigraphy jobs, an A60 unit may be deployed to the field but this may not be necessary as it is commonplace for the analytical and preparation work to be undertaken in a portacabin.

As there are a number of potential setbacks before the wellsite study even begins, it is suggested the analysts arrive at wellsite at least two days before the first set of samples are to be analyzed. Sometimes this is not possible, owing to financial or logistical considerations, but it is always advisable as it can take some time to arrange the set-up of the A60 unit (e.g. arranging power supplied, water, telephone links etc. and the analytical equipment). Furthermore, if any items of equipment are missing or broken, a cushion of 2 days or more may allow enough time to arrange for replacements to be sent to the rig.

The choice of personnel to send to wellsite is absolutely critical to the success of any given study. Most wellsite chemostratigraphy projects are run by two individuals working 'back-to-back' 12 hour shifts, though arrangements may vary according to the working regulations of specific countries and rigs, as well as financial and logistical considerations. In some wellsite jobs, for instance, a high turnover of samples may require three or four individuals to be working at wellsite. Others may only need one person where there is limited 'bed space'. Irrespective of the number of individuals involved in the study, however, the analyst(s) should have the following skills:

(i) Skilled in the use of XRF equipment (assuming this is to be used at wellsite).
(ii) Ability to complete minor repairs to the instrument and be skilled at 'troubleshooting'.
(iii) Full knowledge of all sample preparation procedures.
(iv) Good communication and interpersonal skills as it is often necessary to communicate with numerous persons on the rig such a mudloggers, MWD engineers, wellsite geologists, tool pushers, mud engineers and others.
(v) Ability to work at speed under pressure, without compromising data preparation, analysis and quality.
(vi) Ability to recognize good and bad quality data.
(vii) Good interpretation skills (assuming they are required at wellsite).
(viii) Good basic knowledge of geology and petroleum geology.

The reason these points are mentioned is that there is often a failure to recognize the importance of employing individuals with the correct personal attributes suitable to work on wellsite chemostratigraphy studies. It may be assumed that the analyst only has to know how to set up and run an XRF instrument, something that can be taught in a relatively short space of time. In fact, the knowledge of the analysts should go far beyond this and include all of the aforementioned points. It may be practical to send one inexperienced individual to wellsite to be accompanied by a more experienced person, but there should always be an experienced analyst available at wellsite at any given time. The possession of good interpersonal and technical skills are of equal importance. The former are often overlooked, but should not be, as it is necessary for the analyst to communicate with many individuals working on the rig for the entire duration of the study. For example, it may be necessary to contact an electrician to arrange for the A60 unit to be 'powered up'. It will almost certainly be necessary to open doors of communication with the mudloggers, data engineers, MWD engineers, wellsite geologists, mud engineers and others during the course of drilling.

6.3 Sampling, Sample Preparation and Analytic Requirements at Wellsite

As with any chemostratigraphy study, one of the most important 'keys' to the success of wellsite studies is that the correct procedures are adhered to with respect to sampling, sample preparation and analysis. This is especially true at wellsite, where there is often very little time scheduled for sample preparation/analysis. Unlike conventional laboratory-based studies, it is rarely feasible to re-prep/re-run samples on a regular basis or to waste valuable time trouble-shooting.

Sampling of cuttings may either be done by the analyst, or by 'third party' personnel such as mudloggers. If the latter holds true, it is important for the analyst to check that the samples are taken from the 'correct' depth ranges. In most studies, samples should be analyzed every 10 ft, but the process of taking samples from the wrong depths, either deliberately or by accident, is not entirely uncommon. Another issue that the author has noticed is that, on occasions, the sampling board at the shakers is not washed down after each sample has been taken from a specific depth, resulting in contamination of subsequent samples. For these reasons, it is suggested that the analyst makes periodic checks to ensure that the correct sampling procedures are followed. A common issue noted by the author involves a change of data quality with a change in the shift pattern/change of personnel of mudlogging crews—for example, involving a change from an experienced to an inexperienced mudlogger. Building a good rapport with the mudloggers, or whoever will be sampling the cuttings, is vitally important. It is equally important to obtain information of shift/crew changes of the mudloggers and the levels of experience of each individual involved in the sampling of cuttings. Never overlook the fact that the success of any chemostratigraphy study is dependent on receiving good quality data, and a significant part of that process relates to sampling. It may not be viable to obtain cuttings descriptions from the mudloggers, but this information can be useful and should always be requested. Furthermore, the mudloggers are often a focal point for obtaining information on drilling plans, well trajectories, drilling/sampling rates and other important factors relating to the drilling of the well.

After meeting the mudloggers (or other individuals responsible for the sampling) and establishing a system for the sampling of cuttings material, it is important for the analyst to meet the mud engineers to discuss the type and quantity of additives that will be used, and whether these factors are likely to change during drilling. If possible, request samples of the additives and analyse these by XRF during a 'break' in drilling. In most studies, the use of drilling additives will have little or no effect on the values of key elements and ratios, but this does not always hold true. For example, the introduction of a high proportion of bicarbonate additive may cause certain elements to be 'artificially' diluted more than others. The use of a bentonite additive may result in sharp increases in the some elements (e.g. Nb, Ti, Ni, Cr) rendering them useless for chemostratigraphic purposes. Armed with information on drilling additives, the analyst can make important decisions with respect to sample preparation and the interpretation of data.

In addition to meeting the mudloggers and mud engineers, the analyst should attempt to hold meetings with the wellsite geologist, tool pusher, oilfield data manager (OIM) and drilling supervisor/company representative. These individuals can provide vital information on drilling plans, well trajectories, planned daily/weekly meetings, health and safety concerns, crew changes and other logistical/administrative issues specific to the rig. Meeting the MWD engineers may not be necessary, but is advised in order to obtain wireline log data which can be plotted in conjunction with geochemical profiles.

A minimum turnaround of analysing 2 samples per hour is required in most wellsite chemostratigraphy studies. Consequently, the analyst is usually required to work at speed and under a certain amount of pressure. Ideally, a 'conveyer belt' should be formed—as one sample is being dried in the oven, another is being picked or ground, whilst yet another is being analysed by XRF. Given that the analytical time should be around 20 min, the total sample preparation time should not exceed this, meaning that 3 samples can be analysed per hour. In reality, however, a sample turnaround of 2 samples per hour is more realistic, as the analyst needs to spend some time interpreting the data, analysing standard reference materials to check on data quality and holding brief meetings with the wellsite geologist etc. In spite of the pressures involved to delivered data and interpretations at speed, standard reference materials should be analysed 2–3 times per day in order to check analytical precision, accuracy, detection limits and possible drift. Without these checks, erroneous results and interpretations may be made and, in a worst case scenario, this can result in incorrect decisions being delivered on proposed well trajectories. The reader is referred to Chap. 2 for more details on sampling, drilling additives and sample preparation techniques.

Nearly all wellsite chemostratigraphy studies are now completed using XRF, but the choice of XRF to be utilized is important and should satisfy the following requirements:

(a) Ability to deliver data for 35 elements or more, with detection limits of around 2 ppm for most.
(b) Good levels of analytical accuracy and precision.
(c) Suitable calibration method(s), depending on lithologies that are likely to be encountered.
(d) Robust enough to be transported to wellsite without breakage, except under extreme circumstances.
(e) Easy to use and calibrate.
(f) Good reliability and availability of spare parts.
(g) Small/compact enough to be carried by two individuals with relative ease and to be used within a mudlogging or separate A60 unit.

Hand held XRF instruments have become more popular over that last 5–10 years but are not recommended for wellsite chemostratigraphy as, in the experience of the author, they do not deliver data for a sufficient number of elements, offer good enough detection limits and levels of resolution. Where information is only required on bulk lithology/mineralogy, they are ideal, but this can normally be achieved by the physical examination of cuttings and by obtaining wireline log data. A much higher level of resolution is required in nearly all wellsite chemostratigraphy studies. It is generally accepted that WDXRF instruments deliver data for more elements, and of a higher quality, that their EDXRF counterparts. However, WDXRF instruments are normally too large and cumbersome to take to wellsite and are less robust during transportation. The fact that additional gasses and coolants may be required can cause additional complications when trying to use them in a such an environment. Furthermore, the software and internal workings of WDXRF spectrometers are more complex and their use may require more specialized knowledge. For these reasons, EDXRF is favoured at wellsite but the specific type/model should satisfy all of the aforementioned requirements. The author favours EDXRF instruments operating with polarization as these generally deliver the best quality data for the largest number of elements, particularly with respect to ‘heavy’ elements, though other XRF instruments are also be acceptable.

As technology develops, it may be possible to utilize ICP instruments at wellsite and the data generated from these would certainly be preferred over XRF, owing to the larger number of analysed elements and better detection limits. Unfortunately, the ICP-OES and ICP-MS instruments are simply too large to transport to wellsite with ease and, even if these machines could be utilized at wellsite, they are unstable under non-laboratory conditions, making them an impractical choice for wellsite applications at the time of writing. A further issue may relate to sample preparation as this requires a separate laboratory and the use of harmful acids, which may result in health and safety concerns (see Chap. 2 for more details on ICP sample preparation proceedures).

In summary, EDXRF instruments are favoured for wellsite chemostratigraphy studies but it is rather unfortunate that the words “X-ray” are included in the name of this instrument, as this often gives the impression that the machines are highly dangerous and emit harmful radiation. In fact, XRF instruments are perfectly safe provided that are not modified or misused, and the author has proven on numerous occasions that XRF instruments emit less ration that computer monitors, even when the X-rays are switched on. In spite of this, health and safety officials are often alarmed at the mere thought of a company transporting an XRF instrument to wellsite. For this reason, it is recommended that the analysts brings all the necessary documentation relating to the emission of X-rays which should be provided by the manufacturer. It is also recommend, though it may not be practical, for the analyst to bring a Geiger counter to wellsite to demonstrate the low levels of radiation emitted by XRF. When provided with this information, health and safety officials are usually satisfied that XRF instruments can be used safely at wellsite. In spite of the negligible risks associated with the emission of X-rays from XRF instruments, most rigs require X-ray radiation warnings/signs to be placed on the outside of units in which they are operating in and minicontainers in which they are transported. The author normally prepares a very short PowerPoint presentation on the application of XRF (including XRF safety) that can be presented to any individual that has concerns relating to health and safety and the safe usage of XRF at wellsite.

6.4 Data Interpretation

As with conventional chemostratigraphy studies, checks should be made on data quality in wellsite studies before making interpretations and/or releasing the raw data to the client. The techniques used to assess data quality are discussed in detail in Chap. 2. The analyst should go through the same mental checklist where results are unexpected. For example, if samples consistently yield high Ca concentrations exceeding 30% in quartz arenites, the analyst needs to first check if the samples were collected from the correct depths. If complications relating to sampling can be ruled out, then he/she needs to find out if carbonate additives have been added during the drilling process. If these have not been used, the next step would be to check sample preparation procedures. For example, are there cavings in the samples, and have these been removed prior to analysis. If sampling and sample preparation procedures cannot explain the high Ca values, perhaps the issue is related to malfunction of the XRF instrument and/or related software. Alternatively, the problems may be explained by the calibration method employed during XRF analysis. The ability to think at speed and 'on ones feet' are important attributes for the analyst, particularly at wellsite when there is a requirement to generate data at speed and under pressure. On one particular occasion the author was contacted by a colleague working at wellsite as high levels of Ca and Mg were being produced by sandstone samples. After conducting an extensive investigation, sampling and analytical factors were ruled out. In fact, it transpired that one of the analysts was unwilling/unable to analyse samples and simply 'invented' results by copying and pasting data obtained from an unrelated study. The author instantly recognized the 'invented/copied' data as he had been working on it during the previous week. Such incidents of unwanted data manipulation and invention are extremely rare but should never be completely ruled out. It often helps if the analyst has a highly analytical, and even a cynical, mind when trying to resolve problems relating to data quality.

In most wellsite studies, data corrections are rarely required, but this is not always the case. For example, the presence of a high proportion of bicarbonate additive may result in the 'artificial' dilution of some elements more than others. In order to produce data that makes sense, it may be necessary to make manual corrections to the data under these circumstances. Such corrections, however, should only be done with considerable care. For example, adding specific concentrations of additive to standard reference materials may enable the effects of the drilling additive to be modelled stochastically. It may also be possible to redefine 'cut off' values for particular elements/ratios used to place chemostratigraphic boundaries by employing the same approach.

In some studies, it is only necessary for the analyst to generate data from cuttings samples and supply that to the client. In most projects, however, the data are used to provide stratigraphic control during drilling and/or to set casing points, coring points or total depth (TD). In many basins, the technique is utilized to aid geosteering/well trajectory monitoring. In these circumstances, it is necessary to build a chemostratigraphic framework using the techniques described in Chap. 4. The choice of wells to be used at this stage is crucial and should fulfil the following requirements:

(a) Reasonably close to proposed drilling site.
(b) Good coverage of section penetrating the stratigraphic intervals that are likely to be encountered at wellsite.
(c) Both core and cuttings should be available for analysis, in order to prove that it is possible to acquire good quality data from both.
(d) Ideally, samples from 3 to 4 wells should be analysed, though the minimum number of wells should be 2.

One unfortunate aspect of wellsite chemostratigraphy is that companies are often unable or unwilling to fund such pre-wellsite pilot studies and an inadequate number of samples/wells are analysed to build a chemostratigraphic scheme. Some companies try to use only one well/reference section to build a framework, but this practice should be avoided as it is not possible to determine the lateral extent of chemozones unless two or more wells are considered. Another important consideration is that the same (or very similar) XRF instruments should be used to generate data in this pilot phase and at wellsite (the same callibration method(s) should also be employed in both the pilot and drilling phases of the study). This insures a true 'like-for-like' comparison between data generated in the lab and at wellsite. Another recommendation, one that is often overlooked, is that the persons producing the initial chemostratigraphic scheme should also interpret the data at wellsite. On many occasions this is not possible for practical reasons, but at least one experienced analyst should be present at wellsite for the duration of the study.

It is unfortunate that inexperienced individuals are often sent to wellsite, with interpretations and any trouble shooting done by experienced chemostratigraphys working in a separate office-based location. In the opinion of the author, this practice should be avoided as the required amount of information is often not passed to the interpreter. For example, information on mudlogging crew changes and changes in drilling additives may be difficult to obtain without being present at wellsite. For these reasons, it is always recommended that at least one experienced analyst works at wellsite during the entire duration of a given study.

The number of samples to be analysed will very much depend on the speed of drilling. Where drilling rates are relatively slow (e.g. < 30 ft. per hour), it should be possible for the analyst to analyse every sample (assuming these are taken at standard 10 ft. intervals). More rapid drilling rates, however, may require samples to be analysed at 20–30 ft. spacings or more. Where the stratigraphy is complex, or where a high levels of resolution is required, it may be necessary to advise the company to slow the rate of drilling or even stop drilling to 'circulate bottoms up' until the chemostratigraphic zonation can be clearly established.

6.5 Case Study

As mentioned in Sect. 6.1, very few papers have been published on the applications of wellsite chemostratigraphy and most take the form of conference abstracts lacking in detail (the PowerPoint presentations relating to these abstracts are rarely published). Marsala et al. (2016) published a paper on this subject. Although the potential applications of XRF are mentioned, this work focuses on XRF as an analytical tool, without providing much information on field applications. Arguably, the most informative article on wellsite chemostratigraphy was written by Hildred and Rice (2014). In a study of the Evie, Otter Park and Muskwa members of the Horn River Formation, British Columbia, Canada, the authors produced a chemostratigraphic scheme relating to variations in Si/Zr, Ca, U, EFV (Vanadium enrichment factor—see Chap. 5 for explanation) and Th/U. The details are shown in Fig. 6.1. The King River Formation occurs at the base of both study wells and

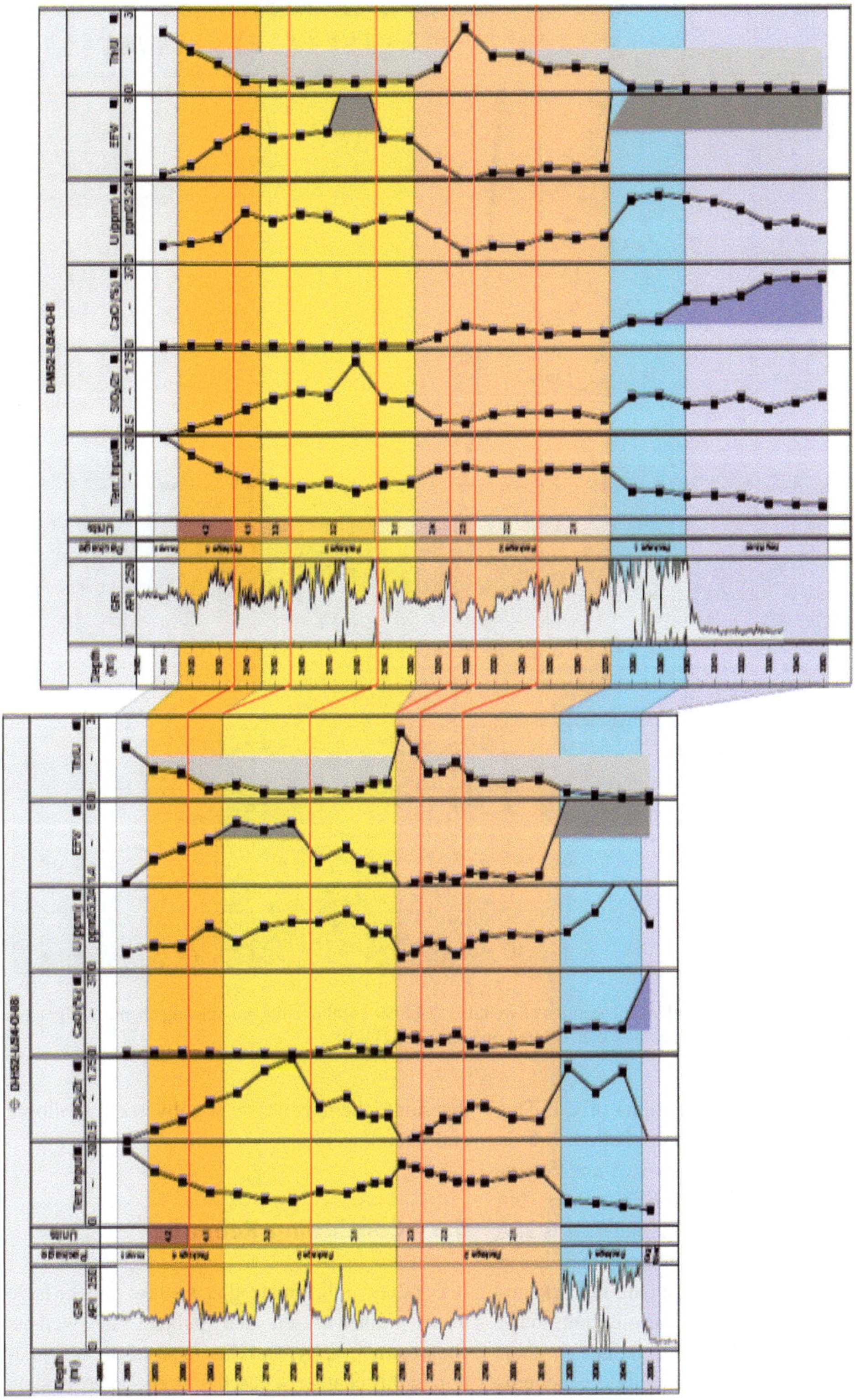

Fig. 6.1 Chemostratigraphic characterization and correlation of the Horn River Formation in two vertical wells (after Hildred and Rice 2014)

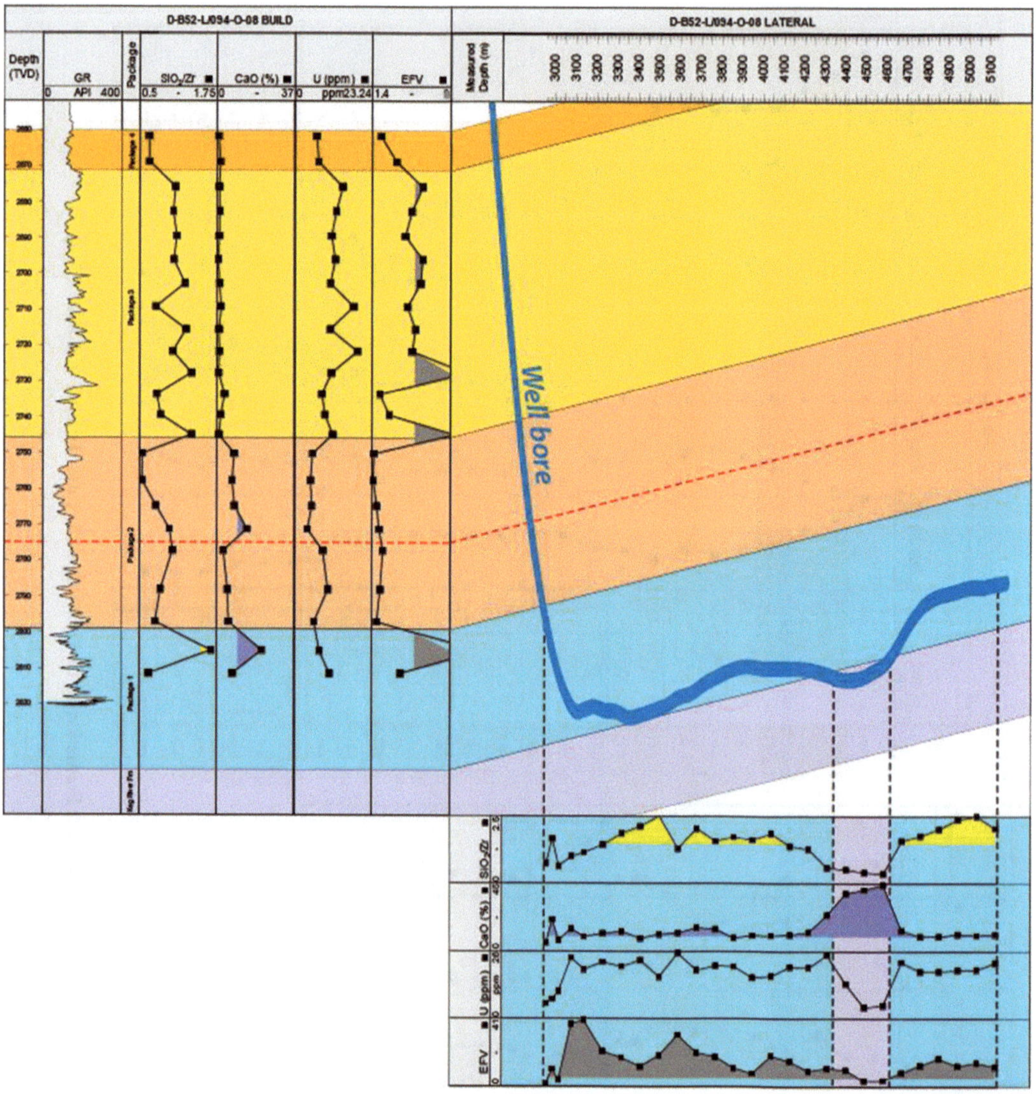

Fig. 6.2 Placement of the build and lateral sections of well d-B52-L/094-O-08 into the pre-existing chemostratigraphic framework for the Horn River Formation (after Hildred and Rice 2014)

produces the highest concentrations of Ca. This is succeeded by five packages labelled 1–5 in ascending order, with packages 1, 3 and 5 generally yielding higher U and EFV values, but lower Th/U, than in Packages 2 and 4. It is also noted that the levels of Ca are very much lower in Packages 3, 4 and 5, than they are in Packages 1 and 2. In addition to this, various units have been identified in Packages 2, 3 and 4, their definition being based on more subtle geochemical criteria. Figure 6.2 shows the definition of the same chemozones in a subvertical wellbore and lateral section.

6.6 Concluding Remarks

It is rather unfortunate that so few papers have been published on the subject of wellsite chemostratigraphy, and that most of the information in this chapter is based entirely on previously unpublished work, relating to the experiences of the author. It is hoped that the reader will find this information useful and that more articles will be published on this subject in the future.

References

Hildred, G., & Rice, C. (2014). Using high resolution chemostratigraphy to determine well-bore pathways in multilateral drilling campaigns: An example from the Horn River, British Columbia, Canada. In *Extended abstract presented at SCPG/CSEG/CWLS GeoConvention 2012*, (vision) Calgary TELUS Convention Centre and ERBC Core Research Centre, Calgary, Alberta, Canada, May 14–18th, 2012.

Marsala, A. F., Loermans, A. M. T., Shen, S., Scheibe, C., & Zereik, R. (2016). Real-time mineralogy, lithology and chemostartigraphy while drilling, using portable energy-dispersive X-ray Fluorescence. *Saudi Aramco Journal of Technology* (2016 Special Edition).

Pilcher, J. (2011). Geosteering in unconventional shale reservoirs has potential. *E&P online magazine website*. www.epmag.com/production-drilling/geosteering-unconventional-shale-reservoirs-potential_80771.

Schmidt, K., Poole, M., & Hildred, G. (2010). A triumvirate of targeting—a three-pronged approach to keeping a horizontal well in the desired Eagle Ford reservoir interval. In *American Association of Petroleum Geologists, International Annual Convention & Exhibition*, September 12–15, 2010.

MIX
Papier aus verantwortungsvollen Quellen
Paper from responsible sources
FSC® C105338

If you have any concerns about our products,
you can contact us on
ProductSafety@springernature.com

In case Publisher is established outside the EU,
the EU authorized representative is:
Springer Nature Customer Service Center GmbH
Europaplatz 3, 69115 Heidelberg, Germany

Printed by Libri Plureos GmbH
in Hamburg, Germany